endorsed for
edexcel

Edexcel A level
CHEMISTRY
2

Cliff Curtis
Jason Murgatroyd
David Scott

ALWAYS LEARNING

PEARSON

Published by Pearson Education Limited, 80 Strand, London, WC2R 0RL.

www.pearsonschoolsandfecolleges.co.uk

Copies of official specifications for all Edexcel qualifications may be found on the website: www.edexcel.com

Text © Pearson Education Limited 2015
Edited by Ashley Craig and Alison Sleigh
Designed by Elizabeth Arnoux for Pearson Education Limited
Typeset by Techset Ltd, Gateshead
Original illustrations © Pearson Education Limited 2015
Illustrated by Techset Ltd and Peter Bull
Cover design by Elizabeth Arnoux for Pearson Education Limited
Picture research by Alison Prior
Cover photo/illustration © *Front:* **Science Photo Library Ltd:** Ramon Andrade 3DCiencia

The rights of Cliff Curtis, Jason Murgatroyd and David Scott to be identified as authors of this work have been asserted by them in accordance with the Copyright, Designs and Patents Act 1988.

First edition published 2009
Second edition published 2015

25
11

British Library Cataloguing in Publication Data
A catalogue record for this book is available from the British Library

ISBN 978 1 447 99117 5

Copyright notice
All rights reserved. No part of this publication may be reproduced in any form or by any means (including photocopying or storing it in any medium by electronic means and whether or not transiently or incidentally to some other use of this publication) without the written permission of the copyright owner, except in accordance with the provisions of the Copyright, Designs and Patents Act 1988 or under the terms of a licence issued by the Copyright Licensing Agency, Saffron House, 6–10 Kirby Street, London EC1N 8TS (www.cla.co.uk). Applications for the copyright owner's written permission should be addressed to the publisher.

Printed and bound by CPI Group (UK) Ltd, Croydon, CR0 4YY

A note from the publisher
In order to ensure that this resource offers high-quality support for the associated Pearson qualification, it has been through a review process by the awarding body. This process confirms that this resource fully covers the teaching and learning content of the specification or part of a specification at which it is aimed. It also confirms that it demonstrates an appropriate balance between the development of subject skills, knowledge and understanding, in addition to preparation for assessment.

Endorsement does not cover any guidance on assessment activities or processes (e.g. practice questions or advice on how to answer assessment questions), included in the resource nor does it prescribe any particular approach to the teaching or delivery of a related course.

While the publishers have made every attempt to ensure that advice on the qualification and its assessment is accurate, the official specification and associated assessment guidance materials are the only authoritative source of information and should always be referred to for definitive guidance.

Pearson examiners have not contributed to any sections in this resource relevant to examination papers for which they have responsibility.

Examiners will not use endorsed resources as a source of material for any assessment set by Pearson.

Endorsement of a resource does not mean that the resource is required to achieve this Pearson qualification, nor does it mean that it is the only suitable material available to support the qualification, and any resource lists produced by the awarding body shall include this and other appropriate resources.

Picture Credits

The publisher would like to thank the following for their kind permission to reproduce their photographs:

(Key: b-bottom; c-centre; l-left; r-right; t-top)

123RF.com: 8-9, angellodeco 233, ensup 96; **Alamy Images:** AC Images 232, agefotostock / Javier Larrea 82-83, Bon Appetit / Herbert Lehmann 125, BSIP SA 228, Chronicle 216; **Corbis:** Brooks Kraft 97; **Courtesy of Toyota (GB) Ltd:** 78; **Dr. Armin Rose:** 136; **Fotolia.com:** ChiccoDodiFC 46, dreamon512 188, Ekaterina Shilova 121, Flaschen 090214 28 124, Gudellaphoto 102, nilanewsom 192, Stefan Merkle 197, stockphoto-graf 18, Subbotina Anna 50-51; **Getty Images:** BSIP 230, Ulrich Baumgarten 225, waldru 201; **Imagestate Media:** John Foxx Collection 185; **Martyn F. Chillmaid:** 122tl, 122bl, 123l, 123r, 135; **Masterfile UK Ltd:** Flowerphotos 140-141; **Pearson Education Ltd:** Trevor Clifford 129, 143br, 193; **Reproduced by permission of The Royal Society of Chemistry:** 24b; **Science Photo Library Ltd:** 24t, Andrew Lambert Photography 22-23, 35, 98, 112, 130, 143tr, 143cl, 179l, 179r, 181, 186, Asrtid & Hanns-Frieder 133, Charles D Winters 183, Cordelia Molloy 209cr, Dick Luria 132, Gustoimages 202, Jiri Loun 177, Look at Science / Vincent Moncorge 168-169, Martyn F. Chillmaid 108, Mauro Fermariello 235, Ramon Andrade 3DCiencia 117; **Shutterstock.com:** Elnur 106-107, Gonul Kokal 234, topseller 209tr, vlastas 66; **Sozaijiten:** 246

All other images © Pearson Education

We are grateful to the following for permission to reproduce copyright material:

Text

Extract on page 18 from *Uncorked: The Science of Champagne*, Princeton University Press (Liger-Belair, G. 2004) pp.37-39, reproduced with permission of Princeton University Press in the format Book via Copyright Clearance Center; Extract on page 78 after How does Toyota's fuel cell vehicle work? by Joe Clifford, 17 November 2014, http://blog.toyota.co.uk/how-does-toyotas-fuel-cell-vehicle-work, with permission from Toyota GB; Extract on page 102 after Battery power by Tony Hargreaves, http://www.rsc.org/education/eic/issues/2008Mar/BatteryPower.asp, with permission from Royal Society of Chemistry; Extract on page 136 from Blue Blood Helps Octopus Survive Brutally Cold Temperatures, posted by Stefan Sirucek in Weird & Wild on July 10, 2013, Stefan Sirucek/National Geographic Creative; Extract on page 164 from Engineered metalloenzyme catalyses Friedel–Crafts reaction by Debbie Houghton, *Chemistry World*, 3 November 2014 http://www.rsc.org/chemistryworld/2014/11/metalloenzyme-friedel-crafts-unnatural-amino-acids, with permission from Royal Society Chemistry; Extract on page 244 from The sweet scent of success by Emma Davies, http://www.rsc.org/chemistryworld/Issues/2009/February/TheSweetScentOfSuccess.asp, with permission from Royal Society of Chemistry.

The publisher would like to thank Chris Curtis, Chris Ryan and Adelene Cogill for their contributions to the Maths skills and Preparing for your exams sections of this book.

The publisher would also like to thank Chris Jennings for his collaboration in reviewing this book.

Contents

How to use this book ... 6

TOPIC 11 Further equilibrium

11.1 Chemical equilibrium ... 8
1. Equilibrium constant, K_c ... 10
2. Equilibrium constant, K_p ... 12
3. Effect of temperature on equilibrium constants ... 14
4. Effect of concentration, pressure and catalysts on equilibrium constants ... 16
Thinking Bigger ... 18
Exam-style questions ... 20

TOPIC 12 Acid–base equilibria

12.1 Strong and weak acids ... 22
1. The Brønsted–Lowry theory ... 24
2. Hydrogen ion concentration and the pH scale ... 27
3. Ionic product of water, K_w ... 30
4. Analysing data from pH measurements ... 32

12.2 Acid–base titrations ... 35
1. Acid–base titrations, pH curves and indicators ... 35
2. Buffer solutions ... 39
3. Buffer solutions and pH curves ... 43
4. Enthalpy changes of neutralisation for strong and weak acids ... 45
Thinking Bigger ... 46
Exam-style questions ... 48

TOPIC 13 Further energetics

13.1 Lattice energy ... 50
1. Lattice energy, $\Delta_{lattice}H$, and Born–Haber cycles ... 52
2. Experimental and theoretical lattice energy ... 56
3. Enthalpy changes of solution and hydration ... 59

13.2 Entropy ... 62
1. Introduction to entropy ... 62
2. Total entropy ... 64
3. Understanding entropy changes ... 67

13.3 Gibbs energy ... 70
1. The Second Law and Gibbs energy ... 70
2. Gibbs energy and equilibrium ... 72
3. Some applications of Gibbs energy: further understanding ... 74
Thinking Bigger ... 78
Exam-style questions ... 80

TOPIC 14 Further redox

14.1 Standard electrode potential ... 82
1. Standard electrode (redox) potentials ... 84
2. Electrochemical cells ... 90
3. Standard electrode potentials and thermodynamic feasibility ... 92

14.2 Redox in action ... 96
1. Storage cells and fuel cells ... 96
2. Redox titrations ... 98
Thinking Bigger ... 102
Exam-style questions ... 104

TOPIC 15 Transition metals

15.1 Principles of transition metal chemistry ... 106
1. Transition metal electronic configurations ... 108
2. Ligands and complexes ... 110
3. The origin of colour in complexes ... 112
4. Common shapes of complexes ... 114
5. Square planar complexes ... 116
6. Multidentate ligands ... 118

15.2 Transition metal reactions ... 122
1. Different types of reaction ... 122
2. Reactions of cobalt and iron complexes ... 124
3. The chemistry of chromium ... 126
4. The chemistry of vanadium ... 130

15.3 Transition metals as catalysts ... 132
1. Heterogeneous catalysis ... 132
2. Homogeneous catalysis ... 134
Thinking Bigger ... 136
Exam-style questions ... 138

TOPIC 16 Further kinetics

16.1 Further kinetics ... 140
1. Methods of measuring the rate of reaction ... 142
2. Rate equations, rate constants and orders of reaction ... 145
3. Determining orders of reaction ... 148
4. Rate equations and mechanisms ... 153
5. Activation energy and catalysis ... 157
6. The effect of temperature on the rate constant ... 160
Thinking Bigger ... 164
Exam-style questions ... 166

TOPIC 17 Further organic chemistry

17.1 Chirality — 168
1. Chirality and enantiomers — 170
2. Optical activity — 172
3. Optical activity and reaction mechanisms — 174

17.2 Carbonyl compounds — 176
1. Carbonyl compounds and their physical properties — 176
2. Redox reactions of carbonyl compounds — 178
3. Nucleophilic addition reactions — 180

17.3 Carboxylic acids — 182
1. Carboxylic acids and their physical properties — 182
2. Preparations and reactions of carboxylic acids — 184
3. Acyl chlorides — 186
4. Esters — 188
5. Polyesters — 190

17.4 Arenes – benzene — 192
1. Benzene – a molecule with two models — 192
2. Some reactions of benzene — 196
3. Electrophilic substitution mechanisms — 198
4. Phenol — 200

17.5 Amines, amides, amino acids and proteins — 202
1. Amines and their preparations — 202
2. Acid–base reactions of amines — 204
3. Other reactions of amines — 206
4. Amides and polyamides — 208
5. Amino acids — 210
6. Peptides and proteins — 212

17.6 Organic structures — 215
1. Principles of organic synthesis — 215
2. Hazards, risks and control measures — 218
3. Practical techniques in organic chemistry – Part 1 — 220
4. Practical techniques in organic chemistry – Part 2 — 222
5. Simple chromatography — 224

17.7 Organic analysis — 226
1. Traditional methods of analysis — 226
2. Determining structures using mass spectra — 228
3. Chromatography – HPLC and GC — 230
4. Chromatography and mass spectrometry — 232
5. Principles of NMR spectroscopy — 234
6. ^{13}C NMR spectroscopy — 236
7. ^{1}H NMR spectroscopy — 240
8. Splitting patterns in ^{1}H NMR spectroscopy — 243

Thinking Bigger — 246
Exam-style questions — 248

Maths skills — 250
Preparing for your exams — 254
Glossary — 261
Periodic Table — 263
Index — 264

How to use this book

Welcome to Book 2 of your A level Chemistry course. This book, covering Topics 11–19 of the specification, follows on from Book 1, which covered Topics 1–10. The exams that you will sit at the end of your A level course will cover content from both books. The following features are included to support your learning:

Topic openers

Each topic starts by setting the context for that topic's learning:

- Links to other areas of chemistry are shown, including previous knowledge that is built on in the topic, and future learning that you will cover later in your course.
- The **All the maths you need** checklist helps you to know what maths skills will be required.

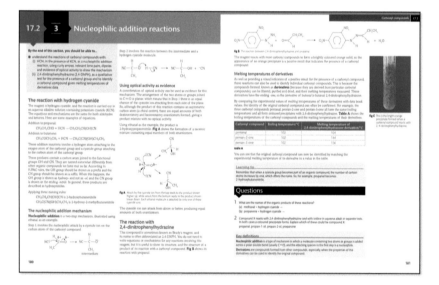

Main content

The main part of each topic covers all the points from the specification that you need to learn. The text is supported by diagrams and photos that will help you understand the concepts.

Within each section, you will find the following features:

- **Learning objectives** at the beginning of each section, highlighting what you need to know and understand.
- **Key definitions** shown in bold and collated at the end of each section for easy reference.
- **Worked examples** showing you how to work through questions, and how your calculations should be set out.
- **Investigations** provide a summary of practical experiments that explore key concepts.
- **Learning tips** to help you focus your learning and avoid common errors.
- **Did you know?** boxes featuring interesting facts to help you remember the key concepts.
- **Questions** to help you check whether you have understood what you have just read, and whether there is anything that you need to look at again. Answers to the questions can be found on Pearson's website as a free resource.

Thinking Bigger

At the end of each topic there is an opportunity to read and work with real-life research and writing about science. The timeline at the bottom of the spreads highlights which other topics the material relates to. These spreads will help you to:

- read real-life material that's relevant to your course
- analyse how scientists write
- think critically and consider the issues
- develop your own writing
- understand how different aspects of your learning piece together.

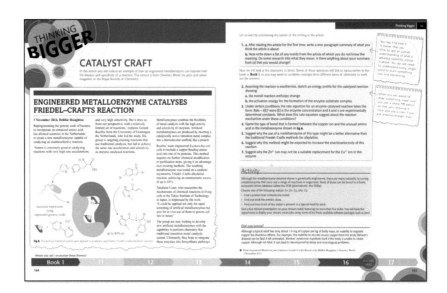

Exam-style questions

At the end of each topic there are also **exam-style questions** to help you to:

- test how fully you have understood the learning
- practise for your exams.

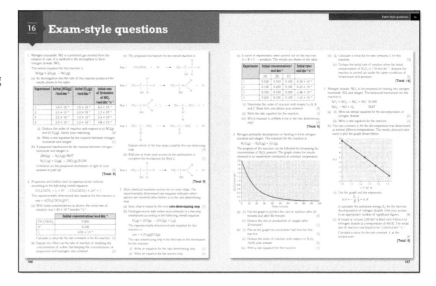

Getting the most from your online ActiveBook

This book comes with 3 years' access to ActiveBook* – an online, digital version of your textbook. Follow the instructions printed on the inside front cover to start using your ActiveBook.

Your ActiveBook is the perfect way to personalise your learning as you progress through your Edexcel A level Chemistry course. You can:

- access your content online, anytime, anywhere
- use the inbuilt highlighting and annotation tools to personalise the content and make it really relevant to you
- search the content quickly using the index.

Highlight tool
Use this to pick out key terms or topics so you are ready and prepared for revision.

Annotations tool
Use this to add your own notes, for example links to your wider reading, such as websites or other files. Or make a note to remind yourself about work that you need to do.

*For new purchases only. If this access code has already been revealed, it may no longer be valid. If you have bought this textbook secondhand, the code may already have been used by the first owner of the book.

TOPIC 11

Further equilibrium

Introduction

Whenever chemical engineers design a chemical plant to manufacture a new chemical they must overcome a problem. They want the best possible yield from reactants, but they find that most chemical reactions do not go to completion because they are reversible and can reach a state of dynamic equilibrium. All reactions have a tendency to approach equilibrium, whether they seem to go to completion or whether they seem to form no products at all. In many cases, such as the combustion of a fuel, the equilibrium composition corresponds to the complete conversion of reactants into products. Reactions that seem to 'not go' at all, like the corrosion of gold in air, also tend to approach equilibrium, but in these cases the equilibrium composition corresponds to virtually unchanged reactants.

Many reactions, however, approach an equilibrium in which there are significant amounts of both reactants and products present. Finding and controlling the position of equilibrium in such reactions is the focus of this topic.

All the maths you need

- Recognise and make use of appropriate units in calculations
- Recognise and use expressions in decimal and ordinary form
- Use ratios, fractions and percentages
- Use an appropriate number of significant figures
- Change the subject of an equation
- Substitute numerical values into algebraic expressions using appropriate units for physical quantities
- Solve algebraic expressions

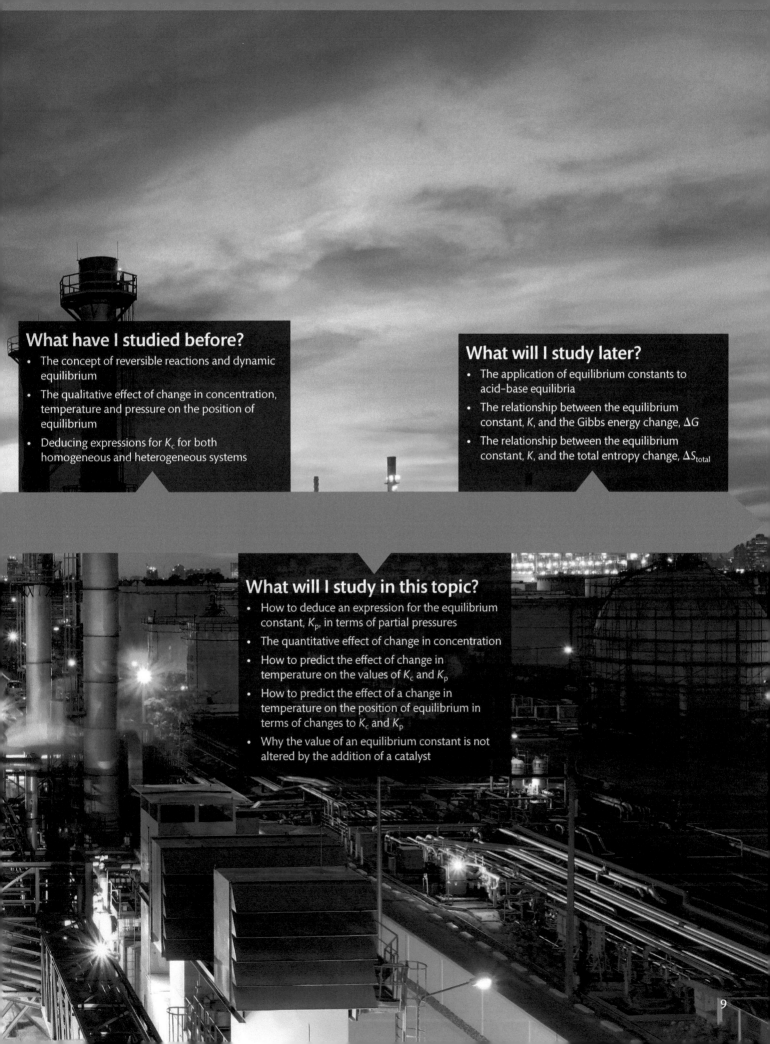

What have I studied before?
- The concept of reversible reactions and dynamic equilibrium
- The qualitative effect of change in concentration, temperature and pressure on the position of equilibrium
- Deducing expressions for K_c for both homogeneous and heterogeneous systems

What will I study later?
- The application of equilibrium constants to acid–base equilibria
- The relationship between the equilibrium constant, K, and the Gibbs energy change, ΔG
- The relationship between the equilibrium constant, K, and the total entropy change, ΔS_{total}

What will I study in this topic?
- How to deduce an expression for the equilibrium constant, K_p, in terms of partial pressures
- The quantitative effect of change in concentration
- How to predict the effect of change in temperature on the values of K_c and K_p
- How to predict the effect of a change in temperature on the position of equilibrium in terms of changes to K_c and K_p
- Why the value of an equilibrium constant is not altered by the addition of a catalyst

11.1 › 1 › Equilibrium constant, K_c

By the end of this section, you should be able to...

- deduce an expression for K_c for homogeneous and heterogeneous reactions, in terms of equilibrium concentrations in mol dm^{-3}
- calculate a value from experimental data, with units where appropriate, for K_c in both homogeneous and heterogeneous reactions

Homogeneous reactions

A homogeneous reaction is one in which all reactants and products are in the same phase. For example, all are gases or all are in aqueous solution.

In **Book 1 Topic 10** we saw how the position of equilibrium of a reversible reaction can change when the conditions are altered.

In this section, we shall study *quantitatively* how changes in concentration affect the equilibrium position.

For the reversible reaction:

$$N_2O_4(g) \rightleftharpoons 2NO_2(g)$$

it can be shown experimentally that at equilibrium and at a given temperature:

$$\frac{[NO_2(g)]^2_{eq}}{[N_2O_4(g)]_{eq}} = \text{a constant}$$

As we mentioned in **Book 1**, this is called the 'equilibrium constant' and is given the symbol K_c.

The subscript 'c' indicates that this equilibrium constant is calculated using concentrations. This is to distinguish it from another equilibrium constant, K_p, which uses partial pressures.

The subscript 'eq' indicates that the concentrations are those *at equilibrium*.

At a given temperature, the value of the equilibrium constant does not change with concentration, unlike the equilibrium position. The only factor that affects the value of the equilibrium constant is temperature.

There are two ways of obtaining a value for the equilibrium constant. The first is by experimentation, and this is the method we shall look at in this section. The second is by calculation using thermodynamic equations, and this will be covered in **Topic 13**.

For a general reaction:

$$wA + xB \rightleftharpoons yC + zD$$

$$K_c = \frac{[C]^y_{eq}[D]^z_{eq}}{[A]^w_{eq}[B]^x_{eq}}$$

For convenience, the 'eq' sign is often not included, but is taken for granted.

Determination of the equilibrium constant

To determine the value of the equilibrium constant for a reaction, known amounts of reactants are allowed to reach equilibrium with their products. At equilibrium, the amount of one of the substances is measured and the amounts of the others can be calculated using the stoichiometry of the equation.

WORKED EXAMPLE 1:

2.00 mol of ethanoic acid and 2.00 mol of ethanol are mixed and allowed to reach equilibrium with ethyl ethanoate and water at 298 K. The amount of ethanoic acid at equilibrium is found to be 0.67 mol. Calculate K_c for the reaction at 298 K.

Answer

Work out the amounts, and hence the concentrations, of each substance at equilibrium:

	$CH_3COOH(l)$ +	$CH_3CH_2OH(l)$ $\rightleftharpoons$	$CH_3COOCH_2CH_3(l)$ +	$H_2O(l)$
amount at start/mol	2.00	2.00	0	0
amount at equilibrium/mol	0.67	0.67	(2.00 − 0.67) = 1.33	(2.00 − 0.67) = 1.33
concentration at equilibrium/mol dm^{-3}	$\frac{0.67}{V}$	$\frac{0.67}{V}$	$\frac{1.33}{V}$	$\frac{1.33}{V}$

where V is the total volume in dm^3 of the reaction mixture.

$$K_c = \frac{[CH_3COOCH_2CH_3(l)][H_2O(l)]}{[CH_3COOH(l)][CH_3CH_2OH(l)]} = \frac{(1.33/V)(1.33/V)}{(0.67/V)(0.67/V)}$$

$$= 3.94 \text{ (no units)}$$

K_c has no units because the same number of moles is present on each side of the equation. For the same reason, V also cancels.

WORKED EXAMPLE 2:

0.100 mol of N_2O_4 is placed into a 0.100 dm³ flask and allowed to reach equilibrium with NO_2 at 398 K. At equilibrium, there is 0.071 mol of N_2O_4. Calculate K_c at this temperature.

Answer

	$N_2O_4(g)$	$\rightleftharpoons$	$2NO_2(g)$	
	0.100		0	amount at start/mol
	0.071		(0.100 − 0.071) × 2 = 0.058	amount at equilibrium/mol
	0.071/0.100 = 0.71		0.058/0.100 = 0.58	concentration at equilibrium/mol dm⁻³

$$K_c = \frac{[NO_2(g)]^2}{[N_2O_4(g)]} = \frac{0.58^2}{0.71} = 0.474 \text{ mol dm}^{-3}$$

Note: this time, K_c has units of mol dm⁻³

$$\frac{\text{mol dm}^{-3} \times \text{mol dm}^{-3}}{\text{mol dm}^{-3}} = \text{mol dm}^{-3}$$

Heterogeneous reactions

A heterogeneous reaction is one in which at least one of the reactants and/or products is in a different phase to the others.

Examples are:

$$CaCO_3(s) \rightleftharpoons CaO(s) + CO_2(g)$$
$$3Fe(s) + 4H_2O(g) \rightleftharpoons Fe_3O_4(s) + 4H_2(g)$$
$$H_2O(l) \rightleftharpoons H^+(aq) + OH^-(aq)$$

Writing equilibrium constants for heterogeneous reactions

If we apply the equilibrium law to the reaction:

$$CaCO_3(s) \rightleftharpoons CaO(s) + CO_2(g)$$

we obtain:

$$\frac{[CaO(s)][CO_2(g)]}{[CaCO_3(s)]} = \text{a constant}$$

However, the concentration of a solid at a given temperature is determined by its density, which has a constant value. Hence, the expression can be simplified to:

$$[CO_2(g)] = \text{a constant}$$

This constant is given the symbol K_c.

So, $K_c = [CO_2(g)]$ for the above equilibrium.

Similarly, for the equilibrium:

$$3Fe(s) + 4H_2O(g) \rightleftharpoons Fe_3O_4(s) + 4H_2(g)$$

$$K_c = \frac{[H_2(g)]^4}{[H_2O(g)]^4}$$

Did you know?

The concentration term of a solid in a heterogeneous equilibrium is constant. It can therefore be absorbed into the equilibrium constant.

Questions

1. A dynamic equilibrium is set up between carbon monoxide, hydrogen and methanol. The equation for the reaction is:

   $$CO(g) + 2H_2(g) \rightleftharpoons CH_3OH(g)$$

   The equilibrium concentrations, in mol dm⁻³, for the three components are shown below:

   $CO(g)$, 3.1×10^{-3}; $H_2(g)$, 2.4×10^{-2}; $CH_3OH(g)$, 2.6×10^{-5}

   (a) Write an expression for K_c for this reaction.
   (b) Calculate a value for K_c. State the units, if any.

2. The reaction between ethanoic acid and ethanol is reversible. The equation for the reaction is:

   $$CH_3COOH(l) + CH_3CH_2OH(l) \rightleftharpoons CH_3COOCH_2CH_3(l) + H_2O(l)$$

   6.0 mol of ethanoic acid and 12.5 mol of ethanol are mixed together with some sulfuric acid to act as a catalyst. The total volume of the reaction mixture is V dm³. The mixture is left for several days to reach equilibrium, after which only 1.0 mol of ethanoic acid remained.

   (a) Write an expression for K_c for the reaction.
   (b) Calculate the amounts of ethanol, ethyl ethanoate and water present in the equilibrium mixture.
   (c) Use your answers to part (b) to calculate a value for K_c. Give your answer to an appropriate number of significant figures and state the units, if any.

3. The reaction between hydrogen and iodine to form hydrogen iodide is reversible.

   $$H_2(g) + I_2(g) \rightleftharpoons 2HI(g)$$

   0.30 mol of $H_2(g)$ is mixed with 0.20 mol of $I_2(g)$ and the mixture is allowed to reach equilibrium. At equilibrium, 0.14 mol of $H_2(g)$ is present.

   Calculate the value of K_c for the reaction. Give your answer to an appropriate number of significant figures and state the units, if any.

4. Nitrogen and oxygen can exist in equilibrium with nitrogen(II) oxide. The equation for the reaction is:

   $$N_2(g) + O_2(g) \rightleftharpoons 2NO(g) \quad \Delta H \text{ is positive}$$

   The equilibrium constant at 289 K for this reaction is 4.8×10^{-31}.

   (a) What does the value of the equilibrium constant tell you about the position of equilibrium of this reaction at 298 K?
   (b) An equilibrium mixture of volume 1.2 dm³ of the three gases contains 1.1 mol of $N_2(g)$ and 4.0×10^{-16} mol of $NO(g)$. Calculate the equilibrium concentration of $O_2(g)$ in this mixture.

5. The equilibrium constant, K_c, for the reaction:

   $$H_2O(g) + C(s) \rightleftharpoons H_2(g) + CO(g)$$

   has a value of 4.92×10^{-5} mol dm⁻³ at 700 K.

   A mixture of water vapour and carbon is heated to 700 K in a closed vessel and allowed to reach equilibrium with hydrogen and carbon monoxide.

   Write an expression for K_c and use it to calculate the equilibrium concentrations of H_2 and CO in the above equilibrium mixture, at 700 K, when the equilibrium concentration of H_2O is 2.00×10^{-2} mol dm⁻³.

11.1 > 2 > Equilibrium constant, K_p

By the end of this section, you should be able to...

- deduce an expression for K_p for both homogeneous and heterogeneous reactions, in terms of equilibrium partial pressures in atmospheres (atm)
- calculate a value for K_p from experimental data, with units where appropriate, for both homogeneous and heterogeneous reactions

Homogeneous reactions

For reversible reactions involving gases we can express the concentrations of the reactants and products in terms of their partial pressures. The **partial pressure** of an individual gas in a mixture of gases is defined as the pressure that the gas would exert if it alone occupied the volume of the mixture. The total pressure of the mixture is then equal to the sum of the partial pressures of all the gases present in the mixture.

If we have a mixture of two gases, A and B, then the total pressure, p, of the mixture is related to the partial pressures of A and B by the following equation:

$$p = p_A + p_B$$

where p_A = partial pressure of A and p_B = partial pressure of B.

For the general reaction:

$$wA(g) + xB(g) \rightleftharpoons yC(g) + zD(g)$$

we can define another equilibrium constant, K_p, in terms of partial pressures:

$$K_p = \frac{(p_C)^y (p_D)^z}{(p_A)^w (p_B)^x}$$

Calculating partial pressures

The partial pressure of an individual gas in a mixture of gases is calculated by multiplying its mole fraction by the total pressure, for example:

$$p_A = x_A p$$

where x_A = the mole fraction of gas A in the mixture.

$$\text{mole fraction of A} = \frac{\text{number of moles of A}}{\text{total number of moles of gas}}$$

WORKED EXAMPLE:

1.00 mol of PCl_5 vapour is heated to 500 K in a sealed vessel. The equilibrium mixture, at a pressure of 6 atm, contains 0.60 mol of chlorine. Calculate K_p for the reaction:

$$PCl_5(g) \rightleftharpoons PCl_3(g) + Cl_2(g)$$

Answer

	$PCl_5(g)$	$\rightleftharpoons$	$PCl_3(g)$	+	$Cl_2(g)$	Total	
	1.00		0		0	1.00	amount at start/mol
	1.00 − 0.60		0.60		0.60	1.60	amount at equilibrium/mol
	$\frac{0.40}{1.60}$		$\frac{0.60}{1.60}$		$\frac{0.60}{1.60}$		mole fraction at equilibrium
	0.25 × 6		0.375 × 6		0.375 × 6		partial pressure/atm

$$K_p = \frac{(p_{PCl_3})(p_{Cl_2})}{(p_{PCl_5})} = \frac{(0.375 \times 6)(0.375 \times 6)}{(0.25 \times 6)} = 3.38 \text{ atm}$$

Heterogeneous reactions

Solids and gases

For solids that are in equilibrium with gases, the partial pressure terms of any solids are not included in the expression for K_p. You will recall that K_c has a similar rule.

Consider the reaction:

$$CaCO_3(s) \rightleftharpoons CaO(s) + CO_2(g)$$

The equilibrium constant in terms of partial pressures is given by:

$$K_p = p_{CO_2(g)}$$

This equation suggests that for a particular temperature, the pressure of carbon dioxide in the equilibrium mixture is constant regardless of the masses of calcium carbonate and calcium oxide present. This prediction can be confirmed experimentally.

At 1073 K, the pressure of carbon dioxide in equilibrium with $CaCO_3(s)$ and $CaO(s)$ is 0.25 atm.

Hence, at 1073 K, $K_p = 0.25$ atm

For this reaction, K_p increases with increasing temperature. It only exceeds 1 atm at temperatures above 1173 K. So, calcium carbonate must be heated to this temperature for it to decompose at 1 atm of pressure.

Chemical equilibrium 11.1

Did you know?

A similar situation exists for liquids that are in equilibrium with gases. For the equilibrium between liquid water and water vapour:

$$H_2O(l) \rightleftharpoons H_2O(g)$$

$$K_p = p_{H_2O(g)}$$

The partial pressure of water (p_{H_2O}) in equilibrium with its liquid is also known as the 'saturated vapour pressure' of water.

At 298 K, K_p, and hence p_{H_2O}, is equal to 0.03 atm; whereas at 373 K, K_p, and hence p_{H_2O}, is equal to 1.00 atm.

Since the boiling temperature of a liquid is defined as the temperature at which the saturated vapour pressure equals the external pressure, the boiling temperature of water at 1 atm is 373 K (100 °C).

Questions

1 One stage in the manufacture of sulfuric acid by the contact process involves the reaction between sulfur dioxide and oxygen to form sulfur trioxide:

$$2SO_2(g) + O_2(g) \rightleftharpoons 2SO_3(g)$$

(a) Write an expression for K_p for this reaction.
(b) An equilibrium is set up for this reaction at 700 K.

At this temperature:
- the partial pressure of SO_2 is 0.100 atm
- the partial pressure of O_2 is 0.500 atm
- K_p for the reaction is 3.00×10^4 atm^{-1}

Calculate the partial pressure of SO_3 in this equilibrium mixture and hence determine the percentage of SO_3 present.

2 The reaction between carbon monoxide and chlorine to form phosgene, $COCl_2$, is reversible. The equation for the reaction is:

$$CO(g) + Cl_2(g) \rightleftharpoons COCl_2(g)$$

Some carbon monoxide and chlorine are allowed to react and reach a position of equilibrium. The equilibrium partial pressures of the mixture are:

$CO(g)$, 2.47×10^{-8} atm; $Cl_2(g)$, 2.47×10^{-8} atm; $COCl_2(g)$, 4.08×10^{-10} atm

(a) What is meant by the term 'partial pressure'?
(b) Write an expression for K_p for this reaction and calculate its value, giving the units, if any.

3 When chlorine gas is heated to a high temperature, the molecules dissociate into chlorine atoms.

$$Cl_2(g) \rightleftharpoons 2Cl(g)$$

Some chlorine gas is placed into a closed container and heated to 1400 K until equilibrium is established. The partial pressure of $Cl_2(g)$ is 0.84 atm and that of $Cl(g)$ is 0.030 atm.

(a) Determine the mole fraction of Cl in the equilibrium mixture.
(b) (i) Write an expression for K_p for this reaction.
    (ii) Calculate the value of K_p for this reaction at 700 K. Give the units, if any.

Key definition

The **partial pressure** of a gas in a mixture of gases is the pressure that the gas would exert if it alone occupied the volume of the mixture.

11.1 — 3 — Effect of temperature on equilibrium constants

By the end of this section, you should be able to...

- determine the effect of changing the temperature on the equilibrium constant (K_c and K_p) for both exothermic and endothermic reactions
- understand that the effect of temperature on the position of equilibrium is explained using a change in the value of the equilibrium constant

Effect of temperature on K_p and K_c

Table A gives the value of the equilibrium constant, K_p, at different temperatures for two gaseous reactions.

$N_2(g) + 3H_2(g) \rightleftharpoons 2NH_3(g)$ $\Delta H^\ominus = -92.2 \text{ kJ mol}^{-1}$		$N_2O_4(g) \rightleftharpoons 2NO_2(g)$ $\Delta H^\ominus = +57.2 \text{ kJ mol}^{-1}$	
T/K	K_p/atm^{-2}	T/K	K_p/atm
298	6.76×10^5	298	1.15×10^{-1}
400	4.07×10^1	400	4.79×10^1
500	3.55×10^{-2}	500	1.70×10^3
600	1.66×10^{-3}	600	1.78×10^4

table A

For the *exothermic* reaction between $N_2(g)$, $H_2(g)$ and $NH_3(g)$, the value of K_p *decreases* with increasing temperature.

The opposite is true for the *endothermic* reaction between $N_2O_4(g)$ and $NO_2(g)$.

Exactly the same trend is observed with the values of K_c.

Table B summarises the effect of changing the temperature on the value of the equilibrium constant.

Thermicity of reaction	Increase in temperature	Decrease in temperature
exothermic	K decreases	K increases
endothermic	K increases	K decreases

table B

Change in temperature alters the equilibrium position

If the equilibrium constant for a reaction changes with a change in temperature, then it follows that the equilibrium position also changes. This is illustrated by the data in **table C**, which shows the value of K_c and the amounts of reactants and products at equilibrium for the reaction between hydrogen and iodine to form hydrogen iodide:

$$H_2(g) + I_2(g) \rightleftharpoons 2HI(g) \qquad \Delta H^\ominus = -9 \text{ kJ mol}^{-1}$$

In each case, one mole of hydrogen is mixed with one mole of iodine and equilibrium is allowed to establish, at the given temperature, in a vessel of volume 1 dm^3.

Temperature	K_p	Amount/mol		
		$H_2(g)$	$I_2(g)$	$HI(g)$
500	160	0.14	0.14	1.72
700	54	0.21	0.21	1.58

table C

As the value of K_p decreases, the percentage of $H_2(g)$ and $I_2(g)$ increases, and the percentage of $HI(g)$ decreases. The equilibrium shifts to the left, in the endothermic direction, as the temperature is increased.

Questions

1. A dynamic equilibrium is set up between carbon monoxide, hydrogen and methanol. The equation for the reaction is:

    $CO(g) + 2H_2(g) \rightleftharpoons CH_3OH(g)$

    When the reaction is carried out at a higher temperature, the value of K_c decreases.
    Explain whether the forward reaction is exothermic or endothermic.

2. The reaction between hydrogen and iodine to form hydrogen iodide is reversible:

    $H_2(g) + I_2(g) \rightleftharpoons 2HI(g)$

    Some $H_2(g)$ is mixed with $I_2(g)$ and the mixture is allowed to reach equilibrium. At equilibrium, 0.14 mol of $H_2(g)$ is present.
    The experiment is repeated with the same amounts of $H_2(g)$ and $I_2(g)$, but at a higher temperature. This time the amount of $H_2(g)$ at equilibrium is greater than 0.14 mol.
    Explain what this information tells you about the reaction.

3. Nitrogen and oxygen can exist in equilibrium with nitrogen(II) oxide. The equation for the reaction is:

    $N_2(g) + O_2(g) \rightleftharpoons 2NO(g)$ ΔH is positive

    The equilibrium constant at 298 K for this reaction is 4.8×10^{-31}.
    Explain the change, if any, in the proportion of $NO(g)$ at a temperature higher than 298 K.

11.1 › 4 › Effect of concentration, pressure and catalysts on equilibrium constants

> **By the end of this section, you should be able to...**
> - understand that the value of the equilibrium constant is not affected by changes in concentration or pressure or by the addition of a catalyst

Effect of concentration

In **Book 1 Topic 10**, we saw that the equilibrium position changes when the concentration of a component of the equilibrium mixture is altered. In this section, we will provide an explanation for this phenomenon in terms of K_c.

The first thing to appreciate is that the only factor that can change the value of K_c is a change in temperature. That is, K_c *is constant at a given temperature*. When the concentration of one of the components of an equilibrium mixture is altered, there is an immediate change in the reaction quotient, Q_c, which is the mathematical relationship between the concentrations of the components. At this point, the reaction quotient is no longer equal to the equilibrium constant, K_c. Therefore, the equilibrium composition changes until the reaction quotient and equilibrium constant become equal.

The following examples illustrate this process.

Reaction quotient, Q_c

For a general reaction:

$$wA + xB \rightleftharpoons yC + zD$$

the reaction quotient, Q_c is given by the expression:

$$Q_c = \frac{[C]^y[D]^z}{[A]^w[B]^x}$$

where the concentration terms are not necessarily those at equilibrium.

At equilibrium $Q_c = K_c$.

WORKED EXAMPLE:

The equation for the reaction between hydrogen and iodine to form hydrogen iodide is:

$$H_2(g) + I_2(g) \rightleftharpoons 2HI(g)$$

Explain the effect on the position of equilibrium of increasing the concentration of hydrogen at a constant temperature and constant volume.

$$K_c = \frac{[HI(g)]^2}{[H_2(g)][I_2(g)]} \quad Q_c = \frac{[HI(g)]^2}{[H_2(g)][I_2(g)]}$$

If $[H_2(g)]$ is suddenly increased, then Q_c decreases and is no longer equal to K_c. In order for equilibrium to be re-established, the magnitude of the denominator in the expression for Q_c has to decrease in value. This will produce a subsequent increase in the value of the numerator. The two values will adjust until the value of Q_c becomes equal to that of K_c and a new equilibrium system is established.

The net result of an increase in $[H_2(g)]$ is that the equilibrium shifts to the right. This is in agreement with the qualitative prediction that we made in **Book 1**.

Effect of pressure

As with concentration, a change of pressure, at constant temperature, on an equilibrium system containing gases has no effect on the value of either K_c or K_p.

Once again, any change in the equilibrium position occurs in order to maintain the equilibrium constant at a constant value.

If the partial pressure of only *one* of the component gases is changed, then the overall effect on K_p can be predicted in the same way the effects of changes in concentration on K_c were described in the previous section.

However, if the *total* pressure of a gaseous system is suddenly increased or decreased, then the partial pressures of *all* the gases will either increase or decrease. The effect on the position of equilibrium can be explained using the equilibrium law. This is shown in the following example.

Consider the reaction:

$$N_2(g) + 3H_2(g) \rightleftharpoons 2NH_3(g)$$

for which:

$$K_p = \frac{(p_{NH_3})^2}{(p_{N_2})(p_{H_2})^3}$$

Suppose the equilibrium partial pressures of nitrogen, hydrogen and ammonia are a, b and c atm, respectively. Then:

$$K_p = \frac{c^2}{ab^3}$$

If the total pressure is suddenly doubled, the partial pressures of nitrogen, hydrogen and ammonia will be doubled to $2a$, $2b$ and $2c$ respectively:

$$Q_p = \frac{(2c)^2}{2a \times (2b)^3} = \frac{(2c)^2}{4ab^3}$$

The value of Q_p is now one-quarter of the value of K_p. In order to re-establish equilibrium the numerator has to increase, with a subsequent decrease in the denominator, until $Q_p = K_p$. This is achieved by some nitrogen and hydrogen reacting to form more ammonia.

Hence, an increase in pressure increases the equilibrium yield of ammonia by shifting the equilibrium to the right.

Effect of adding a catalyst

The expression for an equilibrium constant of a given reaction contains only those substances included in the overall stoichiometric equation for the reaction. A catalyst does *not* appear in the overall equation and so cannot influence the value of the equilibrium constant, and hence the equilibrium position.

We can think about this in another way. You will remember from **Book 1** that a catalyst increases the rate of both the forward reaction and the backward reaction to the same extent. It therefore increases the rate at which equilibrium is established, but has no effect on the final concentrations of reactants and products at equilibrium. Therefore, a catalyst has no effect on the value of the equilibrium constant for a reaction.

Questions

1. A dynamic equilibrium is set up between carbon monoxide, hydrogen and methanol. The equation for the reaction is:

    $CO(g) + 2H_2(g) \rightleftharpoons CH_3OH(g)$

    State the effect that an increase in pressure, at constant temperature, has on the value of K_c and also on the position of equilibrium. Justify your answers.

2. The reaction between ethanoic acid and ethanol is reversible. The equation for the reaction is:

    $CH_3COOH(l) + CH_3CH_2OH(l) \rightleftharpoons CH_3COOCH_2CH_3(l) + H_2O(l)$

    Some ethanoic acid and ethanol are mixed together, and some sulfuric acid is added to act as a catalyst. The mixture is left for several days to reach equilibrium.
    (a) Explain what happens to the composition of the equilibrium mixture if some more ethanol is added.
    (b) Explain what happens to the composition of the equilibrium mixture if some more sulfuric acid is added.

3. The reaction between hydrogen and iodine to form hydrogen iodide is reversible:

    $H_2(g) + I_2(g) \rightleftharpoons 2HI(g)$

    Some $H_2(g)$ is mixed with $I_2(g)$ and the mixture is allowed to reach equilibrium. At equilibrium, 0.14 mol of $H_2(g)$ is present.

    The mixture was compressed to reduce its volume and then left to reach equilibrium at the original temperature. Explain the change, if any, in the composition of the equilibrium mixture.

THINKING BIGGER

BALANCING BUBBLES

A chemical reaction reaches equilibrium when the rates of the forward and reverse reaction are equal so that the system has no further tendency to change. That may sound like a very abstract idea but let's consider a bottle of champagne: from the first pop of the cork to the subtle flavour of the wine on the palate, equilibria are involved at every stage.

WHY DOES CHAMPAGNE MAKE BUBBLES?

The gas responsible for bubble production is carbon dioxide, which is produced by yeast during the second fermentation in the sealed bottle. According to Henry's Law, equilibrium is established between carbon dioxide molecules dissolved in the liquid and carbon dioxide molecules in the vapour phase in the headspace under the cork. Before opening the bottle, the pressure of the carbon dioxide under the cork is about 6 atmospheres. The mass of dissolved carbon dioxide molecules in equilibrium is about 12 grams per litre of champagne. When the bottle is opened, the carbon dioxide pressure in the vapour phase suddenly drops, the thermodynamic equilibrium of the closed bottle is broken, and the liquid is super-saturated, which means that it now contains an excess of carbon dioxide molecules in comparison with the atmosphere outside the bottle. To recover a new and stable thermodynamic state corresponding to the partial pressure of carbon dioxide molecules in the atmosphere, almost all the carbon dioxide molecules dissolved into the champagne must escape.

If we assume that a classic champagne flute contains about 0.1 litre of champagne, we can estimate that approximately 0.7 litre of gaseous carbon dioxide must escape from it in order for equilibrium to be regained. To get an idea of how many bubbles this involves, we can divide the volume of the gaseous carbon dioxide in the flute (0.7 litre) by the volume of an average bubble (about 500 micrometres in diameter), and if we do so, we discover that a huge number of bubbles need to escape a flute of champagne before the liquid can reach equilibrium: almost 11 million bubbles – more than the population of New York City!

fig A The gas responsible for bubbles in champagne is carbon dioxide.

Where else will I encounter these themes?

Let us start by considering the nature of the writing in the article.

1. This article has been written with a general audience in mind rather than scientists. Go back through the article and highlight the incidents where scientific terms are used. How could a reader find out more information about these terms?

> Think of all of the sources of information that you might use to do research. Are they all equally reliable?

Now we will look at the chemistry in detail. Some of these questions will link to **Book 1**, so you may need to combine concepts from different areas of chemistry to work out the answers. Some questions will also link to topics covered later in this book. Don't worry if you are not ready to give answers to these questions yet. You may like to return to the questions later.

2. Consider the two equilibria shown below:

   $CO_2(g) \rightleftharpoons CO_2(aq)$

   $H_2O(l) + CO_2(aq) \rightleftharpoons HCO_3^-(aq) + H^+(aq)$

   $K_a = 4.5 \times 10^{-7}$ mol dm^{-3}

   a. How would you expect an increase in the partial pressure of carbon dioxide to affect both equilibria shown above?

   b. What happens to the position of both equilibria when the champagne cork is popped?

3. a. If 0.5 g of CO_2 is dissolved in 1.0 dm^3 of water, calculate the pH of this solution (use the K_a value given above) stating any assumptions made.

   b. The pressure inside a bottle of champagne can reach about 6 atm. How would this affect the pH of the champagne?

   c. With reference to the equations above suggest why it is better to open a bottle of champagne that is chilled rather than warm.

> It's often necessary to make assumptions in science in order to make sense of complex problems. In fact, assumptions and models are a very important part of the scientific method. The important thing is to be aware of the assumptions you are making and be able to justify why they are reasonable.

4. The concept of equilibrium is also an important one when it comes to the flavour and aroma of champagne. For example, ethanol can react with ethanoic acid to give an ester.

   a. Write an equation for the reaction between ethanol and ethanoic acid to give an ester and name the ester formed.

   b. The enthalpy change for this reaction is −3.5 kJ mol^{-1}. Explain why a change in temperature will have very little effect on the position of the equilibrium in question **4a**.

5. The structure shown below has the common name malic acid. Malic acid is one of the compounds that contributes to the acidity of champagne.

   a. Give the IUPAC name for malic acid.

   b. Explain what is meant by a chiral centre and identify the chiral centre in malic acid.

   c. How many esters could be formed between ethanol and malic acid? Explain your answer and draw one of the structures.

   fig B The structure of malic acid.

Activity

In many cases foods have additives called E numbers that are introduced for reasons of preservation, improving colour, flavour or texture. Some E numbers can be used to regulate pH. Such combinations include the following: E200/E202, E210/E211, E221/E222 and E280/E281. Choose one pair of E numbers and complete the following tasks:

- Identify the chemical compounds giving their structures.
- Find out how they are used in the food industry.
- Using appropriate equations, show how they can act as pH buffers. Is this the main reason why they are used?

Did you know?

It is surprising what can be added to food. For example, E173, E174 and E175 are the elemental forms of aluminium, silver and gold, respectively! India alone is reported to use about 11 tonnes of 'culinary' gold per year for use in religious feasts.

● From *Uncorked: The Science of Champagne Revised Edition* by Gérard Liger-Belair

11 Exam-style questions

1. Nitrogen and hydrogen react together to form ammonia in a reversible reaction that can reach a position of equilibrium. The equation for the reaction is:

   $N_2(g) + 3H_2(g) \rightleftharpoons 2NH_3(g)$ $\Delta_r H = -92 \text{ kJ mol}^{-1}$

   (a) Which change would affect **both** the value of the equilibrium constant, K_p, and the proportion of ammonia present in the equilibrium mixture? [1]

   A Adding a catalyst of finely divided iron.

   B Adding more nitrogen at the same temperature and volume.

   C Carrying out the reaction at a lower temperature at the same pressure and volume.

   D Carrying out the reaction at a higher pressure at the same temperature and volume.

   (b) At a fixed temperature and a total pressure of 5.00 atm, a sealed vessel of volume 20.0 dm³ contained 1.00 mol of $N_2(g)$, 2.00 mol of $H_2(g)$ and 1.00 mol of $NH_3(g)$.

   (i) Calculate the value, stating units, of the equilibrium constant, K_c, for this reaction. [5]

   (ii) Calculate the value, stating units, of the equilibrium constant, K_p, for this reaction. [5]

   [Total: 11]

2. When phosphorus(V) chloride, PCl_5, is heated to a constant temperature in a sealed vessel, it forms a gaseous equilibrium mixture with phosphorus(III) chloride, PCl_3, and chlorine. The equation for the reaction is:

   $PCl_5(g) \rightleftharpoons PCl_3(g) + Cl_2(g)$

   (a) The equilibrium constant may be expressed in terms of either concentration, K_c, or partial pressure, K_p.

   (i) Write an expression for K_p. [1]

   (ii) If the units of partial pressure are atm, state the units of K_p for the above equilibrium. [1]

   (b) Explain, in terms of the expression for K_p, the effect on the position of equilibrium of increasing the pressure at constant temperature. [4]

   (c) The relationship between K_c and K_p is given by the expression:

   $K_p = K_c (0.0821\ T)^{\Delta n}$

   where T = the absolute temperature in Kelvin (K)

   Δn = the change in number of moles from reactants to products

   Calculate a value for K_c if the value for K_p at 500 K is 0.810 atm. [3]

   [Total: 9]

3. Ethanol is manufactured in industry by the direct hydration of ethene in the presence of a phosphoric acid catalyst. The equation for the reaction is:

   $C_2H_4(g) + H_2O(g) \rightleftharpoons C_2H_5OH(g)$

   The table gives some details of various experiments carried out under different conditions in order to determine the percentage of ethane converted to ethanol at equilibrium.

Experiment	Molar ratio of ethene : steam	Temperature /K	Pressure /atm	Percentage of ethanol at equilibrium
1	1:1	573	50	32
2	1:2	573	50	40
3	1:3	573	50	50
4	1:2	573	60	46
5	1:2	573	70	55
6	1:2	523	50	42
7	1:2	623	50	38

   (a) Deduce, using the data in the table, how the position of equilibrium varies with changes in concentration of reactants, temperature and pressure. [6]

   (b) (i) Write an expression for the equilibrium constant, K_p, for this reaction. [1]

   (ii) Calculate a value for K_p under the conditions used in Experiment 2 in the table. State the units. [4]

   [Total: 11]

4. A chemist has discovered a method for making a commercially important chemical, **R**. The reaction involves the following reversible reaction:

   $2\mathbf{P}(g) + \mathbf{Q}(g) \rightleftharpoons \mathbf{R}(g)$

   The process is normally carried out in industry at a pressure of 500 atm and a temperature of 573 K.

   (a) (i) Write an expression for K_p for this reaction. [1]

   (ii) Give the units of K_p in this expression. [1]

   (b) During their initial research the chemist carried out several experiments, all at 500 atm pressure, using mixtures of **P** and **Q** starting with 2.0 mol of **P** and 1.0 mol of **Q**. In each case they allowed the reaction mixture to reach equilibrium before determining the percentage of **Q** converted.

The table shows the results of two of his experiments.

	Experiment 1	Experiment 2
Temperature/K	423	573
Percentage of **Q** converted	50	70

(i) Calculate, using the results from experiment 1, the value of K_p. [4]

(ii) Deduce the conditions of temperature and pressure that would give the highest yield of **R**. [2]

(iii) Explain why the conditions you have suggested in part (ii) may be different from those used industrially. [2]

[Total: 10]

5 Hydrogen and iodine react together, in the presence of a suitable catalyst, in a reversible reaction. The equation for the reaction is:

$H_2(g) + I_2(g) \rightleftharpoons 2HI(g)$ $\Delta H^\ominus = -9\,kJ\,mol^{-1}$

(a) The graph shows the changes in concentration when 2.0×10^{-3} mol of H_2 were mixed with 2.0×10^{-3} mol of I_2 in a sealed container of volume 1 dm³ and left to reach equilibrium.

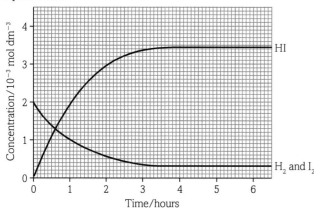

(i) Write the expression for the equilibrium constant, K_c, for this reaction. [1]

(ii) Use the graph to determine the time at which equilibrium was established. [1]

(iii) Use the graph to determine the equilibrium concentrations of HI(g), H_2(g) and I_2(g). [2]

(iv) Calculate a value for the equilibrium constant, K_c. State the units. [2]

(b) Explain how the value of K_c changes when:

(i) the total pressure at equilibrium is increased at constant temperature. [2]

(ii) the temperature is increased at constant pressure. [2]

(c) Which statement about the effect of a catalyst on a reversible reaction is correct? [1]

A It increases the value of the equilibrium constant of the reaction.

B It increases the equilibrium yield of product.

C It increases the rate of both the forward reaction and backward reaction.

D It increases the rate of the forward reaction, but not that of the backward reaction.

[Total: 11]

6 One of the stages in the contact process for the manufacture of sulfuric acid involves a reversible reaction between SO_2(g), O_2(g) and SO_3(g). The equation for this reaction is:

$2SO_2(g) + O_2(g) \rightleftharpoons 2SO_3(g)$

The equilibrium constant, K_p, for this reaction is given by the expression:

$$K_p = \frac{(p_{SO_3})^2}{(p_{SO_2})^2(p_{O_2})}$$

The table shows three values of K_p at different temperatures.

Temperature/K	K_p/atm^{-1}
298	4.0×10^{24}
500	2.5×10^{10}
700	3.0×10^{4}

(a) Use the data to comment on the percentage of SO_3(g) present in the equilibrium mixture at 298 K. [2]

(b) Use the data to deduce the sign of the enthalpy change for the forward reaction. [3]

(c) Explain, using the expression for K_p, why an increase in pressure increases the percentage of SO_3(g) present in the equilibrium mixture. [4]

[Total: 9]

TOPIC 12
Acid-base equilibria

Introduction

Acids and bases play an important part in our lives. They are of fundamental importance in industry, and life depends on the complex networks of acid-base reactions taking place inside living organisms. To understand life, and death, we need to understand acids and bases and the reactions they undergo. In this topic we shall build on the knowledge you have gained so far on the properties of acids and bases. In particular, we shall use the concept of equilibrium constants to look at their properties quantitatively. We shall also see how the strengths of acids are related to their molecular structures.

When your body is working normally, the pH of your blood plasma is kept fairly constant at around 7.4. You are likely to die if it rises or falls by more than 0.4 from this normal value. The pH of your blood plasma can fall as a result of disease or shock, both of which can generate acidic conditions in your body. You are also likely to die if the pH of your blood plasma rises; this could happen during the recovery from severe burns. To survive, your body must control its own pH. If your natural control systems fail, then medicine must come immediately to the rescue by, for example, intravenously administering electrolyte solutions.

In this topic we will look at how different ions affect pH and how they can be used to control it.

All the maths you need

- Recognise and make use of appropriate units in calculations
- Recognise and use expressions in decimal and ordinary form
- Use ratios, fractions and percentages
- Use calculators to find and use power, exponential and logarithmic functions
- Use an appropriate number of significant figures
- Change the subject of an equation
- Substitute numerical values into algebraic expressions using appropriate units for physical quantities
- Solve algebraic expressions
- Use logarithms in relation to quantities that range over several orders of magnitude

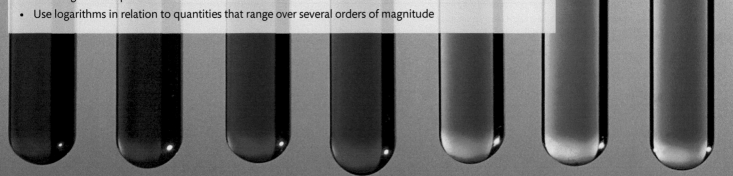

What have I studied before?
- Reactions of acids and bases
- A qualitative appreciation of the significance of pH of aqueous solutions
- Calculation of equilibrium constants based on concentrations (i.e. K_c)
- An understanding of the effect of changes of temperature on the value of equilibrium constants

What will I study later?
- The effect of pH change on some reactions of transition metal ions
- The effect of pH change on the equilibrium between an amino acid and its zwitterion
- Acid and base hydrolysis of organic compounds such as esters
- Acid catalysed reactions such as the iodination of propanone

What will I study in this topic?
- Acid–base reactions in terms of proton transfer
- The relationship between hydrogen ion concentration and pH
- How to calculate the pH of aqueous solutions
- The difference between strong and weak acids
- How to draw and interpret titration curves
- How to select a suitable indicator for an acid-base titration
- The concept of buffer solutions

12.1 1 The Brønsted–Lowry theory

By the end of this section, you should be able to...

- understand that a Brønsted-Lowry acid is a proton donor and a Brønsted-Lowry base is a proton acceptor
- recognise that acid-base reactions involve the transfer of protons
- identify Brønsted-Lowry conjugate acid-base pairs
- understand the difference between a strong and a weak acid in terms of degree of dissociation

Brønsted-Lowry acids and bases

In 1923 physical chemists Johannes Nicolaus Brønsted in Denmark and Thomas Martin Lowry in England independently proposed the theory that carries their names. In the Brønsted–Lowry theory acids and bases are defined by the way they react with each other.

fig A Johannes Nicolaus Brønsted.

fig B Thomas Martin Lowry.

They defined an acid as a substance that can donate a proton, i.e. a **proton donor** (hydrogen ion, H^+), and a base as a substance that can accept a proton, i.e. a **proton acceptor**.

We might expect that any substance containing hydrogen could act as an acid. In practice, a substance behaves as an acid only if the hydrogen carries a slight positive charge. As an example, this is the case when it is bonded to a highly electronegative atom to the right of the Periodic Table, e.g. oxygen or a halogen.

In order to accept a proton, a base has to contain a lone pair of electrons that it can use to form a dative covalent bond with the proton. So, a base must contain an atom to the right-hand side of the Periodic Table, and this is often oxygen.

Conjugate acid-base pairs

When hydrogen chloride dissolves in water an equilibrium is established that can be represented by the following equation:

$$HCl(aq) + H_2O(l) \rightleftharpoons H_3O^+(aq) + Cl^-(aq)$$

In the forward reaction:

- HCl is acting as an acid because it is donating a proton to H_2O
- H_2O is behaving as a base as it is accepting a proton from HCl.

In the reverse reaction:

- H_3O^+ is behaving as an acid because it is donating a proton to Cl^-
- Cl^- is behaving as a base because it is accepting a proton from H_3O^+.

When the acid HCl loses a proton it forms a base, Cl^-. These two species are called a **conjugate acid–base pair**.

When the acid H_3O^+ loses a proton it forms the base H_2O. These two species also form a conjugate acid–base pair.

So the equilibrium mixture above contains two conjugate acid-base pairs.

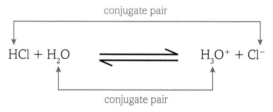

fig C Conjugate acid-base pairs formed from hydrochloric acid and water.

Cl^- is the **conjugate base** of HCl; H_2O is the conjugate base of H_3O^+.

Hydrochloric acid can donate *one* proton – it is called a *monoprotic* or monobasic acid.

Sulfuric acid can donate *two* protons and is therefore called a *diprotic* or dibasic acid.

Step 1: $H_2SO_4 \rightarrow H^+ + HSO_4^-$
Step 2: $HSO_4^- \rightarrow H^+ + SO_4^{2-}$

Similarly, some bases such as the carbonate ion can accept more than one proton:

Step 1: $CO_3^{2-} + H^+ \rightarrow HCO_3^-$
Step 2: $HCO_3^- + H^+ \rightarrow H_2CO_3$

The carbonate ion is therefore described as a *diprotic* or diacidic base.

Did you know?

- Acids that donate a maximum of one, two or three protons are called monoprotic, diprotic or triprotic, respectively. They are also called monobasic, dibasic or tribasic.
- Bases that can accept one, two or three protons are called monoprotic, diprotic or triprotic respectively. They are also called monoacidic, diacidic or triacidic.

We will use the terms monobasic, dibasic etc. for acids throughout the book.

Another example of a conjugate acid–base pair

When ammonia dissolves in water an equilibrium is established that can be represented by the following equation:

$$NH_3(aq) + H_2O(l) \rightleftharpoons NH_4^+(aq) + OH^-(aq)$$

In the forward reaction:

- NH_3 is acting as a base because it is accepting a proton from H_2O
- H_2O is acting as an acid because it is donating a proton to NH_3.

In the reverse reaction:

- NH_4^+ is behaving as an acid because it is donating a proton to OH^-
- OH^- is behaving a base because it is accepting a proton from NH_4^+.

So the equilibrium contains two acid–base conjugate pairs:

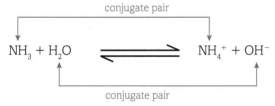

fig D Conjugate acid–base pairs formed from ammonia and water.

Amphoteric substances

In its reaction with HCl, H_2O behaves as a base by accepting a proton. However, in its reaction with NH_3, H_2O behaves as an acid by donating a proton.

A substance that can act as either an acid or a base is described as being **amphoteric**.

Did you know?

You may also come across the term amphiprotic. An amphiprotic substance is one that can both donate and accept protons. So water is described as being amphiprotic. Other examples of amphiprotic substances are amino acids and the hydrogensulfate ion, HSO_4^-.

All amphiprotic substances are also amphoteric, but the reverse is not true. There are amphoteric substances, like aluminium oxide, that do not donate or accept protons when they act as acids and bases respectively. These substances fall into the category of acting as acids and bases according to the Lewis theory, which states that an acid is an electron pair acceptor and a base is an electron pair donor.

The advantage of the Lewis theory is that it shows the similarity between acid–base reactions that involve proton transfer and other acid–base reactions that do not. It is a useful way of explaining why some reactions in organic chemistry that are catalysed by acids can also be catalysed by some substances that can accept a pair of electrons. For example, the nitration of benzene is catalysed by sulfuric acid, but the reaction also takes place if nitric acid is used in conjunction with boron trifluoride, BF_3.

Neither the term 'amphiprotic' nor the Lewis theory of acids and bases is required learning for your A level course.

Nitric acid as a base

The most common method to nitrate benzene (see **Section 17.4.2**) is to use the 'nitrating mixture' of concentrated nitric acid and concentrated sulfuric acid. This mixture forms the equilibrium:

$$H_2SO_4 + HNO_3 \rightleftharpoons HSO_4^- + H_2NO_3^+$$
Acid 1   Base 2   Base 1   Acid 2

In this reaction:

- H_2SO_4 is an acid; its conjugate base is HSO_4^-
- HNO_3 is a base; its **conjugate acid** is $H_2NO_3^+$.

It seems strange to refer to nitric acid as a base, but in this reaction that is exactly what it is behaving as.

Strong and weak acids

A strong acid is defined as one that is almost completely dissociated in aqueous solution.

In a dilute solution of hydrochloric acid, virtually all the hydrogen chloride molecules are dissociated. We usually represent this by using a single direction arrow in the equation for the dissociation:

$$HCl(aq) \rightarrow H^+(aq) + Cl^-(aq)$$

Learning tip

From now on we are going to use $H^+(aq)$ to represent the acid protons in aqueous solution, rather than H_3O^+. It is debatable how many water molecules are bonded to a given hydrogen ion. The formula could just as easily be $H_5O_2^+$ or $H_7O_3^+$. For this reason, $H^+(aq)$ is possibly the best representation of acid protons in aqueous solution. It makes no difference to your understanding of the concepts involved, whichever formula you use.

Learning tip

The conversion of acid molecules into ions when an acid dissolves in water is also called ionisation, and you may see this term used in other books.

Did you know?

Although, strictly speaking, an equilibrium exists between HCl molecules, H^+ ions and Cl^- ions, the acid is considered to be completely dissociated in dilute solution. So strong acids are often said to be 100% dissociated.

By contrast, a weak acid is defined as one that is only partially dissociated (often less than 10%) in aqueous solution. Organic acids such as ethanoic acid are typically weak acids.

We represent partial dissociation by using the reversible arrow sign ($\rightleftharpoons$) in the equation for the dissociation:

$$CH_3COOH(aq) \rightleftharpoons CH_3COO^-(aq) + H^+(aq)$$

Questions

1. Give the formula of the conjugate acid of each of the following species:
   (a) CH_3COO^-
   (b) CH_3NH_2
   (c) HSO_4^-

2. Give the formula of the conjugate base of each of the following species:
   (a) $HClO_4$
   (b) H_3O^+
   (c) HSO_4^-

3. In each of the following examples, identify the two conjugate acid–base pairs. In each case identify the species that are acting as a Brønsted–Lowry acid.
   (a) $H_2CO_3 + H_2O \rightleftharpoons HCO_3^- + H_3O^+$
   (b) $HCO_3^- + H_2O \rightleftharpoons CO_3^{2-} + H_3O^+$
   (c) $CH_3COOH + HNO_3 \rightleftharpoons CH_3COOH_2^+ + NO_3^-$

4. Explain why the following reaction may be described as an acid–base reaction:
   $$NH_4^+ + NH_2^- \rightarrow 2NH_3$$

Key definitions

An acid is a **proton donor**.
A base is a **proton acceptor**.
A **conjugate acid–base pair** consists of either a base and its conjugate acid or an acid and its conjugate base.
When a base accepts a proton, the species formed is the **conjugate acid** of the base.
When an acid donates a proton, the species formed is the **conjugate base** of the acid.
An **amphoteric** substance is one that can act both as an acid and as a base.

12.1 — 2 — Hydrogen ion concentration and the pH scale

By the end of this section, you should be able to...
- define the term pH
- calculate pH from hydrogen ion concentration, and vice versa
- calculate the pH of an aqueous solution of a strong acid
- deduce the expression for the acid dissociation constant, K_a, for a weak acid and carry out relevant calculations
- calculate the pH of a weak acid making relevant assumptions
- define the term pK_a

Hydrogen ion concentration and pH

Strong acids

As already mentioned, strong acids are assumed to be dissociated completely when they are dissolved in water. This means that the hydrogen ion concentration is related directly to the concentration of the acid.

For example, a solution of HCl of concentration $0.100 \text{ mol dm}^{-3}$ will produce a hydrogen ion concentration of $0.100 \text{ mol dm}^{-3}$.

The **pH** of an aqueous solution is related to the hydrogen ion concentration by the following equation:

$$pH = -\lg [H^+] \quad \text{or} \quad pH = \lg \frac{1}{[H^+]}$$

The hydrogen ion concentration, $[H^+]$, is measured in mol dm^{-3}.

Which equation you decide to use is purely a matter of personal preference.

Did you know?

Logarithms can only be taken of a number, not a quantity with a unit. So, strictly speaking, the hydrogen ion concentration has to be divided by the standard concentration, $c^\ominus$, which has a value of 1 mol dm^{-3}.

Hence, the correct expression for calculating pH is $-\lg [H^+]/c^\ominus$. This, however, is not something you need to consider at this level.

Learning tip

Do not worry if you are unfamiliar with the use of logarithms. They are merely a way to convert a scale of numbers in powers of 10 to a linear scale. For example:

$\lg 100 = 2$ (as 100 is 10^2)
$\lg 10 = 1$ (as 10 is 10^1)
$\lg 1 = 0$ (as 1 is 10^0)
$\lg 0.01 = -2$ (as 0.01 is 10^{-2})

The lg of a number can easily be found by using the 'lg' or 'log' button on your calculator.

Although the accepted abbreviation for logarithm to the base 10 is lg, you can also use $\log_{10}$ or even log.

WORKED EXAMPLES

1. Calculate the pH of aqueous solutions of the following monobasic, strong acids. In each case assume the acid is completed dissociated. Give your answers to two decimal places.
   (a) $0.00100 \text{ mol dm}^{-3}$ HCl
   (b) $0.0500 \text{ mol dm}^{-3}$ HNO_3
   (c) $0.150 \text{ mol dm}^{-3}$ HBr

   Answer
   (a) $pH = -\lg [H^+]$
       $= -\lg (0.00100)$
       $\lg (0.00100) = -3$
       so, $-\lg (0.00100) = +3.00$
       $pH = 3.00$
   (b) $pH = -\lg (0.0500)$
       $\lg (0.0500) = -1.30$
       $pH = 1.30$
   (c) $pH = -\lg (0.150)$
       $\lg (0.150) = -0.82$
       $pH = 0.82$

2. Calculate the pH of an aqueous solution of $10.00 \text{ mol dm}^{-3}$ of HCl. In this solution the HCl is 55% dissociated. Give your answer to two decimal places.

   Answer
   $[H^+] = 0.55 \times 10.00 \text{ mol dm}^{-3} = 5.50 \text{ mol dm}^{-3}$
   $pH = -\lg (5.50)$
   $\lg (5.50) = 0.74$
   so, $pH = -0.74$

Did you know?

pH and activity

The measured pH of a concentrated acid is never as low as the calculated value, even though the acid may be completely dissociated. This is because the ions that are close together in solution interact with one another. This makes their effective concentration less than the actual concentration. This effective concentration is called the activity. For example, a solution of HCl of 1.00 mol dm^{-3} has an effective hydrogen ion concentration of 0.81 mol dm^{-3}. This makes the pH of this solution 0.09, not 0.00, as calculated from a hydrogen ion concentration of 1.00 mol dm^{-3}.

Only at concentrations below 0.10 mol dm^{-3} do activities and concentrations have similar values.

This activity effect means that the minimum pH and maximum pH that concentrated solutions can have are 0.30 and 14.30, respectively. This activity effect can be ignored at A level, hence the calculation of -0.74 above.

Calculating hydrogen ion concentration from pH

It is a simple matter to calculate the hydrogen ion concentration from a given pH value.

The following equation should be used:

$$[H^+(aq)] = 10^{-pH}$$

WORKED EXAMPLE

Calculate the hydrogen ion concentration of a solution with a pH of 4.8:

$$[H^+(aq)] = 10^{-4.8}$$
$$= 1.58 \times 10^{-5} \text{ mol dm}^{-3}$$

The key to solving this calculation is to press the 10^X or $10^\blacksquare$ button on your calculator and then entering the negative pH value, followed by the equals button. The answer will then be displayed.

Weak acids

Determining the hydrogen ion concentration of an aqueous solution of a weak acid is more complicated because a significant amount of undissociated acid is present in solution. It is necessary to refer to the acid dissociation constant, K_a, for the acid.

Acid dissociation constant

If we use HA to represent a weak acid, then the equation for its dissociation in aqueous solution is:

$$HA(aq) \rightleftharpoons H^+(aq) + A^-(aq)$$

By applying the equilibrium law to this reaction we obtain:

$$\frac{[H^+(aq)][A^-(aq)]}{[HA(aq)]} = \text{a constant}$$

This constant is called the acid dissociation constant and is given the symbol K_a.

We shall now calculate the hydrogen ion concentration of an aqueous solution of ethanoic acid of concentration $0.0500 \text{ mol dm}^{-3}$. The value of K_a for ethanoic acid is $1.74 \times 10^{-5} \text{ mol dm}^{-3}$ at 298 K.

$$CH_3COOH(aq) \rightleftharpoons CH_3COO^-(aq) + H^+(aq)$$

$$K_a = \frac{[CH_3COO^-(aq)][H^+(aq)]}{[CH_3COOH(aq)]}$$

Every time a molecule of CH_3COOH dissociates, a CH_3COO^- ion and a H^+ ion are formed.

This means that $[CH_3COO^-(aq)] = [H^+(aq)]$.

So, the expression for K_a can be simplified to:

$$K_a = \frac{[H^+(aq)]^2}{[CH_3COOH(aq)]}$$

It is important to recognise at this stage that the concentrations in the expression for K_a are the *equilibrium* concentrations. However, if the value of K_a is very small (as it is in this case) then the concentration of the undissociated acid at equilibrium is very similar to the initial concentration of the acid. It is therefore reasonable to take the concentration at equilibrium as being the same as the initial concentration.

Therefore,

$$K_a = \frac{[H^+(aq)]^2}{0.0500} = 1.74 \times 10^{-5} \text{ mol dm}^{-3}$$

$$[H^+(aq)] = (0.0500 \times 1.74 \times 10^{-5})^{1/2} = 9.33 \times 10^{-4} \text{ mol dm}^{-3}$$

This gives a pH of 3.03.

Learning tip

If you are concerned that the approximation we made in the calculation is not justified, then it is possible to solve the precise equation:

$$K_a = [H^+(aq)]^2/(0.0500 - [H^+(aq)]) = 1.74 \times 10^{-5} \text{ mol dm}^{-3}$$

The solution to this equation gives a value for $[H^+(aq)]$ of $9.327 \times 10^{-4} \text{ mol dm}^{-3}$

The difference is significant only for very accurate work.

There is also one other approximation that we made in both the original calculation and this calculation. As we shall see in **Section 12.1.3**, water ionises very slightly to form hydrogen ions and hydroxide ions. We have ignored the contribution to the total hydrogen ion concentration from the dissociation of water. However, unless the acid is *very* dilute, this approximation is also justified.

If you are good at remembering equations, then the following equation can be used to calculate the hydrogen ion concentration of an aqueous solution of a weak monobasic acid:

$$[H^+(aq)] = \sqrt{(K_a \times [\text{acid}])}$$

where K_a is the dissociation constant for the weak acid.

K_a and pK_a values

Table A (next page) lists some weak monobasic organic acids together with their K_a and **pK_a** values at 298 K, where:

$$pK_a = -\lg K_a$$

You will notice that the *larger* the value of K_a the stronger the acid. By contrast, the *smaller* the value of pK_a the stronger the acid.

Calculating the pH of a dibasic acid

Sulfuric acid is the most common dibasic acid. It dissociates in two stages:

$$H_2SO_4(aq) \rightarrow H^+(aq) + HSO_4^-(aq)$$
$$HSO_4^-(aq) \rightleftharpoons H^+(aq) + SO_4^{2-}(aq)$$

H_2SO_4 is a strong acid and is therefore fully dissociated. HSO_4^- is a weak acid ($K_a = 0.0100 \text{ mol dm}^{-3}$.)

In a $0.500 \text{ mol dm}^{-3}$ aqueous solution of H_2SO_4 the contribution to the $[H^+(aq)]$ from the H_2SO_4 will be $0.500 \text{ mol dm}^{-3}$. If we assume that the contribution to the $[H^+(aq)]$ from the HSO_4^- ion is $x \text{ mol dm}^{-3}$, then we have the following relationship:

$$K_a (HSO_4^-) = 0.0100 = (0.500 - x)x/0.500$$

Solving this quadratic equation gives $x = 0.0098$.

This gives a total $[H^+(aq)]$ of $0.05098 \text{ mol dm}^{-3}$ with a subsequent pH of 0.293.

Strong and weak acids | 12.1

Name of acid	Formula of acid	K_a/mol dm^{-3}	pK_a
propanoic acid	CH_3CH_2COOH	1.35×10^{-5}	4.87
ethanoic acid	CH_3COOH	1.74×10^{-5}	4.76
benzoic acid	C_6H_5COOH	6.31×10^{-5}	4.20
methanoic acid	$HCOOH$	1.60×10^{-4}	3.80
chloroethanoic acid	$CH_2ClCOOH$	1.38×10^{-3}	2.86
dichloroethanoic acid	$CHCl_2COOH$	5.13×10^{-2}	1.29
trichloroethanoic acid	CCl_3COOH	2.24×10^{-1}	0.65

table A

↓ increasing acid strength

It is interesting to perform this calculation in order to recognise that the contribution to the [H^+(aq)] from the HSO_4^- ions is negligible. This is because its dissociation is significantly reduced owing to the high [H^+(aq)] from the full first ionisation of the H_2SO_4.

A second interesting reason for performing this calculation is because many books state that 0.5 mol dm^{-3} H_2SO_4(aq) can be used as the acid solution in a standard hydrogen electrode (see **Section 14.1.1** on standard electrode potentials). As the standard hydrogen electrode requires a [H^+(aq)] of 1.00 mol dm^{-3} (pH = 0.00), this is clearly incorrect.

Did you know?

It is interesting to note that the chloro-substituted ethanoic acids are stronger than ethanoic acid itself, and that the acid strength increases with the number of chlorine atoms present. This will be explained in Section **13.3.3**.

Questions

1 Calculate the pH of each of the following aqueous solutions of strong monobasic acids. Assume the acid is fully dissociated in each case. Give your answers to two decimal places.
   (a) 0.0100 mol dm^{-3} HI
   (b) 0.500 mol dm^{-3} HNO$_3$
   (c) 0.00405 mol dm^{-3} HCl

2 Calculate the pH of a mixture of 20.0 cm^3 of 1.00 mol dm^{-3} HCl(aq) and 5 cm^3 of 1.00 mol dm^{-3} NaOH(aq).

3 Calculate the pH of each of the following aqueous solutions of weak monobasic acids. Give your answers to two decimal places.
   (a) 0.100 mol dm^{-3} HCOOH [K_a(HCOOH) = 1.60 × 10^{-4} mol dm^{-3}]
   (b) 1.00 mol dm^{-3} HF [K_a(HF) = 5.62 × 10^{-4} mol dm^{-3}]
   (c) 0.505 mol dm^{-3} NH$_4$Cl [K_a(NH$_4^+$) = 5.62 × 10^{-10} mol dm^{-3}]

4 The pH of an aqueous solution of a weak acid, HA, of concentration 0.0305 mol dm^{-3} is 4.97. Calculate the dissociation constant, K_a, for this weak acid.

Key definition

The **pH** of an aqueous solution is defined as the reciprocal of the logarithm to the base 10 of the hydrogen ion concentration measured in moles per cubic decimetre, pH = –lg [H^+]. This definition is difficult to remember, so either of the two equations given on page 27 can be used to define pH.

pK_a = –lg K_a

12.1　3　Ionic product of water, K_w

By the end of this section, you should be able to...
- define the ionic product of water, K_w
- define the term pK_w
- calculate the pH of an aqueous solution of a strong base using K_w

Dissociation of water

Pure water has a slight electrical conductivity, so it must contain some ions. It self-ionises according to the following equation:

$$H_2O(l) \rightleftharpoons H^+(aq) + OH^-(aq)$$

If we apply the equilibrium law to this reaction we obtain:

$$\frac{[H^+(aq)][OH^-(aq)]}{[H_2O(l)]} = \text{a constant}$$

As $[H_2O(l)]$ is constant at a given temperature, the expression may be simplified to:

$$[H^+(aq)][OH^-(aq)] = \text{a constant}$$

This constant is called the ionic product of water and is given the symbol K_w.

The value of K_w at 298 K is $1.00 \times 10^{-14}\,mol^2\,dm^{-6}$.

A neutral solution is defined as one in which the hydrogen ion concentration is equal to the hydroxide ion concentration. This is the case for pure water.

At 298 K, $[H^+(aq)] = 1.00 \times 10^{-7}\,mol\,dm^{-3}$ [$(1.00 \times 10^{-14})^{½}$]

So, the pH of pure water at 298 K is 7.00 [$-lg(1.00 \times 10^{-7})$]

K_w and pK_w

The relationship between K_w and pK_w is given by the following equation:

$$pK_w = -lg\,K_w$$

At 298 K, when $K_w = 1.00 \times 10^{-14}\,mol^2\,dm^{-6}$, $pK_w = 14.00$

Table A gives the values of K_w and pK_w at various temperatures.

Temperature/K	273	283	293	303	313
$K_w/mol^2\,dm^{-6}$	1.14×10^{-15}	2.93×10^{-15}	6.81×10^{-15}	1.47×10^{-14}	2.92×10^{-14}
pK_w	14.94	14.53	14.17	13.83	13.53

table A

> **Learning tip**
>
> The pH of a neutral solution is often quoted a being 7.00. However, this is true only for a solution that has a temperature of 298 K.
>
> As with all equilibrium constants, K_w varies with temperature.
>
> At 288 K it has a value of $4.52 \times 10^{-15}\,mol^2\,dm^{-6}$.
>
> At this temperature, $[H^+(aq)] = 6.72 \times 10^{-8}\,mol\,dm^{-3}$, giving a pH of 7.17.
>
> So, at 288 K a neutral solution has a pH of 7.17.
>
> Similarly it can be shown that at 308 K, the pH of a neutral solution is 6.84.

> **Learning tip**
>
> The equation should strictly be $pK_w = -lg\,K_w/(c^\ominus)^2$ in order to be able to take the logarithm of a dimensionless quantity. However we will use the simplified version of the equation.

pH of aqueous solutions of strong bases

When an acid is dissolved in water, it produces so many hydrogen ions that the small contribution from the water is insignificant, unless the acid concentration is very small.

However, the fact that water ionises is the reason why even the most alkaline solutions contain some hydrogen ions.

Sodium hydroxide is a strong base, so in dilute aqueous solutions we can consider its ions to be completely dissociated.

A sodium hydroxide solution of concentration $0.100 \text{ mol dm}^{-3}$ therefore has a hydroxide ion concentration of $0.100 \text{ mol dm}^{-3}$.

If $[OH^-(aq)] = 0.100 \text{ mol dm}^{-3}$ and $[H^+(aq)][OH^-(aq)] = 1.00 \times 10^{-14} \text{ mol}^2 \text{ dm}^{-6}$, then

$[H^+(aq)] = 1.00 \times 10^{-14}/0.100 \text{ mol dm}^{-3}$

$= 1.00 \times 10^{-13} \text{ mol dm}^{-3}$

The pH of this solution is therefore 13.0.

WORKED EXAMPLE

Calculate the pH, at 298 K, of an aqueous solution of potassium hydroxide of concentration $0.0200 \text{ mol dm}^{-3}$.

K_w (298 K) $= 1.00 \times 10^{-14} \text{ mol}^2 \text{ dm}^{-6}$

Answer

As potassium hydroxide is a strong base we may assume that its ions are completely dissociated.

$[OH^-(aq)] = 0.0200 \text{ mol dm}^{-3}$

So, $[H^+(aq)] = 1.00 \times 10^{-14}/0.0200 = 5.00 \times 10^{-13} \text{ mol dm}^{-3}$

$pH = -\lg (5.00 \times 10^{-13}) = 12.3$

Learning tip

If you are not happy working with logarithms of small numbers, you can use the logarithmic form of the equation $[H^+(aq)][OH^-(aq)] = 1.00 \times 10^{-14}$, which is $pH + pOH = 14.0$, where $pOH = -\lg [OH^-(aq)]$.

So, in the worked example above:

$pH = 14.0 - pOH = 14.0 - (-\lg (0.0200)) = (14.0 - 1.70) = 12.3$

Questions

1. The ionic product of water, K_w, has a value of $1.00 \times 10^{-14} \text{ mol}^2 \text{ dm}^{-6}$ at 298 K and a value of $6.81 \times 10^{-14} \text{ mol}^2 \text{ dm}^{-6}$ at 293 K. Use this information, where relevant, to answer the following questions.
   (a) Calculate the pH of water at (i) 298 K and (ii) 293 K.
   (b) Even though pure water at 298 K and at 293 K has different pH values, both samples of water are said to be neutral. Explain why.
   (c) Is the following reaction exothermic or endothermic?

   $H_2O(l) \rightarrow H^+(aq) + OH^-(aq)$

   Explain how you arrived at your answer.

2. Calculate the pH at 298 K of each of the following aqueous solutions of strong bases. Assume the base is fully dissociated in each case. Give your answers to two decimal places.
   (a) $0.0100 \text{ mol dm}^{-3}$ NaOH
   (b) $0.0500 \text{ mol dm}^{-3}$ Ca(OH)$_2$
   (c) $0.0315 \text{ mol dm}^{-3}$ KOH

   [$K_w = 1.00 \times 10^{-14} \text{ mol}^2 \text{ dm}^{-6}$ at 298 K]

Key definition

$K_w = [H^+(aq)][OH^-(aq)]$

12.1 | 4 | Analysing data from pH measurements

By the end of this section, you should be able to...

- analyse data from the following experiments:
  - (i) measuring the pH of equimolar aqueous solutions of strong and weak acids
  - (ii) measuring the pH of equimolar aqueous solutions of strong and weak bases
  - (iii) measuring the pH of equimolar aqueous solutions of various salts
  - (iv) comparing the pH of aqueous solutions of strong and weak acids after dilution
- calculate K_a for a weak acid from experimental data given the pH of an aqueous solution containing a known mass of acid

Comparing solutions through pH measurement

Strong and weak acids

The relative strengths of different acids can be determined by measuring the pH of equimolar aqueous solutions of the acids, at the same temperature.

Table A shows the pH of $0.100\,mol\,dm^{-3}$ aqueous solutions of various acids at 298 K.

Formula of acid	HCl	$CHCl_2COOH$	$CH_2ClCOOH$	HCOOH	CH_3COOH	CH_3CH_2COOH
pH of $0.100\,mol\,dm^{-3}$ aqueous solution	1.00	1.14	1.93	2.38	2.87	2.93

→ decreasing acid strength

table A

The higher the value of the pH, the weaker the acid.

Strong and weak bases

The same method can be used to determine the relative strengths of bases.

Table B shows the pH of $0.100\,mol\,dm^{-3}$ aqueous solutions of various bases at 298 K.

Formula of base	NH_3	CH_3NH_2	$(CH_3)_2NH$	$CH_3CH_2NH_2$	$CH_3CH_2CH_2NH_2$	NaOH
pH of $0.100\,mol\,dm^{-3}$ aqueous solution	11.13	11.82	11.83	11.84	11.86	13.00

→ increasing base strength

table B

The higher the value of the pH, the stronger the base.

Salts

Table C shows the pH of aqueous solutions of various salts of concentration $0.100\,mol\,dm^{-3}$ at 298 K.

Formula of salt	NaCl	KNO_3	CH_3COONa	NH_4Cl	CH_3COONH_4
pH of aqueous solution	7.00	7.00	8.88	5.13	7.00

table C

The pH of NaCl is 7.00 because the salt is made from a strong acid (HCl) and a strong base (NaOH).

The same is true for KNO_3, which is a product of the strong acid HNO_3, and the strong base KOH.

An aqueous solution of CH_3COONa is alkaline because it is a product of a weak acid (CH_3COOH) and a strong base (NaOH).

An aqueous solution of NH_4Cl is acidic because it is a product of a strong acid (HCl) and a weak base (NH_3).

An aqueous solution of CH_3COONH_4 is neutral (pH = 7.00) because it is a product of a weak acid (CH_3COOH) and a weak base (NH_3), and the relative strengths of the acid and base are the same. This is shown by their dissociation constants.

CH_3COOH: $K_a = 1.74 \times 10^{-5}$ mol dm^{-3}

NH_3: $K_b = 1.74 \times 10^{-5}$ mol dm^{-3}

Learning tip
K_a is a measure of acid strength and K_b is a measure of base strength.

Summary table for aqueous solutions at 298 K

salt of a strong acid and a strong base	pH = 7 (solution is neutral)
salt of a weak acid and a strong base	pH > 7 (solution is alkaline)
salt of a strong acid and a weak base	pH < 7 (solution is acidic)
salt of a weak acid and a weak base	pH depends on relative strength of acid and base; if $K_a = K_b$, pH = 7; if $K_a > K_b$, pH < 7; if $K_a < K_b$, pH > 7

table D

Effect of dilution on the pH of aqueous solutions of acids

Strong acids

Table E shows the pH of five aqueous solutions of hydrochloric acid. In each case the acid has been diluted by a factor of ten from 1.00 (1.00×10^0) to 0.000100 (1.00×10^{-4}) mol dm^{-3}. All solutions have a temperature of 298 K.

Concentration/mol dm^{-3}	1.00 (1.00×10^0)	0.100 (1.00×10^{-1})	0.0100 (1.00×10^{-2})	0.00100 (1.00×10^{-3})	0.00100 (1.00×10^{-4})
pH	0.00	1.00	2.00	3.00	4.00

table E

Did you know?
The reason why some salts form alkaline or acidic solutions is that they undergo hydrolysis.

To learn more about this, research 'salt hydrolysis'.

It is important to recognise that different salts have different pH values because this will explain why, for example, the pH of the solution formed when a strong acid reacts with an equivalent amount of a weak base is less than 7. We shall meet this concept when we look at acid–base titration curves in **Section 12.2.1**.

You will notice that the pH increases by a factor of one unit for each 10-fold decrease in concentration.

If we follow this to its logical conclusion, then an aqueous solution of hydrochloric acid of concentration 1.00×10^{-8} mol dm^{-3} should have a pH of 8.00. This is clearly nonsense because this would mean that a solution of an acid was alkaline.

Earlier in this topic we mentioned the fact that the contribution to the hydrogen ion concentration from the dissociation of water can usually be ignored. When solutions are as dilute as 10^{-8} mol dm^{-3} this is no longer the case. The pH of 10^{-8} mol dm^{-3} hydrochloric acid is very close to 7, as the contribution to the concentration of hydrogen ions from the water (10^{-7} mol dm^{-3}) is now greater than that of the acid.

Weak acids

Table F shows the pH of five aqueous solutions of ethanoic acid (CH_3COOH). In each case the acid has been diluted by a factor of ten from 1.00 (1.00×10^0) to 0.000100 (1.00×10^{-4}) mol dm^{-3}. Once again, all solutions have a temperature of 298 K.

Concentration/mol dm^{-3}	1.00 (1.00×10^0)	0.100 (1.00×10^{-1})	0.0100 (1.00×10^{-2})	0.00100 (1.00×10^{-3})	0.000100 (1.00×10^{-4})
pH	2.38	2.88	3.38	3.88	4.38

table F

With a weak acid the pH value increases by a factor of about 0.50 for each 10-fold decrease in concentration.

12.1 4

Determining K_a of a weak acid from experimental data

The following experiment can be performed to determine K_a of benzoic acid (C_6H_5COOH) – a weak monobasic acid.

- Accurately weigh between 0.40 and 0.50 g of benzoic acid and then dissolve it in a small volume (say 50 cm³) of deionised water contained in a beaker. (Benzoic acid is not very soluble, so it may be necessary to warm the water to get it to dissolve. If so, allow the solution to cool before performing the next stage.)
- Transfer the solution to a 250 cm³ volumetric flask. Add several washings from the beaker using deionised water, and then make up to the mark with deionised water.
- Mix the solution by inverting the flask several times.
- Withdraw a sample of the solution and place it in a small beaker.
- Using a calibrated pH meter, measure the pH of the solution.

WORKED EXAMPLE

Sample results

Mass of benzoic acid = 0.49 g
pH of solution = 3.00

Analysis of results

$C_6H_5COOH(aq) \rightleftharpoons C_6H_5COO^-(aq) + H^+(aq)$
molar mass of benzoic acid (C_6H_5COOH) = 122 g mol⁻¹
$n(C_6H_5COOH)$ in 250 cm³ of solution = 0.49/122 mol
so, $[C_6H_5COOH(aq)] = (0.49/122) \times 4$ mol dm⁻³
pH = 3.00
so, $[H^+(aq)] = 10^{-3.00} = [C_6H_5COO^-(aq)]$
$K_a = (10^{-3.00})^2 \div ((0.49/122) \times 4)$
 = 6.22 × 10⁻⁵ mol dm⁻³
The Data Book value is 6.32 × 10⁻⁵ mol dm⁻³ at 298 K.

The result obtained is close to the accepted value, but that does not necessarily mean that we have performed an accurate experiment. There may have been a number of errors that have, by chance, cancelled out one another.

To start with, we have assumed that the $[C_6H_5COOH(aq)]$ at equilibrium is identical to the original concentration of benzoic acid. Obviously this is not correct because it must have dissociated slightly to produce a solution of pH = 3.00.

The major uncertainty in this experiment is the measurement of the pH value. Even a small error leads to a large discrepancy in the final answer. A pH of 3.10 would give a final answer of 3.92 × 10⁻⁵, whereas a value of 2.90 would give 9.87 × 10⁻⁵ for the value of K_a.

Transfer errors may also be significant when weighing out small amounts.

Questions

1. Predict whether aqueous solutions of the following salts will be neutral, acidic or alkaline. Justify your answers.
   (a) Ammonium nitrate, NH_4NO_3
   (b) Potassium propanoate, CH_3CH_2COOK
   (c) Sodium nitrate, $NaNO_3$

2. What information is required in order to make a prediction about the pH of an aqueous solution of ammonium methanoate, $HCOONH_4$?

3. Calculate K_a for chloroethanoic acid from the following data.
   1.89 g of chloroethanoic acid was dissolved in 50 cm³ of water and the solution was diluted to 250 cm³ in a volumetric flask. The pH of this solution was 1.99.

12.2 1 Acid–base titrations, pH curves and indicators

By the end of this section, you should be able to...

- draw and interpret titration curves using any combination of strong and weak monobasic acids and monoacidic bases
- select a suitable indicator, using a titration curve and suitable data

Acid–base titrations

End point and equivalence point

When you carry out a simple acid–base titration, you usually use an indicator to tell you when the acid and base are mixed in exactly the right proportions to react in equivalent amounts, as dictated by the stoichiometric equation. When the indicator changes colour, this is often described as the 'end point' of the titration.

The 'equivalence point' is when the acid and base have reacted together in the exact proportions as dictated by the stoichiometric equation. When titrating an aqueous solution of a monobasic acid with an aqueous solution of a monoacidic base of the same concentration, 25 cm³ of acid will react exactly with 25 cm³ of base.

The pH at the equivalence point depends on the combination of acid and base used. For example, if you are titrating aqueous sodium hydroxide with dilute hydrochloric acid then the pH at the equivalence point is 7.00 (at 298 K), as both the base and the acid are strong. The solution at equivalence point will contain the salt sodium chloride.

If, however, ethanoic acid (a weak acid) is titrated against sodium hydroxide (a strong base) the solution at the equivalence point will contain the salt sodium ethanoate, and the pH will be greater than 7 (see **Section 12.1.4**).

If hydrochloric acid (a strong acid) is titrated against aqueous ammonia (a weak base), the solution at equivalence point will contain the salt ammonium chloride, and the pH will be less than 7.

Learning tip

The term 'neutralisation point' should *not* be used to describe the point at which the acid and base have reacted in the exact proportions as dictated by the stoichiometric equation. As seen by the examples listed, the pH of the solution formed is not always 'neutral' (i.e. does not always have a pH of 7.00 at 298 K).

Remember that the term 'end point' refers to when the colour of the indicator just changes colour – this does not always occur at the equivalence point (see discussion of indicator choice later in this section).

Titration of a strong acid with a strong base

As you add an aqueous solution of an acid to an aqueous solution of a base, you might expect there to be a gradual change in the pH of the solution formed. This is not the case. The shape of the curve when the pH of the solution is plotted against the volume of acid added depends on the nature of the acid and base used. The curves so produced are called 'pH titration curves'.

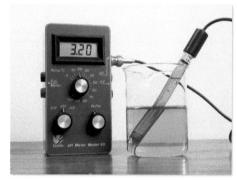

fig A A pH meter.

We shall start by looking at the pH titration curve when a strong acid is added to a strong base.

The following graph is produced when adding 1.00 mol dm⁻³ HCl(aq) to a 25 cm³ sample of 1.00 mol dm⁻³ NaOH(aq). The pH is measured using a pH meter such as the one shown in **fig A**.

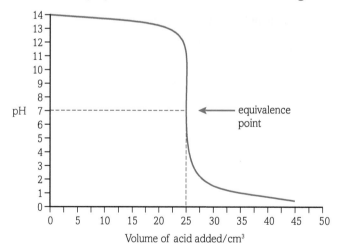

fig B pH curve for a strong acid–strong base titration.

You can see that the pH only falls a very small amount until quite near the equivalence point. Then there is a really steep plunge. If you calculate the values, the pH falls all the way from 11.30 when you have added 24.90 cm³ to 2.70 when you have added 25.10 cm³.

There is a large 'steep section' to the curve. As we shall see later in this section, this is an important point to consider when choosing an appropriate acid–base indicator to determine the end point of this titration.

35

Titration of a weak acid with a strong base

For this example we shall add 1.00 mol dm^{-3} ethanoic acid to 25 cm^3 of 1.00 mol dm^{-3} sodium hydroxide.

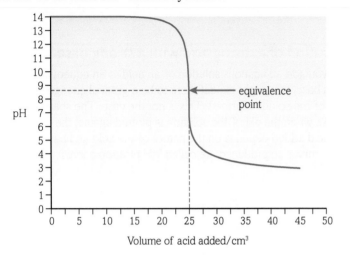

fig C pH curve for a weak acid–strong base titration.

The curve is the same as that for a strong acid–strong base up to the equivalence point, but there is a difference once the acid is present in excess. Past the equivalence point the solution contains a mixture of ethanoic acid and sodium ethanoate. This mixture acts as a buffer solution and hence resists any large change in pH upon addition of further acid. (See **Section 12.2.2**).

Note that the pH at the equivalence point is between 8 and 9; it is *not* 7.

Titration of a strong acid with a weak base

For this example we shall add 1.00 mol dm^{-3} hydrochloric acid to 25 cm^3 of 1.00 mol dm^{-3} aqueous ammonia.

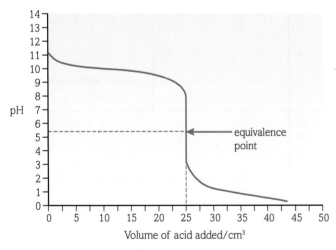

fig D pH curve for a strong acid–weak base titration.

When the acid is first added the pH starts to fall quite sharply, but the curve quickly levels out. This is because a buffer solution has been formed, containing ammonia and ammonium chloride (again, see **Section 12.2.2**).

Notice that the pH at the equivalence point is less than 7 because the salt formed, ammonium chloride, is composed of a strong acid and a weak base.

Titration of a weak acid with a weak base

For this example we shall add 1.00 mol dm^{-3} ethanoic acid to 25 cm^3 of 1.00 mol dm^{-3} aqueous ammonia.

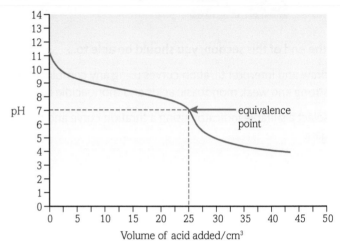

fig E pH curve for a weak acid–weak base titration.

Notice that there is not any steep section to this graph. Instead, there is what is known as a 'point of inflexion'. The lack of a steep section means that it is difficult to do a titration of a weak acid against a weak base using an indicator. The reason for this will be explained in **Section 12.2.3**.

Acid–base indicators

Earlier in this section, we looked at the pH titration curves obtained when an aqueous solution of an acid is added to an aqueous solution of a base. The four cases we considered were:

- strong acid–strong base
- weak acid–strong base
- strong acid–weak base
- weak acid–weak base

We are now going to use these four curves to help us understand why different indicators are required for different types of titrations.

An acid–base indicator is either a weak acid or a weak base. Most indicators are weak acids (HIn). For an indicator that is a weak acid, its dissociation in aqueous solution can be represented as:

$$HIn(aq) \rightleftharpoons H^+(aq) + In^-(aq)$$

The molecule, HIn, and its conjugate base, In$^-$ have different colours in aqueous solution. For methyl orange these are red and yellow respectively:

$$HIn(aq) \rightleftharpoons H^+(aq) + In^-(aq)$$
$$\text{red} \qquad\qquad\qquad \text{yellow}$$

When [H$^+$(aq)] is sufficiently large the equilibrium will shift far enough to the left for the red colour to predominate. If [H$^+$(aq)] is very low then the equilibrium will lie far over to the right and the yellow colour will predominate. Hence, the indicator changes colour according to the pH of the solution.

There will be a stage at which [HIn(aq)] = [In⁻(aq)], and the indicator will appear orange. The exact pH at which this stage is reached can be determined using the equilibrium constant, K_{In}, for methyl orange.

$$K_{In} = \frac{[H^+(aq)][In^-(aq)]}{[HIn(aq)]} = 2.00 \times 10^{-4}\, mol\, dm^{-3}$$

When [HIn(aq)] = [In⁻(aq)], the expression becomes:

$$[H^+(aq)] = 2.00 \times 10^{-4}\, mol\, dm^{-3}$$

This gives a pH of 3.70 for the 'half-way' stage. So, methyl orange will change colour at a pH of 3.70.

Note this pH value is also the same as the value of pK_{In} for methyl orange. Therefore, the pH at which different indicators will change colour can be determined from their pK_{In} values.

pH range of indicators

As a 'rule of thumb', the red colour of methyl orange will first predominate when [HIn(aq)] is ten times [In⁻(aq)], and the yellow colour will predominate when [In⁻(aq)] is ten times [HIn(aq)].

The approximate pH at which each colour predominates can be calculated as follows.

When [HIn(aq)] = 10[In⁻(aq)]:

$$\frac{[H^+(aq)][In^-(aq)]}{10[In^-(aq)]} = 2.00 \times 10^{-4}\, mol\, dm^{-3}$$

$$[H^+(aq)] = 2.00 \times 10^{-3}\, mol\, dm^{-3}$$

So, the pH at which the red colour first predominates is 2.70.

A similar calculation will show that the pH at which the yellow colour first predominates is 4.70.

The 'pH range' of methyl orange is therefore approximately 2.70 to 4.70.

The *exact* pH range of methyl orange is 3.10 to 4.40. This means that at pH below 3.10, methyl orange will appear red. At pH above 4.40, methyl orange will appear yellow. Between 3.10 and 4.40, methyl orange will be a shade of orange.

Table A shows the pK_{In} values, pH ranges and colours of several common indicators.

Indicator	pK_{In}	pH range	Colour HIn(aq)	Colour In⁻(aq)
methyl orange	3.70	3.10–4.40	red	yellow
bromophenol blue	4.00	2.80–4.60	yellow	blue
bromothymol blue	7.00	6.00–7.60	yellow	blue
phenol red	7.90	6.80–8.40	yellow	red
phenolphthalein	9.30	8.20–10.00	colourless	red

table A

Choice of indicator

A good indicator shows a complete colour change upon the addition of one drop of acid from the burette. This is necessary in order to accurately determine the end point of the titration.

Therefore, there has to be a minimum pH change equivalent to the pH range of the indicator in order for the indicator to successfully determine the end point.

Strong acid–strong base titration

Let us first of all consider the use of methyl orange and phenolphthalein as indicators for a strong acid–strong base titration.

Fig F shows the pH titration curve for 25 cm³ of 1.00 mol dm⁻³ NaOH(aq) titrated with 1.00 mol dm⁻³ HCl(aq). The pH ranges of methyl orange and phenolphthalein have also been included.

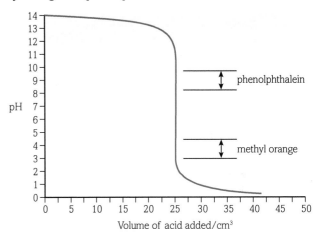

fig F Strong acid–strong base pH curve.

You will notice that the pH range of each indicator falls within the steep section of the curve, where a large pH change is occurring upon the addition of just one drop of acid. This means that both indicators will change colour at the end point and therefore both are suitable indicators to use in this titration.

Weak acid–strong base titration

We shall now consider the suitability of each indicator for a weak acid–strong base titration.

Fig G shows the pH titration curve for 25 cm³ of 1.00 mol dm⁻³ NaOH(aq) titrated with 1.00 mol dm⁻³ CH₃COOH(aq). Once again, the pH ranges of methyl orange and phenolphthalein have also been included.

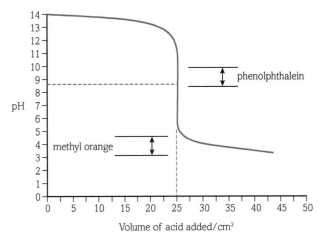

fig G Weak acid–strong base pH curve.

This time, only the pH range of phenolphthalein falls within the steep section of the curve. So, phenolphthalein is suitable, but methyl orange is not.

Strong acid–weak base titration

Fig H shows the pH titration curve for 25 cm³ of 1.00 mol dm⁻³ NH₃(aq) titrated with 1.00 mol dm⁻³ HCl(aq). Again, the pH ranges of methyl orange and phenolphthalein have also been included.

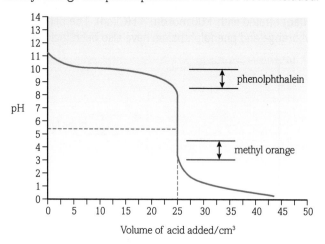

fig H Strong acid–weak base pH curve.

This time, only the pH range of methyl orange falls within the steep section of the curve. So, methyl orange is suitable, but phenolphthalein is not.

Weak acid–weak base titration

Fig I shows the pH titration curve for 25 cm³ of 1.00 mol dm⁻³ NH₃(aq) titrated with 1.00 mol dm⁻³ CH₃COOH(aq). Again, the pH ranges of methyl orange and phenolphthalein have also been included.

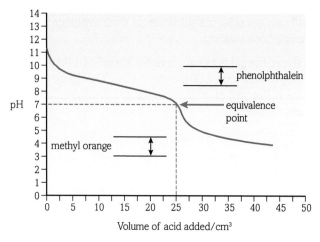

fig I Weak acid–weak base pH curve.

As there is no prominent steep section to the curve, neither indicator is suitable.

In fact, the end point of a titration of a weak acid and a strong base cannot be determined using an acid–base indicator. The end point of such a titration is best determined by measuring the temperature changes (thermometric titration) or electrical conductivity changes (conductometric titration).

Choosing the best indicator

The best indicator to choose for a particular titration is the one whose pK_{In} value is as close as possible to the pH at the equivalence point.

Bromothymol blue (pK_{In} = 7.00) is a particularly good indicator for a strong acid–strong base titration.

Questions

1. The equation for the reaction between hydrochloric acid and ammonia is:

$$HCl(aq) + NH_3(aq) \rightarrow NH_4Cl(aq)$$

A 25.0 cm³ sample of 0.0200 mol dm⁻³ HCl(aq) was placed in a conical flask. Aqueous ammonia was added gradually from a burette and the pH was measured after each addition, until the pH no longer changed.
The pH curve for this titration is shown below.

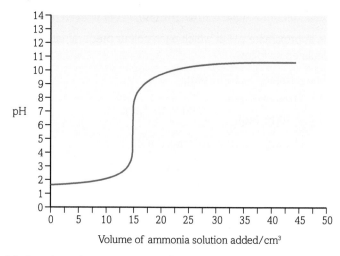

(a) State how the curve suggests that ammonia is a weak base.
(b) Use the information given to calculate the concentration of the ammonia solution.
(c) The pH ranges for three indicators are:
Thymol blue: 1.2 to 2.8
Methyl red: 4.2 to 6.3
Thymolphthalein: 9.3 to 10.5
Explain which of these three indicators is the most suitable for this titration.

2. (a) Calculate the pH at 298 K of an aqueous solution of CH₃COOH of concentration 0.100 mol dm⁻³. [K_a = 1.74 × 10⁻⁵ mol dm⁻³ at 298 K]
(b) Sketch the pH titration curve for the addition of 50.0 cm³ of 0.100 mol dm⁻³ NaOH(aq) to 25.0 cm³ of 0.100 mol dm⁻³ CH₃COOH(aq).
(c) State two differences in the pH curve that would be obtained if the titration were repeated using 25.0 cm³ of 0.0500 mol dm⁻³ CH₃COOH(aq) instead of 25.0 cm³ of 0.100 mol dm⁻³ CH₃COOH(aq).

12.2　2　Buffer solutions

By the end of this section, you should be able to...
- understand what is meant by the term 'buffer solution'
- understand the action of a buffer solution
- calculate the pH of a buffer solution given appropriate data
- calculate the concentrations of solutions required to prepare a buffer solution of a given pH
- understand the roles of carbonic acid molecules and hydrogencarbonate ions in controlling the pH of blood

What is a buffer solution?

Many experiments, particularly in biochemistry, have to be performed in aqueous solutions of fairly constant pH. Unfortunately it is impossible to make a solution whose pH is totally unaffected by the addition of even small amounts of acid or base. However, it is possible to make a solution whose pH remains *almost* unchanged when small amounts of acid or base are added. Such a solution is called a **buffer solution**.

There are many ways of making a buffer solution but two of the most common are:
- to mix a weak acid with its conjugate base
- to mix a weak base with its conjugate acid.

A buffer made from a weak acid and its conjugate base

The simplest example of this is ethanoic acid and sodium ethanoate. The salt of the weak acid has to be soluble in water, which is why sodium and potassium salts are commonly used to make buffer solutions.

In this mixture the acid is partially dissociated, whereas the salt is fully dissociated. The relevant equations are:

$$CH_3COOH(aq) \rightleftharpoons CH_3COO^-(aq) + H^+(aq)$$
$$CH_3COONa(aq) \rightarrow CH_3COO^-(aq) + Na^+(aq)$$

This mixture will produce a buffer solution with a pH less than 7. The exact pH depends on the concentration of both acid and its conjugate base, and can be calculated as follows.

Calculating the pH of a buffer solution

For our example we shall use a solution that has been made by mixing equal volumes of $1.00\,mol\,dm^{-3}$ ethanoic acid and $1.00\,mol\,dm^{-3}$ sodium ethanoate at 298 K.

If we assume that the extent of dissociation of the acid is negligible, then the concentration of CH_3COOH at equilibrium will be $0.500\,mol\,dm^{-3}$.

Also, again because the dissociation of the acid is negligible, the concentration of the ethanoate ions at equilibrium can be considered to be entirely made up from the sodium salt. So, the concentration of ethanoate ions at equilibrium is also $0.500\,mol\,dm^{-3}$.

Note the concentration of both acid and salt in the mixture is half of the concentrations used because equal volumes of each solution were mixed. The total volume of the mixture is twice that of the original volume of each solution used.

At 298 K,

$$K_a = \frac{[CH_3COO^-(aq)][H^+(aq)]}{[CH_3COOH(aq)]} = 1.74 \times 10^{-5}\,mol\,dm^{-3}$$

Rearranging this equation and substituting values for $[CH_3COOH(aq)]$ and $[CH_3COO^-(aq)]$ gives:

$$[H^+(aq)] = (1.74 \times 10^{-5} \times 0.500)/0.500\,mol\,dm^{-3}$$
$$= 1.74 \times 10^{-5}\,mol\,dm^{-3}$$

This gives a pH for the buffer solution of 4.76.

Learning tip

A potential trap when calculating the pH of a buffer solution is not appreciating that when two solutions are mixed there is a dilution. If $50\,cm^3$ of one solution is mixed with $50\,cm^3$ of another solution, then the total volume of the mixture will be $100\,cm^3$. So, the concentration of each solution will be halved on mixing.

How does the buffer action work?

When a small amount of acid is added to the buffer solution, the majority of the H^+ ions added react with the CH_3COO^- ions to form CH_3COOH molecules:

$$CH_3COO^-(aq) + H^+(aq) \rightarrow CH_3COOH(aq)$$

When a little base is added, the majority of the OH^- ions added react with the CH_3COOH molecules:

$$CH_3COOH(aq) + OH^-(aq) \rightarrow CH_3COO^-(aq) + H_2O(l)$$

A new equilibrium mixture will be established in which the concentrations of both CH_3COOH and CH_3COO^- will have changed slightly from their original values. So, there will be a change in pH, but this will be minimal. This is shown by the following argument.

To show that the pH has changed very little, we once again need to make use of the expression for K_a for the acid:

$$K_a = \frac{[CH_3COO^-(aq)][H^+(aq)]}{[CH_3COOH(aq)]}$$

Rearranging this equation gives:

$$[H^+(aq)] = K_a \times \frac{[CH_3COOH(aq)]}{[CH_3COO^-(aq)]}$$

As there are a relative large number of CH_3COOH molecules in the solution, as the extent of dissociation of the acid is very small, the change in $[CH_3COOH(aq)]$ will be negligible.

Also, as there are a relatively large number of CH_3COO^- ions present, resulting from the total dissociation of the CH_3COONa, the change in $[CH_3COO^-(aq)]$ will be negligible.

This means that the ratio $\dfrac{[CH_3COOH(aq)]}{[CH_3COO^-(aq)]}$ will remain fairly constant.

If this ratio remains fairly constant, then $[H^+(aq)]$ remains fairly constant because, at a given temperature, K_a is also constant.

If $[H^+(aq)]$ remains fairly constant, then the pH remains fairly constant.

To summarise, a buffer solution of a weak acid and its conjugate base maintains a fairly constant pH because the ratio of $[CH_3COOH(aq)]$ to $[CH_3COO^-(aq)]$ remains fairly constant when *small* amounts of either acid or base are added.

The best way to demonstrate the effect of adding a small amount of acid on the pH of a buffer solution is to calculate the pH of the buffer both before and after adding the acid.

We have already calculated the pH of a buffer solution made by mixing equal volumes of $1.00\ mol\ dm^{-3}$ ethanoic acid and $1.00\ mol\ dm^{-3}$ sodium ethanoate at 298 K. It is 4.76.

Let us imagine that we have mixed $500\ cm^3$ of each solution to make $1\ dm^3$ of buffer solution. The amount CH_3COOH and CH_3COO^- in this solution will both be equal to $0.500\ mol$.

Let us imagine we have $1\ dm^3$ of this solution and we add 1.00×10^{-2} (0.0100) mol of HCl to it. To make the mathematics easier, we shall assume that the volume of the solution does not change.

0.0100 mol of HCl will provide 0.0100 mol of H^+ ions. These will react with the CH_3COO^- ions in the buffer in a 1:1 molar ratio:

$$CH_3COO^-(aq) + H^+(aq) \rightarrow CH_3COOH(aq)$$

The amount of CH_3COOH present will now have increased from 0.500 to 0.510 mol.

The amount of CH_3COO^- present will have decreased from 0.500 to 0.490 mol.

So, the new concentrations of acid and base present are:

$[CH_3COOH(aq)] = 0.510\ mol\ dm^{-3}$
and $[CH_3COO^-(aq)] = 0.490\ mol\ dm^{-3}$

The new hydrogen ion concentration is given by:

$$[H^+(aq)] = 1.74 \times 10^{-5} \times \dfrac{0.510}{0.490} = 1.81 \times 10^{-5}\ mol\ dm^{-3}$$

The new pH $= -\lg(1.81 \times 10^{-5}) = 4.74$ (to 3 significant figures).

The pH has changed by 0.02 units from 4.76 to 4.74.

To appreciate how effective the buffer solution is in controlling the pH, let us consider adding 0.0100 mol of H^+ ions to $1\ dm^3$ of deionised water.

The pH, at 298 K, of deionised water is 7.00 so:

$[H^+(aq)] = 1.00 \times 10^7\ mol\ dm^{-3}$

Adding 0.0100 mol of H^+ ions gives:

$[H^+(aq)] = (1.00 \times 10^{-7} + 0.0100) = 1.00001 \times 10^{-2}\ mol\ dm^{-3}$

The new pH is given by:

$pH = -\lg(1.00001 \times 10^{-2}) = 2.00$

The pH has dropped by 5 units as opposed to 0.02 units with the buffer solution. A considerable difference!

The Henderson–Hasselbalch equation

If you are convinced that you can remember equations, then the Henderson–Hasselbalch equation can be used to calculate the pH of a buffer solution.

For a weak acid, HA, and its conjugate base, A^-, the following equation applies:

$$[H^+(aq)] = K_a \times \dfrac{[HA(aq)]}{[A^-(aq)]}$$

The concentration terms in the equation are, technically speaking, the concentrations at equilibrium. However, for reasons we have already discussed, it is reasonable to take the original concentrations.

The original concentration of A^- will be the same as that of the salt, providing a sodium or potassium salt has been used. The equation can now be rewritten as:

$$[H^+(aq)] = K_a \times \dfrac{[acid]}{[salt]}$$

If we take the logarithm to base ten of both sides of this equation we get:

$$\lg[H^+(aq)] = \lg K_a + \lg \dfrac{[acid]}{[salt]}$$

Or:

$$-\lg[H^+(aq)] = -\lg K_a - \lg \dfrac{[acid]}{[salt]}$$

Or:

$$pH = pK_a - \lg \dfrac{[acid]}{[salt]}$$

Or:

$$pH = pK_a + \lg \dfrac{[salt]}{[acid]}$$

This last equation is the most common form of the Henderson–Hasselbalch equation.

Learning tip

Do not worry if you did not follow the derivation of the Henderson–Hasselbalch equation. You do not need to perform this derivation but you may use this equation if you wish. The equation is also given in another form in which [base] replaces [salt], since A^- is the base in the buffer.

If we use the Henderson–Hasselbalch equation to calculate the pH of our solution containing ethanoic and ethanoate ions, both of concentration $0.500\ mol\ dm^{-3}$ we obtain:

$pH = 4.76 + \lg(0.500/0.500) = 4.76$
$[pK_a = -\lg K_a = -\lg(1.74 \times 10^{-5}) ; \lg 1 = 0]$

This is the same answer as we obtained earlier.

A buffer made from a weak base and its conjugate acid

The most common example of this type of buffer is ammonia and the ammonium ion. The ammonium ion is usually supplied in the form ammonium chloride. This mixture will provide a buffer solution with a pH greater than 7.

The most convenient equilibrium to consider is:

$$NH_4^+(aq) \rightleftharpoons NH_3(aq) + H^+(aq)$$

This mixture provides a relatively high concentration of both NH_3 molecules and NH_4^+ ions.

The buffer works in a similar manner to the weak acid–conjugate base system and the addition of acid results in the added H^+ ions reacting with NH_3 molecules:

$$NH_3(aq) + H^+(aq) \rightarrow NH_4^+(aq)$$

whereas the addition of base results in the added OH^- ions reacting with NH_4^+ ions:

$$NH_4^+(aq) + OH^-(aq) \rightarrow NH_3(aq) + H_2O(l)$$

As there is a relative high concentration of both NH_3 molecules and NH_4^+ ions, the ratio of $[NH_3(aq)]$ to $[NH_4^+(aq)]$ remains relatively constant upon the addition of small amounts of either acid or base. This results in the pH remaining fairly constant because it is given by the equation:

$$pH = pK_a + \lg \frac{[NH_3(aq)]}{[NH_4^+(aq)]}$$

As pK_a is constant at a given temperature, the pH of the solution depends on the ratio of $[NH_3(aq)]$ to $[NH_4^+(aq)]$.

How to make a buffer solution with a required pH

To make a buffer solution with a pH less than 7, you need to use a mixture of a weak acid and its conjugate base.

Conversely, to make a buffer solution with a pH greater than 7, you need to use a mixture of a weak base and its conjugate acid.

WORKED EXAMPLE 1

Imagine we are faced with making a buffer solution of pH 5.00 at a temperature of 298 K.

To make this solution, we need a hydrogen ion concentration, $[H^+(aq)]$, of 1.00×10^{-5} mol dm^{-3}.

The hydrogen ion concentration of a buffer solution of a weak acid and its conjugate base is calculated using the formula:

$$[H^+(aq)] = K_a \times \frac{[acid]}{[salt]}$$

If we use ethanoic acid as the weak acid, then:

$$K_a = 1.74 \times 10^{-5} \text{ mol dm}^{-3}$$

If we now substitute our known values into the equation we obtain:

$$1.00 \times 10^{-5} = 1.74 \times 10^{-5} \times \frac{[acid]}{[salt]}$$

This gives a value for:

$\frac{[acid]}{[salt]}$ of $(1.00 \times 10^{-5} \div 1.74 \times 10^{-5}) = 0.574$

(to 3 significant figures)

So, if we were supplied with an ethanoic acid solution of concentration 0.574 mol dm^{-3} and a sodium ethanoate solution of 1.00 mol dm^{-3}, we could make a buffer solution of pH 5.00 by mixing equal volumes of the two solutions. This would give us a solution in which the acid concentration was 0.287 mol dm^{-3} and the salt concentration was 0.500 mol dm^{-3}.

$[(0.287 \div 0.500) = 0.574]$

WORKED EXAMPLE 2

In what proportions should 0.100 mol dm^{-3} solutions of ammonia and ammonium chloride be mixed to obtain a buffer of pH 9.80?

[K_a for NH_4^+ is 5.62×10^{-10} mol dm^{-3}]

$$pH = pK_a + \lg \frac{[NH_3(aq)]}{[NH_4^+(aq)]}$$

$$\lg \frac{[NH_3(aq)]}{[NH_4^+(aq)]} = 0.55$$

$$\frac{[NH_3(aq)]}{[NH_4^+(aq)]} = 3.55$$

The solutions, therefore, have to be mixed in a ratio by volume of 3.55 NH_3(aq) to 1 NH_4Cl(aq).

Controlling the pH of blood

The human body operates within a narrow range of pH values. For example, the pH of arterial blood plasma needs to be in the range of 7.35 to 7.45. If the pH of this blood plasma were to change significantly, particularly if it were to fall, the way that the whole body functions would be affected.

The pH of blood is controlled by a mixture of buffers, the most important of which is the carbonic acid–hydrogencarbonate buffer mixture.

In this mixture the carbonic acid molecule, H_2CO_3, acts as the weak acid. The hydrogencarbonate ion, HCO_3^-, is the conjugate base of H_2CO_3.

The equilibrium that exists is represented by the equation:

$$H_2CO_3(aq) \rightleftharpoons HCO_3^-(aq) + H^+(aq)$$

Under normal circumstances the amount of HCO_3^- ion present is approximately 20 times that of H_2CO_3. As normal metabolism produces more acids than bases, this is consistent with the needs of the body.

Any increase in the concentration of hydrogen ions in the blood, by for example the production of lactic acid in the muscles, results in the equilibrium shown above moving to the *left* as the added H^+ ions react with the HCO_3^- ions.

The pH of the blood can also be raised by a variety of respiratory and metabolic causes. For example, the overuse of diuretics increases the amount of urine excreted from the body. If the urine contains large amounts of acids then the pH of the blood will increase. If this happens, the equilibrium shown above will move to the *right* as the H_2CO_3 molecules ionise to increase the H^+ concentration and restore the pH to its normal level.

This is considered to be the most important buffer because it is coupled with the respiratory system of the body. Carbonic acid is not particularly stable and in aqueous solution it decomposes to form carbon dioxide and water:

$$H_2CO_3(aq) \rightleftharpoons CO_2(aq) + H_2O(aq)$$

It is the respiratory system that is responsible for removing carbon dioxide from the body. Aqueous carbon dioxide exists in equilibrium with gaseous carbon dioxide:

$$CO_2(g) \rightleftharpoons CO_2(aq)$$

Combining these three reactions gives us:

$$CO_2(g) + H_2O(aq) \rightleftharpoons CO_2(aq) + H_2O(aq) \rightleftharpoons H_2CO_3(aq) \rightleftharpoons HCO_3^-(aq) + H^+(aq)$$

When these equilibria shift to the left as a result of an increase in hydrogen ion concentration, the concentration of carbon dioxide in the blood increases. The carbon dioxide leaves the blood in the lungs and is then exhaled, thus maintaining the normal pH of the blood.

Questions

1. (a) Explain what is meant by the term 'buffer solution'.
   (b) Explain how an aqueous solution containing a mixture of methanoic acid, HCOOH and potassium methanoate, HCOOK acts as a buffer.
   (c) A buffer solution contains equal concentrations of methanoic acid and potassium methanoate. Explain the effect on the pH of this solution of adding some solid potassium methanoate.
   (d) Calculate the pH, at 298 K, of a buffer solution made by mixing equal volumes of $1.00\ mol\ dm^{-3}$ methanoic acid and $0.500\ mol\ dm^{-3}$ potassium methanoate.
   [$K_a(HCOOH) = 1.79 \times 10^{-4}\ mol\ dm^{-3}$]

2. A student prepares two solutions.
   Solution **A** is prepared by mixing $50\ cm^3$ of $0.100\ mol\ dm^{-3}$ $CH_3COOH(aq)$ with $25\ cm^3$ of $0.100\ mol\ dm^{-3}$ NaOH(aq).
   Solution **B** us prepared by mixing $25\ cm^3$ of $0.200\ mol\ dm^{-3}$ $CH_3COOH(aq)$ with $50\ cm^3$ of $0.100\ mol\ dm^{-3}$ NaOH(aq).
   Explain why solution **A** is a buffer solution, but solution **B** is not.

3. A buffer solution was made by mixing $50\ cm^3$ of $0.200\ mol\ dm^{-3}$ aqueous ammonia, $NH_3(aq)$, with $50\ cm^3$ of aqueous ammonium chloride, $NH_4Cl(aq)$. The pH of the resulting solution was 9.55. Calculate the concentration of the $NH_4Cl(aq)$ used.
   [$K_a(NH_4^+) = 5.62 \times 10^{-10}\ mol\ dm^{-3}$]

4. Calculate the pH of a buffer solution containing 12.20 g of benzoic acid (C_6H_5COOH) and 7.20 g of sodium benzoate (C_6H_5COONa) in $1.00\ dm^3$ of solution.
   [$pK_a(C_6H_5COOH) = 4.20$]

5. Like water, liquid ammonia undergoes self dissociation:

   $$2NH_3 \rightleftharpoons NH_4^+ + NH_2^-$$

   (a) Explain why ammonia can be classified as an amphoteric substance.
   (b) For each of the following substances, indicate whether a solution of liquid ammonia will be 'acidic', 'basic' or 'neutral'.
      (i) Ammonium chloride, NH_4Cl
      (ii) Sodium amide, $NaNH_2$
      (iii) Potassium hydroxide, KOH

Key definition

A **buffer solution** is a solution that *minimises* the change in pH when a *small* amount of either acid or base is added.

12.2 3 Buffer solutions and pH curves

By the end of this section, you should be able to...

- understand how to use a weak acid–strong base titration curve to:
  (i) demonstrate buffer action
  (ii) determine K_a from the point at where half the acid is neutralised

Buffer action during a titration

Fig A shows a typical pH curve obtained when a weak acid is titrated against a strong base.

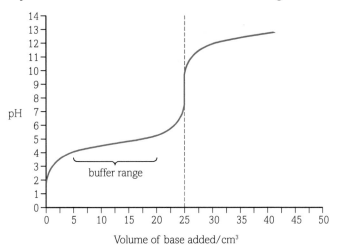

fig A Strong base–weak acid pH curve showing buffer range.

In the region marked 'buffer range', the change in pH as the base is added is gradual. Over this range there is a considerable concentration of both acid and conjugate base molecules. This mixture is displaying buffer action.

Determining K_a from a pH titration curve

This experiment involves performing a titration with an aqueous solution of a weak acid in the conical flask. A standard solution of a strong base, such as sodium hydroxide, is added from a burette and the pH of the solution is measured after each addition.

A graph of pH against volume of base added is then plotted. From the graph, the minimum volume of base required to completely react with all of the acid is determined. This is the volume at the equivalence point.

The graph is then used to determine the pH at the half-equivalence point. This pH value is equal to the pK_a value of the weak acid. From this it is a simple matter to calculate K_a for the acid.

A typical pH titration curve is shown in **fig B**.

The volume at the equivalence point is 25 cm³, so the volume at the half-equivalence point is 12.5 cm³. The pH when 12.5 cm³ of base is added is 4.80.

  The pK_a of the acid = 4.80
    K_a for the acid = $10^{-4.80}$ = 1.58×10^{-5} mol dm⁻³

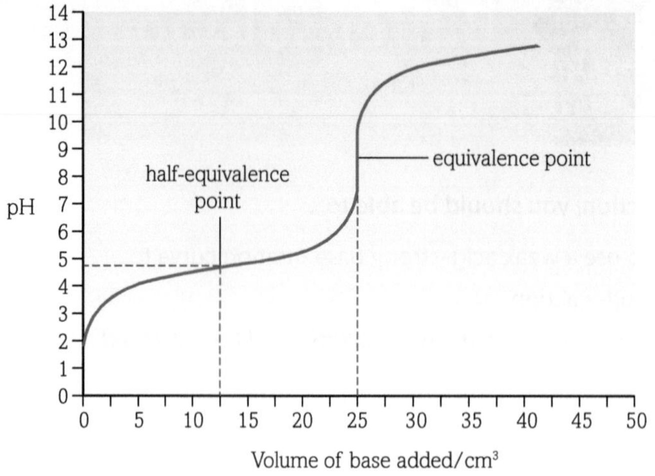

fig B Strong base–weak acid pH curve showing equivalence and half-equivalence points.

Supporting theory

The theoretical justification for determining K_a for a weak acid by the method shown above is quite straightforward.

The mixture at the half-equivalence point is a buffer solution (as stated above). The pH of a buffer solution is calculated using the following equation:

$$pH = pK_a + \lg \frac{[\text{salt}]}{[\text{acid}]}$$

At the half-equivalence point:

[salt] = [acid], so [salt]/[acid] = 1

The logarithm to the base 10 of 1 (lg 1) = 0.

So, the equation becomes:

$$pH = pK_a$$

Alternative method

This method has been called the 'half-volume method'.

- Using a volumetric pipette, place 25.0 cm³ of an aqueous solution of the weak acid into a conical flask.
- Add a few drops of phenolphthalein indicator.
- Titrate against a solution of aqueous sodium hydroxide until the end point colour is obtained.
- Note the volume of sodium hydroxide required. This is the minimum volume required to completely react with the acid.
- Use a fresh 25.0 cm³ sample of the same aqueous solution of the weak acid and the same aqueous solution of sodium hydroxide, but this time do not add the phenolphthalein.
- Add only *half* the volume of sodium hydroxide required to react with the acid.
- Measure the pH of this solution. This pH value is equal to the pK_a value of the acid.

12.2 — 4 — Enthalpy changes of neutralisation for strong and weak acids

By the end of this section, you should be able to...

- understand why there is a difference in the standard enthalpy changes of neutralisation values for strong and weak acids

Standard enthalpy change of neutralisation

In **Book 1** we defined the standard enthalpy change of neutralisation as follows:

> The standard enthalpy change of neutralisation ($\Delta_{neut}H^{\ominus}$) is the enthalpy change measured at 100 kPa and a specified temperature, usually 298 K, when one mole of water is produced by the neutralisation of an acid with an alkali.

We also explained that the standard enthalpy change of neutralisation of a strong acid with a strong base will always have a very similar value, because strong acids and strong bases are almost fully dissociated in aqueous solution. This value is around $-57.6 \text{ kJ mol}^{-1}$.

Table A shows the values of the standard enthalpy changes of neutralisation for various weak acids with the strong base, sodium hydroxide.

Formula of weak acid	$\Delta_{neut}H^{\ominus}$/kJ mol^{-1}
CH_3COOH	−55.2
HCN	−11.2
HF	−68.6

table A

It is interesting to note that for two of the weak acids, CH_3COOH and HCN, the magnitude of the value of $\Delta_{neut}H^{\ominus}$ is less than that for a strong acid–strong base (i.e. less than 57.6), but the magnitude of the value for HF is greater than 57.6.

The equation for the reaction between a weak acid, HA, and a strong base such as sodium hydroxide can be represented as:

$$HA(aq) + OH^-(aq) \rightarrow A^-(aq) + H_2O(l)$$

The major difference in this reaction, compared to that of a strong acid with a strong base, is that the majority of the acid molecules are undissociated. Obviously energy will be required to dissociate them, and this may explain why less heat energy is given out when both CH_3COOH and HCN are neutralised by NaOH.

However, it does not explain why more heat energy is given out when HF is neutralised by NaOH. There must be at least one other energy term involved.

In the above argument, we have ignored the fact that, once the acid molecules are dissociated (which is an endothermic process), the ions formed will then be hydrated by the water molecules present in the aqueous solution. Hydration of ions is an exothermic process, so heat energy will be generated. The two processes can be represented by the following equations for a weak acid of formula HA:

$$HA(aq) \rightarrow H^+(g) + A^-(g) \quad \Delta H = +x \text{ kJ mol}^{-1}$$
DISSOCIATION (endothermic)

$$H^+(g) + A^-(g) \xrightarrow{aq} H^+(aq) + A^-(aq) \quad \Delta H = -y \text{ kJ mol}^{-1}$$
HYDRATION (exothermic)

If the magnitude of x is greater than that of y, then the overall heat energy change involved in dissociating the acid molecules and then hydrating the ions will be endothermic. This is the case with CH_3COOH and HCN.

However, with HF, the enthalpy change of hydration of the ions releases more heat energy than the dissociation of the molecules takes in. Hence the overall effect is that the standard enthalpy change of neutralisation of HF by NaOH is greater than 57.6 kJ mol^{-1}.

Questions

1. Explain why reactions between a strong acid and a strong base all have values for the standard enthalpy change of neutralisation of about -57 kJ mol^{-1}.

2. Explain why the standard enthalpy change of neutralisation for the reaction between aqueous ammonia and aqueous hydrochloric acid is $-52.2 \text{ kJ mol}^{-1}$ and not -57 kJ mol^{-1}.

3. If the standard enthalpy change of neutralisation of HCN(aq) with KOH(aq) is $-11.7 \text{ kJ mol}^{-1}$, predict a value for the standard enthalpy change of neutralisation of HCN(aq) with NH_3(aq).

THINKING BIGGER

A GROWING PROBLEM

Ocean acidification impairs mussels' ability to attach to surfaces. This is an alarming prospect for commercial mussel growers who farm in the waters around Puget Sound, USA.

MUSSELS LOSE FOOTING IN MORE ACIDIC OCEAN

fig A Carbon dioxide from greenhouse gas emissions has steadily turned seawater more acidic, disrupting organisms accustomed to the slightly alkaline waters of the past 20 million years.

PENN COVE, Wash. – Cookie tray in hand and lifejacket around chest, Laura Newcomb looks more like a confused baker than a marine biologist. But the University of Washington researcher is dressed for work. Her job: testing how mussels in this idyllic bay, home to the nation's largest harvester of mussels, are affected by changing ocean conditions, especially warmer and more acidic waters. It's a question critical to the future of mussel farmers in the region. More important, it's key to understanding whether climate change threatens mussels around the world, as well as the food chains mussels support and protect in the wild.

'Along the West Coast, mussels are well-known ecosystem engineers,' said Bruce Menge, an Oregon State University researcher who studies how climate impacts coastal ecosystems. 'They provide habitat for dozens of species, they provide food for many predators and occupy a large amount of space, so are truly a dominant species.'

20 million years

Carbon dioxide from greenhouse gas emissions has steadily turned seawater more acidic, disrupting organisms accustomed to the slightly alkaline waters of the past 20 million years. In the case of mussels, an earlier University of Washington lab study found that increased carbon dioxide weakens the sticky fibers, called byssus, that mussels use to survive by clinging to objects like shorelines or the ropes used by commercial harvesters.

'If byssal thread weakening does eventually become important,' Menge added, 'the consequences would be major if not catastrophic.'

Newcomb's goal now is to apply in the real world what was learned in the lab. 'Instead of spending a lot of time tightly controlling the temperature and pH conditions mussels grow in, I use the natural seasonal variation to try to answer the same questions,' Newcomb said.

Newcomb's field office is the rear deck of a harvesting boat – right between the toilet and the microwave. The quarters are cramped, but the view is grand: The blue waters of Penn Cove on Washington state's Whidbey Island are set against rolling bluffs and snow-capped mountains.

30 percent increase in acidity

The University of Washington marine biologist is there courtesy of Penn Cove Shellfish, which is also the oldest and best known mussel operation in the United States. If you're a mussel fan, you've probably had a few – they're sold at Costco as well as upscale restaurants across the country.

Placing just-harvested mussels on her tray, Newcomb samples for size, thickness and strength. The mussels are grown on 21-foot-long ropes hanging from several dozen rafts in the bay, and Newcomb takes samples from two depths: 3 feet and 21 feet. She also samples water temperature and pH levels at those depths.

Prior to the Industrial Revolution and the explosion of manmade CO_2, ocean pH averaged 8.2. Today it's 8.1, a 30 percent increase in acidity on the logarithmic scale. Computer models peg ocean acidity at 7.8 to 7.7 by the end of the century at the current rate of greenhouse gas emissions.

Washington State is a bit ahead of that curve because ancient carbon stores in the deep ocean are periodically churned up by local currents. The surprising lab discovery was that mussel byssus weakened by 40 percent when exposed to a pH of 7.5. At Penn Cove, low pH levels are not uncommon – Newcomb has even seen 7.4 in the year that she's been sampling.

'We're worried they're going to see it more frequently,' said Emily Carrington, Newcomb's graduate adviser and leader of the University of Washington team that published the earlier lab results.

Where else will I encounter these themes?

Let us start by considering the nature of the writing in the article.

1. This article is not taken from a scientific journal but is instead designed to draw a wider audience's attention to a pressing environmental problem. Identify writing techniques in the text that the author uses to make the science more accessible.

> Being able to communicate scientific ideas to a wide audience is an important skill, particularly when the science has implications for decision making in society.

Now we will look at the chemistry in detail. Some of these questions will link to topics earlier in this book or **Book 1**, so you may need to combine concepts from different areas of chemistry to work out the answers. Some questions will also link to topics covered later in this book. Don't worry if you are not ready to give answers to these questions yet. You may like to return to the questions later.

2. Write an equation for the reaction of carbon dioxide with water to form hydrogencarbonate ions and H⁺(aq) ions.
3. a. Using the relationship $pH = -\lg [H^+]$ show that a change in pH from 8.2 to 8.1 is approximately equivalent to a 30% increase in acidity.
   b. Calculate the pH of the ocean if the acidity increases by 100% from a starting pH of 8.2.
4. Many sea organisms form shells made of calcium carbonate. Use the equation below to explain why increasing levels of carbon dioxide in the air are making this process increasingly difficult.
   $$CO_2(aq) + CO_3^{2-}(aq) + H_2O(l) \rightleftharpoons 2HCO_3^-(aq)$$

> Think about how an increase in carbon dioxide would affect the equilibrium of this reaction, and therefore the availability of carbonate ions.

5. Many organisms use intracellular HCO_3^- ions to buffer changes in pH. Use equations to show how HCO_3^- ions can buffer small changes in H⁺(aq) and OH⁻(aq).
6. The equation for the dissociation of carbonic acid in water can be represented as:
   $$H_2CO_3(aq) \rightleftharpoons H^+(aq) + HCO_3^-(aq)$$

   a. Write an expression for K_a for carbonic acid.
   b. Given that the value of this pK_a is 6.3 and that the pH of blood is 7.4, calculate the ratio of hydrogencarbonate ions to non-dissociated carbonic acid molecules in blood plasma.
   c. What assumptions have you made your answer to question **6b**?

Activity

The molecule histidine is one of the 20 amino acids that make up all proteins. Histidine residues in haemoglobin molecules are also involved in buffering the blood's pH. Research and create a presentation suitable for GCSE Chemistry students addressing the questions below:
- What does a histidine molecule look like?
- Which part of the molecule acts as the proton donor and which as the proton acceptor?
- Histidine can frequently be found at the catalytic centres of enzymes. Can you suggest why?

● From *The Daily Climate* by Miguel Llanos http://www.dailyclimate.org/tdc-newsroom/2014/09/acidification-mussels

Did you know?

March 2014 was the first month in which the average atmospheric carbon dioxide levels reached the 400 ppm milestone. But carbon dioxide is not alone in causing climate change. If oxides of nitrogen and methane and minute trace amounts of other gases such as SF_6 are factored in, the 'equivalent CO_2' atmospheric concentration is about 480 ppm.

12 Exam-style questions

1. A fruit juice contains a monobasic acid HA.
   (a) The fruit juice has a hydrogen ion concentration of 2.50×10^{-4} mol dm^{-3}. Calculate the pH of the fruit juice. [2]
   (b) A 25.0 cm^3 sample of the fruit juice reacted exactly with 26.70 cm^3 of 0.0100 mol dm^{-3} sodium hydroxide.
       (i) Write a chemical equation for the reaction taking place. State symbols are not required. [1]
       (ii) Calculate the concentration, in mol dm^{-3}, of HA in the fruit juice. [3]
       (iii) Compare your answer in (b)(ii) to the hydrogen ion concentration and hence make a deduction about the strength of the acid, HA, in the fruit juice. [2]
   (c) (i) Write an equation to represent the dissociation of HA into its ions in aqueous solution. [1]
       (ii) Write an expression for the acid dissociation constant, K_a, of HA(aq). [1]
       (iii) The value of K_a for HA(aq) is 6.00×10^{-5} mol dm^{-3}. Calculate the concentration of the undissociated acid under these conditions. [2]
   [Total: 12]

2. A mixture of ethanoic acid, CH$_3$COOH, and its sodium salt CH$_3$COONa, can act as a buffer solution.
   (a) State what is meant by the term 'buffer solution'. [2]
   (b) Explain how a mixture of ethanoic acid and sodium ethanoate acts as a buffer. [4]
   (c) A buffer solution is made by adding a solution of sodium hydroxide, of concentration 1.00 mol dm^{-3}, to a sample of 1.00 mol dm^{-3} ethanoic acid until half of the amount of acid present has reacted. Calculate the pH of this buffer solution.
       [K_a of ethanoic acid at the temperature used is 1.70×10^{-5} mol dm^{-3}] [3]
   [Total: 9]

3. The values of the ionic product of water, K_w, are 1.00×10^{-14} mol^2 dm^{-6} at 298 K and 5.48×10^{-14} mol^2 dm^{-6} at 323 K.
   (a) Calculate the pH of water at each of these two temperatures. [4]
   (b) Using your answers to (a), comment of the validity of the following statement:
       'Pure water is neutral because it has a pH of 7'. [2]
   (c) Show that the data supplied can be used to deduce the sign of ΔH for the dissociation of water into ions. [2]
   [Total: 8]

4. In 1923 Johannes Brønsted and Thomas Lowry proposed independently a theory that when an acid reacts with a base the acid forms its conjugate base. The theory is known as the 'Brønsted–Lowry' theory.
   (a) State what is meant by the term **conjugate base**. [1]
   (b) (i) In each of the two equations, identify the species on the left-hand side of the equation that is behaving as a Brønsted–Lowry acid. [2]
       Equation 1 $C_6H_5COO^- + HF \rightleftharpoons C_6H_5COOH + F^-$
       Equation 2 $C_6H_5COOH + CN^- \rightleftharpoons C_6H_5COO^- + HCN$
       (ii) Explain the relative strengths of the three acids involved in the two equilibria. [Assume that both equilibria lie well to the right-hand side.] [4]
   (c) Liquid ammonia, like water, undergoes self-ionisation, according to the following equation:
       $NH_3 + NH_3 \rightleftharpoons NH_4^+ + NH_2^-$
       For each of the two substances listed, explain whether a solution in liquid ammonia would be 'acidic', 'basic' or 'neutral'.
       (i) Ammonium chloride, NH$_4^+$Cl$^-$
       (ii) Sodium amide, Na$^+$NH$_2^-$ [4]
   [Total: 11]

5. Sulfur dioxide reacts with water to produce sulfurous acid, H$_2$SO$_3$, which is a weak, dibasic acid.
   The equation for the first dissociation into ions is:
   $H_2SO_3(aq) \rightleftharpoons H^+(aq) + HSO_3^-(aq)$
   $K_{a(1)}(298\ K) = 1.20 \times 10^{-2}$ mol dm^{-3}
   (a) Calculate the value of p$K_{a(1)}$ for H$_2$SO$_3$. [1]
   (b) Use $K_{a(1)}$ to calculate the approximate pH of an aqueous solution of 0.500 mol dm^{-3} H$_2$SO$_3$ at 298 K. [3]
   (c) The measured pH of 0.500 mol dm^{-3} H$_2$SO$_3$ is slightly lower than that calculated in part (b). Comment on a possible reason for this difference. [1]
   (d) The constant K_w has a value of 1.00×10^{-14} mol^2 dm^{-6} at 298 K.
       (i) Give the name of the constant K_w. [1]
       (ii) Write the expression for K_w. [1]
   (e) Potassium hydroxide, KOH, is a strong base in aqueous solution. Calculate the pH of 0.500 mol dm^{-3} KOH. [3]
   [Total: 10]

6 Hydrochloric acid, HCl(aq), is usually sold as a solution. The solution is made by dissolving hydrogen chloride gas in water. A chemistry technician bought 25.0 dm³ of 10.00 mol dm⁻³ hydrochloric acid.

(a) (i) Calculate the volume of hydrogen chloride gas that is required to make this solution. [Assume that the molar volume of hydrogen chloride is 24.0 dm³ mol⁻¹ under the conditions used] [2]

(ii) Describe how the technician could make 5.00 dm³ of 0.0200 mol dm⁻³ hydrochloric acid from the 10.00 mol dm⁻³ stock solution of hydrochloric acid. [3]

(b) Calculate the pH of 0.0200 mol dm⁻³ hydrochloric acid. [1]

(c) Hydrochloric acid reacts with ammonia to form ammonium chloride.

$$NH_3(aq) + HCl(aq) \rightleftharpoons NH_4Cl(aq)$$

A 25.0 cm³ sample of 0.0200 mol dm⁻³ hydrochloric acid is placed into a conical flask. The pH of this solution is measured using a pH meter. Aqueous ammonia was gradually added and the pH was measured after each addition, until the pH no longer changed.

The diagram is a graph of the results obtained.

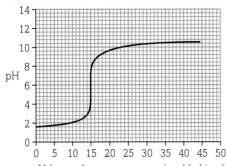

(i) Explain how the pH at the equivalence point shows that ammonia is a weak base. [2]

(ii) Calculate the concentration of the aqueous ammonia. [2]

(iii) The table shows the pK_{In} values for three indicators.

Indicator	pK_{In}
thymol blue	1.7
methyl red	5.1
thymolphthalein	9.7

Explain which of these three indicators is the most suitable for the titration. [2]

[Total: 12]

TOPIC 13

Further energetics

Introduction

People lived very simply in the Stone Age, with their energy needs being supplied by the Sun and by plants and trees that could be burned. Because the plants and trees were continuously renewed, and populations were small, energy supplies during the Stone Age period were plentiful. However, this simple lifestyle has almost completely vanished, and people today cannot survive without abundant fuel.

Today, most of the world's energy is supplied from fossil fuels, which consist of natural gas, coal and products from crude oil, such as gasoline, diesel and kerosene. Fossil fuels are not renewable and will one day run out. To manage these energy resources and develop new fuels we need to understand how energy is released or used in chemical reactions, including the reactions in plants, animals and our own bodies.

Energy is the very essence of chemistry as well as of civilisation as we know it. In chemistry it determines which reactions can occur and which compounds can exist. You have already met a large number of chemical reactions. All of these reactions either absorb or release energy. In this topic we shall develop an understanding of how to measure and report these energy changes.

Have you ever wondered why water evaporates? Why hot objects cool? Why hydrogen combines with oxygen? Why green leaves turn red in the autumn? Why anything happens? Part of the answer is related to energy. We need energy to think, to move and to live. Every chemical reaction makes use of energy to rearrange the bonding between elements in compounds. Thermodynamics deals with questions like these. In the second half of this topic we shall deal with the second law of thermodynamics, which governs the direction of natural change. The second law enables us to predict whether or not a reaction has a tendency to occur, and to what extent it will occur. This law is of fundamental importance in chemistry since it provides a basis for discussing, explaining and predicting equilibria, the subject of **Topics 11** and **12** in this book. It is also the foundation of the whole field of electrochemistry, the subject of **Topic 14**.

All the maths you need

- Recognise and make use of units in calculations
- Recognise and use expressions in decimal and ordinary form
- Carry out calculations using numbers in ordinary form
- Use calculators to find and use power, exponential and logarithmic functions
- Use an appropriate number of significant figures
- Change the subject of an equation
- Substitute numerical values into algebraic equations using appropriate units for physical quantities
- Solve algebraic equations
- Use logarithms in relation to quantities that range over several orders of magnitude

What have I studied before?
- Standard conditions of temperature and pressure for thermodynamic measurements
- Enthalpy changes and Hess's Law
- Energy level diagrams and enthalpy profile diagrams
- Bond enthalpies and mean bond enthalpies

What will I study later?
- The relationship between Gibbs energy and cell potentials (emf)

What will I study in this topic?
- Lattice energies and Born–Haber cycles
- Enthalpy changes of atomisation, solution and hydration
- Electron affinity
- Polarisation of anions by cations to explain the degree of covalent character of ionic compounds
- Entropy
- Gibbs energy
- The relationship between entropy, Gibbs energy and equilibrium constants

13.1 1 Lattice energy, $\Delta_{lattice}H$, and Born–Haber cycles

By the end of this section, you should be able to:
- define lattice energy, $\Delta_{lattice}H$
- understand that lattice energy provides a measure of the strength of ionic bonding
- define the terms:
  (i) enthalpy change of atomisation, $\Delta_{at}H$
  (ii) electron affinity, E_{ea}
- construct Born–Haber cycles and carry out related calculations
- understand the effect of ionic charge and ionic radius on the value of lattice energy

Lattice energy

In **Book 1**, we saw that bond enthalpies can be used as a measure of the strength of the covalent bonding in molecules. The equivalent energy change for ionic bonding in ionic compounds is lattice energy, $\Delta_{lattice}H$ (or lattice enthalpy).

The lattice energy of a compound is the energy change when one mole of the ionic solid is formed from its gaseous ions. If standard conditions of 100 kPa and a stated temperature (usually 298 K) are applied (as indicated by ⦵), then the energy change is called the **standard lattice energy**.

The equation that represents the standard lattice energy of sodium chloride is:

$$Na^+(g) + Cl^-(g) \rightarrow NaCl(s) \qquad \Delta_{lattice}H^\ominus = -780 \text{ kJ mol}^{-1}$$

For magnesium chloride it is:

$$Mg^{2+}(g) + 2Cl^-(g) \rightarrow MgCl_2(s) \qquad \Delta_{lattice}H^\ominus = -2526 \text{ kJ mol}^{-1}$$

Learning tip

The terms 'lattice energy' and 'lattice enthalpy' are commonly used as if they mean exactly the same thing – you will often find both terms used within the same textbook, article or website, including some university websites.

In fact, there is a difference between them that relates to the conditions under which they are calculated. However, the difference is small. In fact, it is negligible compared with the differing values for lattice energies that you will find from different data sources.

Unless you go on to study chemistry at degree level, the difference between the two terms need not concern you.

Did you know?

There are two ways of defining lattice energy. The one mentioned in the specification is the one we have already defined. This is sometimes called lattice energy of *formation*, since the compound is being formed from its ions. Using this definition, the energy change will always be negative.

However, some books define it as lattice energy of *dissociation*, in which it is the energy change when one mole of the compound is broken down (or dissociated) into its ions. In this case, the energy change will be positive.

Factors affecting the magnitude of lattice energy

You will have noticed that the lattice energy of magnesium chloride, $Mg^{2+}(Cl^-)_2$, is much larger (i.e. more negative) than that of sodium chloride, Na^+Cl^-. A number of factors are responsible for this difference.

The first thing to realise is that a magnesium ion carries twice the charge of a sodium ion.

Secondly, there are more cation-to-anion interactions in magnesium chloride because there are twice as many chloride ions per cation than in sodium chloride.

A third factor that determines the magnitude of the lattice energy is the distance between the centres of the cations and their neighbouring anions, which is equal to the sum of their ionic radii. This distance is determined in part by the relative sizes of the ions involved (the Mg^{2+} ion is smaller than Na^+, thus reducing the sum of the ionic radii) and also by the type of lattice structure the compound has. In fact, the relative ion sizes determine the type of lattice structure.

The effects that both the inter-ionic distance and the charges on the ions have on the lattice energy are illustrated in **table A**.

Compound	Inter-ionic distance/nm	Charges on the ions	Lattice energy/ kJ mol^{-1}
LiF	0.207	+1, −1	−1031
NaF	0.235	+1, −1	−918
CaF$_2$	0.233	+2, −1	−2630
Li$_2$O	0.214	+1, −2	−2814
MgO	0.212	+2, −2	−3791
Al$_2$O$_3$	0.193	+3, −2	−15 504

table A

Comparison of LiF and NaF, where the charges on the ions are the same, shows that a decrease in the distance between the centres of the two ions (the inter-ionic distance) results in a more negative value for the lattice energy.

Comparison of NaF and CaF$_2$, where the inter-ionic distances are almost the same, shows that an increase in charge of even one of the ions (Ca^{2+} as opposed to Na$^+$) results in a more negative value for the ionisation energy. Similar comparisons can be made between Li$_2$O and MgO and also between Li$_2$O and Al$_2$O$_3$.

Finally, as we shall see later in this topic, there are also covalent interactions between the ions, and these affect the magnitude of the lattice energy. The values quoted in **table A** take these interactions into consideration since they are calculated values using a Born–Haber cycle (see later in this section).

Learning tip

Lattice energies are determined by:
- the magnitudes of the charges on the ions
- the sum of the ionic radii
- the type of lattice structure
- the extent of covalent interactions between the ions.

At A level, you are not required to know the different types of lattice structure. In addition, the extent of the covalent interactions is already taken into consideration in the quoted values for lattice energy. Therefore, you need only consider the magnitude of charges on the ions and the size of the ions.

Standard enthalpy change of atomisation, $\Delta_{at}H^\ominus$

The enthalpy change measured at a stated temperature (usually 298 K) and 100 kPa when one mole of gaseous atoms is formed from an element in its standard state is called the **standard enthalpy change of atomisation** of the element. It is given the symbol $\Delta_{at}H^\ominus$.

Equations representing some standard enthalpy changes of atomisation at 298 K are given below.

$$C(s) \rightarrow C(g) \qquad \Delta_{at}H^\ominus = +717 \text{ kJ mol}^{-1}$$
$$Na(s) \rightarrow Na(g) \qquad \Delta_{at}H^\ominus = +107 \text{ kJ mol}^{-1}$$
$$\tfrac{1}{2}H_2(g) \rightarrow H(g) \qquad \Delta_{at}H^\ominus = +218 \text{ kJ mol}^{-1}$$
$$\tfrac{1}{2}Cl_2(g) \rightarrow Cl(g) \qquad \Delta_{at}H^\ominus = +122 \text{ kJ mol}^{-1}$$

Learning tip

Note that the standard enthalpy change of atomisation is the enthalpy change when one mole of gaseous *atoms* is formed. In the case of elements that exist as polyatomic molecules, it is *not* the enthalpy change when one mole of gaseous *molecules* is atomised.

Electron affinity

The first electron affinity of an element, $E_{ea(1)}$, is the energy change when each atom in one mole of atoms in the gaseous state gains an electron to form a -1 ion.

The equations below represent the first electron affinities of some elements.

$$Cl(g) + e^- \rightarrow Cl^-(g) \qquad E_{ea(1)} = -349 \text{ kJ mol}^{-1}$$
$$Br(g) + e^- \rightarrow Br^-(g) \qquad E_{ea(1)} = -325 \text{ kJ mol}^{-1}$$
$$O(g) + e^- \rightarrow O^-(g) \qquad E_{ea(1)} = -141 \text{ kJ mol}^{-1}$$

The first electron affinity has a negative value for many elements, including the alkali metals. As examples, the $E_{ea(1)}$ of Li is -60 kJ mol^{-1} and the $E_{ea(1)}$ of Na is -53 kJ mol^{-1}. There is a notable exception with the noble gases. For these, repulsion caused by the electrons already present in the valence shell results in a positive value for the first electron affinity as the additional electron would have to occupy a new valence shell.

In contrast, second electron affinities, $E_{ea(2)}$, tend to be positive. For example:

$$O^-(g) + e^- \rightarrow O^{2-}(g) \qquad E_{ea(2)} = +798 \text{ kJ mol}^{-1}$$

Therefore, the formation of the oxide ion, O^{2-}, in the gaseous state from its atom in the gaseous state is an endothermic process overall:

$$O(g) + 2e^- \rightarrow O^{2-}(g) \qquad E_{ea(1)} + E_{ea(2)} = +657 \text{ kJ mol}^{-1}$$

This raises an interesting question: why does oxygen form O^{2-} ions in its binary ionic compounds rather than O$^-$ ions, as the O$^-$ ion would appear to be the more energetically favourable state? We will attempt to answer this question later in this topic.

Born–Haber cycles

We are now in a position to consider the overall energy changes that take place when an ionic compound is made from its elements. These energy changes are summarised in an energy level diagram called the Born–Haber cycle. **Fig A** shows the Born–Haber cycle for sodium chloride.

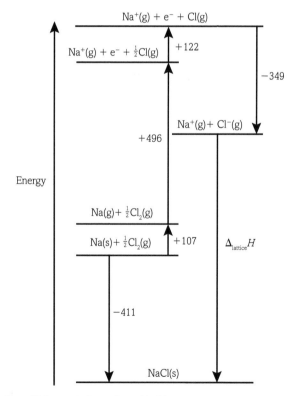

fig A Born–Haber cycle for sodium chloride.

The cycle includes the following energy changes, all of which can be determined experimentally:
- The enthalpy change of formation of NaCl(s), $\Delta_f H[\text{NaCl(s)}] = -411$ kJ mol^{-1}
- $\Delta_{at}H[\text{Na(s)}] = +107$ kJ mol^{-1}

- The first ionisation energy, $IE_{(1)}[Na(g)] = +496 \text{ kJ mol}^{-1}$
- $\Delta_{at}H[Cl_2(g)] = +122 \text{ kJ mol}^{-1}$
- $E_{ea}[Cl(g)] = -349 \text{ kJ mol}^{-1}$

Applying Hess's Law to the cycle gives us:

$$+107 + 496 + 122 + (-349) + \Delta_{lattice}H = -411$$

Hence:

$$\Delta_{lattice}H[NaCl(s)] = -411 - 107 - 496 - 122 + 349 \text{ kJ mol}^{-1}$$
$$= -787 \text{ kJ mol}^{-1}$$

> **Learning tip**
>
> Note: see **Book 1 Section 1.1.4** and **8.1** for a reminder of these definitions.

Did you know?

Why is calcium oxide $Ca^{2+}O^{2-}$ and not Ca^+O^-?

You may be wondering why we have asked this question and think that the answer is obvious. Surely, you might say, it is $Ca^{2+}O^{2-}$ because that way both ions will have a stable noble gas electronic configuration.

If this is what you believe, then the following may come as a surprise. Acquiring a noble gas electronic configuration is not a reason for electronic changes to take place when atoms and molecules react together. After all, many ions exist in stable compounds, in particular cations of the transition metals, in which they do not have the electronic configuration of a noble gas. In fact, compounds in which the ions do have a noble gas electronic configuration are in the minority.

So, why does calcium oxide form as $Ca^{2+}O^{2-}$ and not as Ca^+O^-, and why are we asking the question?

You may remember that we have mentioned that the formation of $O^-(g)$ from $O(g)$ is exothermic:

$$O(g) + e^- \rightarrow O^-(g) \qquad E_{ea(1)} = -141 \text{ kJ mol}^{-1}$$

However, the formation of $O^{2-}(g)$ from $O(g)$ is endothermic:

$$O(g) + 2e^- \rightarrow O^{2-}(g) \qquad E_{ea(1)} + E_{ea(2)} = +657 \text{ kJ mol}^{-1}$$

Since more energy is required, why is the formation of the O^{2-} ion preferred over the formation of O^-?

A similar pattern is observed with the formation of $Ca^+(g)$ and $Ca^{2+}(g)$ from $Ca(g)$:

$$Ca(g) \rightarrow Ca^+(g) + e^- \qquad IE_{(1)} = +590 \text{ kJ mol}^{-1}$$
$$Ca(g) \rightarrow Ca^{2+}(g) + 2e^- \qquad IE_{(1)} + IE_{(2)} = +1735 \text{ kJ mol}^{-1}$$

Similarly, since more energy is required, why is the formation of the Ca^{2+} ion preferred over the formation of the Ca^+ ion?

To answer this question it is necessary to consider *all* of the energy changes involved in the formation of an ionic compound from its elements – not just those involved in the formation of the gaseous ions from the gaseous atoms. In other words, we need to look at the information supplied by the Born–Haber cycle for each compound.

Using the equation:

$$\Delta_f H[Ca^+O^-(s)] = \Delta_{at}H[Ca(s)] + IE_{(1)}[Ca(g)] + \Delta_{at}H[\tfrac{1}{2}O_2(g)] + E_{ea(1)}[O(g)] + \Delta_{lattice}H[Ca^+O^-(s)]$$

we can calculate a value for $\Delta_f H[Ca^+O^-(s)]$ using a theoretical value for $\Delta_{lattice}H[Ca^+O^-(s)]$ of -650 kJ mol^{-1}.

In this way, the value calculated for $\Delta_f H[Ca^+O^-(s)]$ is:

$$178 + 590 + 249 + (-141) + (-650) = +226 \text{ kJ mol}^{-1}.$$

Similarly, the value calculated for $\Delta_f H[Ca^{2+}O^{2-}(s)]$ is -635 kJ mol^{-1}

From these values, it is clear that the formation of $Ca^{2+}O^{2-}$ is energetically more favourable than the formation of Ca^+O^-.

The extra energy required to form the 2+ and 2– ions is more than compensated for by the much larger (i.e. more negative) lattice energy of $Ca^{2+}O^{2-}$ ($-3401 \text{ kJ mol}^{-1}$) compared to that of Ca^+O^- (-650 kJ mol^{-1}).

It is the *overall* energy change involved, and not the incorrect notion of desirability to obtain a noble gas electronic configuration, that determines how atoms and molecules will interact with one another to form a compound.

However, it is important to recognise that the conditions under which a reaction takes place will influence the overall energy change involved, and hence the nature of the compound formed.

For example, if iron is heated in chlorine gas, iron(III) chloride is the product. However, if hydrogen chloride gas is used in place of chlorine, iron(II) chloride is formed.

Therefore, changing the conditions of temperature and/or pressure can also have an effect on the exact compound formed.

Questions

1. Lattice energies can be used to compare the strength of bonding present in an ionic compound.
   (a) State what is meant by the term 'lattice energy'.
   (b) Explain the effect of ionic charge and ionic radius on the magnitude of the lattice energy of an ionic compound.

2. The Born–Haber cycle can be used to calculate the lattice energy, $\Delta_{lattice}H$, for magnesium oxide.

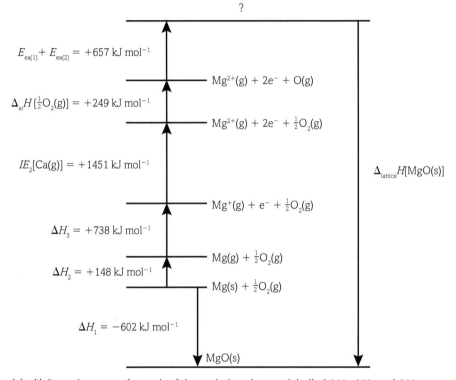

   (a) (i) State the names for each of the enthalpy changes labelled ΔH_1, ΔH_2 and ΔH_3.
       (ii) Give the formula missing at the top of the cycle, indicated by a question mark (?). Give the state symbols.
   (b) The equations representing the first and second electron affinities of oxygen are:
       $O(g) + e^- \rightarrow O^-(g)$ $E_{ea(1)} = -141$ kJ mol^{-1}
       $O^-(g) + e^- \rightarrow O^{2-}(g)$ $E_{ea(2)} = +798$ kJ mol^{-1}
       Suggest why the second of these processes is endothermic.
   (c) Use the information in the Born–Haber cycle shown above to calculate the lattice energy of magnesium oxide, $\Delta_{lattice}H[MgO(s)]$.
   (d) Explain how the lattice energy of barium oxide differs from that of magnesium oxide.

Key definition

The **standard enthalpy change of atomisation** of an element is the enthalpy change measured at a stated temperature, usually 298 K, and 100 kPa when one mole of gaseous atoms is formed from an element in its standard state.

13.1 2 Experimental and theoretical lattice energy

By the end of this section, you should be able to:

- understand that a comparison of the experimental lattice energy value (obtained from a Born–Haber cycle) with the theoretical value (obtained from electrostatic theory) in a particular compound indicates the degree of covalent bonding
- understand the meaning of polarisation as applied to ions
- understand that the polarising power of a cation depends on its radius and charge
- understand that the polarisability of an anion depends on its radius and charge

Experimental lattice energy

The Born–Haber cycle allows us to calculate a value for the lattice energy of an ionic compound from knowledge of the other energy changes, all of which can be determined experimentally. The value of the lattice energy calculated this way is called the 'experimental lattice energy'.

Theoretical lattice energy

The type of lattice structure and the inter-ionic distance can be found by X-ray crystallography. Using this information, it is possible for us to calculate a value for the lattice energy of an ionic compound. However, we first need to make the following assumptions.

- The ions are in contact with one another.
- The ions are perfectly spherical.
- The charge on each ion is evenly distributed around the centre so that each ion can be considered as point charges.

A value for the lattice energy can be calculated using the principles of electrostatics. There are three main methods for performing such calculations. If you are interested, you can research them under the headings of 'The Born–Landé equation', 'The Born–Mayer equation' and 'The Kapustinskii equation'.

Table A shows a comparison of the experimental lattice energies (obtained by using a Born–Haber cycle) with the theoretical lattice energies (calculated using the principles of electrostatics) for various compounds.

Compound	Experimental lattice energy/kJ mol^{-1}	Theoretical lattice energy/kJ mol^{-1}
NaF	−918	−912
NaCl	−780	−770
NaBr	−742	−735
AgF	−958	−920
AgCl	−905	−833
AgBr	−891	−816

table A

You will notice that there is good agreement between the experimental values and the theoretical values for the halides of sodium. However, the agreement is not so good for the halides of silver.

An agreement between the experimental and theoretical values of lattice energy for a compound indicates that the ionic model is a good one for that compound. A significant difference suggests that the ionic model needs to be modified. In such compounds, the bonding in the lattice has a considerable covalent character, which makes the experimental value for the lattice energy more negative than the theoretical value.

Lattice energy 13.1

The covalency in bonding is caused by polarisation of the anion by the cation. Polarisation results in distortion of the electron density within the anion, resulting in a higher electron density near the cation. This means that there is some electron density existing between the two ions. That is to say, there will be a degree of covalent bonding in the compound.

The extent of covalent character: polarisation of the anion

In an ionic lattice, the positive ion (cation) will attract the electrons of the anion. If the electrons are pulled towards the cation, the anion is said to be *polarised* because the even distribution of its electron density has been distorted.

The extent to which an anion is polarised by a cation depends on several factors. The two main factors are summarised below. These are known as Fajan's Rules.

Polarisation will be increased by:

- a high charge and small size of the cation (i.e. a high charge density of the cation)
- a high charge and large size of the anion.

High charge and small size of cation

The ability of a cation to attract electrons from the anion towards itself is called its 'polarising power'. A cation with a high charge and a small radius has a large polarising power. An approximate value for the polarising power of a cation can be obtained by calculating its charge density (sometimes called surface charge density). The charge density of a cation is the charge divided by the surface area of the ion. If the ion is assumed to be a sphere, its surface area is equal to $4\pi r^2$, where r is the ionic radius.

A rough approximation of the charge density can be determined by dividing the charge by the square of its ionic radius. This calculation is beyond the scope of your A level course.

$$\text{charge density} \sim \frac{\text{charge}}{r^2}$$

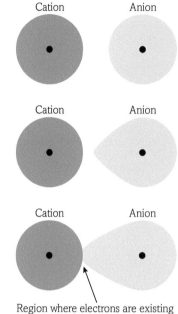

fig A A representation of a cation attracting the electrons of an anion in an ionic lattice.

Table B compares the extent of covalent bonding in sodium chloride and magnesium chloride.

Compound	Charge density of cation	Experimental lattice energy/kJ mol⁻¹	Theoretical lattice energy/kJ mol⁻¹	Percentage difference	Extent of covalent bonding
NaCl	$\frac{1}{(0.095)^2} = 111$	−780	−770	0.13	very little
MgCl$_2$	$\frac{2}{(0.060)^2} = 556$	−2526	−2326	7.92	more than in NaCl

table B

The charge density of the magnesium ion is larger than that of the sodium ion, resulting in greater polarisation of the chloride ion and increased covalent bonding in MgCl$_2$ than in NaCl.

High charge and large size of the anion

The ease with which an anion is polarised depends on its charge and its size. Anions with a large charge and a large size are polarised most easily.

Table C compares the extent of covalent bonding in silver fluoride and silver iodide.

Compound	Charge of anion	Radius of anion/nm	Experimental lattice energy/kJ mol⁻¹	Theoretical lattice energy/kJ mol⁻¹	Percentage difference	Extent of covalent bonding
AgF	−1	0.133	−958	−920	4.13	fairly large
AgI	−1	0.215	−889	−778	12.49	greater than in AgF

table C

The larger iodide ion is more easily polarised, which leads to a greater degree of covalent bonding in silver iodide.

Did you know?

The interesting case of silver compounds

Table D compares the extent of covalent bonding in sodium chloride and silver chloride.

Compound	Charge density of cation	Experimental lattice energy/ kJ mol^{-1}	Theoretical lattice energy/ kJ mol^{-1}	Percentage difference	Extent of covalent bonding
NaCl	$\frac{1}{(0.095)^2} = 111$	−771	−766	0.65	very little
AgCl	$\frac{1}{(0.126)^2} = 63$	−905	−770	14.92	much greater than in NaCl

table D

Despite the polarising power of the Ag$^+$ ion being less than that of the Na$^+$ ion, there is considerably more covalent bonding in AgCl.

This is explained by considering the valence shell electronic configurations of the two ions concerned.

$\quad$ Na$^+$ 1s^2 2s^2 2p^6

$\quad$ Ag$^+$ [Kr] 3d^{10}

A d^{10} configuration offers less shielding than a p^6 electronic configuration, so compounds which have a d^{10} configuration show a greater tendency toward a covalent character.

For the same reason, zinc compounds also contain a significant degree of covalent bonding.

Questions

1. The experimental and theoretical lattice energies, in kJ mol^{-1}, of calcium fluoride, CaF$_2$, and silver fluoride, AgF, are given below.

	experimental value	theoretical value
CaF$_2$	−2630	−2609
AgF	−958	−920

   Suggest why there is good agreement between the two values for CaF$_2$, but there is a significant difference between the two values for AgF.

2. Anions in an ionic lattice can be polarised by the cations adjacent to them. The extent of polarisation depends on the nature of both the anion and cation involved.
   (a) Explain what is meant by the term 'polarised' in this context.
   (b) State whether the oxide ion, O^{2-}, or the sulfide ion, S^{2-}, is more easily polarised.
   (c) Place the following cations (with ionic radii shown in parenthesis) in order of increasing polarising power:

   $\quad$ Mg^{2+} (0.072 nm) Al^{3+} (0.053 nm) Li$^+$ (0.074 nm)
   $\quad$ Na$^+$ (0.102 nm) Ca^{2+} (0.100 nm) K$^+$ (0.138 nm)

   Support your conclusion by suitable calculations.

3. Suggest why the oxide Na^{2+}O^{2-} does not exist. State what further energy change, other than those quoted in the data books, you would require in order to confirm your suggestion.

13.1 3 Enthalpy changes of solution and hydration

By the end of this section, you should be able to...

- define the terms enthalpy change of solution, $\Delta_{sol}H$, and enthalpy change of hydration, $\Delta_{hyd}H$
- use energy cycles and energy level diagrams to carry out calculations involving enthalpy change of solution, enthalpy change of hydration and lattice energy
- understand the effect of ionic charge and ionic radius on the value of enthalpy change of hydration

Enthalpy change of solution, $\Delta_{sol}H$

The solubilities of ionic solids in water show a very wide variation and there is no obvious pattern.

One of the factors that determines solubility is the value of the **enthalpy change of solution**, $\Delta_{sol}H$, the enthalpy change when one mole of an ionic solid dissolves in water to form an infinitely dilute solution.

The enthalpy change of solution for sodium chloride is the energy change associated with the following process:

$$NaCl(s) \xrightarrow{aq} Na^+(aq) + Cl^-(aq)$$

It is important to specify the extent of dilution of the final solution when quoting a value for the enthalpy change. Upon dilution, the ions in the solution move further apart (an endothermic process) and also become more hydrated (an exothermic process). The relative importance of these two processes changes with dilution and they affect the value of $\Delta_{sol}H$ in a complicated way. For this reason the quoted values for $\Delta_{sol}H$ refer to an *infinitely dilute solution*. This value cannot be determined experimentally and is found by a process of extrapolation. In practice, there comes a point when further dilution has no measurable effect on $\Delta_{sol}H$. This is known as the point of infinite dilution.

Enthalpy changes of solution can be either negative or positive, as is shown in **table A**.

Ionic solid	Equation	$\Delta_{sol}H$/kJ mol^{-1}
NaCl	$NaCl(s) \xrightarrow{aq} Na^+(aq) + Cl^-(aq)$	+11.0
NaOH	$NaOH(s) \xrightarrow{aq} Na^+(aq) + OH^-(aq)$	−44.5
NH$_4$NO$_3$	$NH_4NO_3(s) \xrightarrow{aq} NH_4^+(aq) + NO_3^-(aq)$	+25.7
MgSO$_4$	$MgSO_4(s) \xrightarrow{aq} Mg^{2+}(aq) + SO_4^{2-}(aq)$	−91.3

table A

Enthalpy change of hydration, $\Delta_{hyd}H$

The **enthalpy change of hydration**, $\Delta_{hyd}H$, is the enthalpy change when one mole of an ion in its gaseous state is completely hydrated by water. In practice, complete hydration is said to have occurred when the solution formed is at infinite dilution (see the definition of enthalpy change of solution).

For the sodium and chloride ions, the enthalpy change of hydration is the enthalpy change for the following processes:

$$Na^+(g) \xrightarrow{aq} Na^+(aq)$$

and $$Cl^-(g) \xrightarrow{aq} Cl^-(aq)$$

When an ion is placed in water it immediately interacts with the water molecules. Water molecules are polar (see **Book 1 Section 2.2.4**) and are attracted to both positive and negative ions. **Fig A** shows the hydration of sodium and chloride ions.

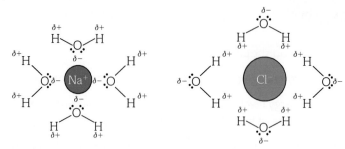

fig A Hydration of sodium and chloride ions.

In the case of the sodium ion, the interaction is the result of the attraction between the $\delta-$ oxygen atom of the water molecule and the cation. Such an interaction is often referred to as an *ion–dipole* interaction. With some other positive ions, notably those of the transition metals, a dative covalent bond is formed between the water molecule and the cation using one of the lone pairs of electrons on the oxygen atom.

Ion–dipole interactions exist in the hydrated chloride ion, but some hydrogen bonds are also formed between the $\delta+$ hydrogen atoms of the water molecules and the chloride ion, making use of the lone pairs of electrons on the chloride ion (see **Book 1 Section 2.2.5**).

Enthalpy changes of hydration are always negative. Some examples are given in **table B**.

Ion	Ionic radius/nm	Equation	$\Delta_{hyd}H$/kJ mol^{-1}
Na$^+$	0.102	Na$^+$(g) $\xrightarrow{aq}$ Na$^+$(aq)	−406
K$^+$	0.138	K$^+$(g) $\xrightarrow{aq}$ K$^+$(aq)	−322
Rb$^+$	0.149	Rb$^+$(g) $\xrightarrow{aq}$ Rb$^+$(aq)	−301
Mg^{2+}	0.072	Mg^{2+}(g) $\xrightarrow{aq}$ Mg^{2+}(aq)	−1920
Ca^{2+}	0.100	Ca^{2+}(g) $\xrightarrow{aq}$ Ca^{2+}(aq)	−1650
Sr^{2+}	0.113	Sr^{2+}(g) $\xrightarrow{aq}$ Sr^{2+}(aq)	−1480
Cl$^-$	0.180	Cl$^-$(g) $\xrightarrow{aq}$ Cl$^-$(aq)	−363
Br$^-$	0.195	Br$^-$(g) $\xrightarrow{aq}$ Br$^-$(aq)	−335
I$^-$	0.215	I$^-$(g) $\xrightarrow{aq}$ I$^-$(aq)	−293

table B

Factors affecting the magnitude of the hydration enthalpy

The first thing you will notice from the values in **table B** is that $\Delta_{hyd}H$ is much more negative for 2+ ions than for 1+ ions. This makes sense because we would expect a doubly charged ion to have a stronger interaction with the water molecules compared with a singly charged ion. That is to say, the electrostatic force of attraction between a doubly charged ion and water molecules will be greater than that between a singly charged ion and water molecules.

As we go down a group (e.g. Na$^+$ to Rb$^+$ or Cl$^-$ to I$^-$) the magnitude of $\Delta_{hyd}H$ becomes less negative. This seems to correlate with an increase in ionic radius. This trend can be explained using a simple electrostatic model for hydration, similar to that used to explain the variation in lattice energies for ionic solids. As the ions become larger, the electrostatic force of attraction between them and the water molecules decreases, and hence the energy released upon hydration decreases.

As with lattice energy, there is a strong correlation between $\Delta_{hyd}H$ and the charge densities of the ions. **Table C** shows this for four cations.

Cation	Charge density of cation	$\Delta_{hyd}H$/kJ mol^{-1}
Na$^+$	$\frac{1}{(0.095)^2} = 111$	−406
K$^+$	$\frac{1}{(0.133)^2} = 57$	−322
Mg^{2+}	$\frac{2}{(0.060)^2} = 556$	−1920
Ca^{2+}	$\frac{2}{(0.099)^2} = 204$	−1650

table C

The greater the charge density of the cation, the more negative the value of $\Delta_{hyd}H$.

Relationship between $\Delta_{sol}H$, $\Delta_{hyd}H$ and $\Delta_{lattice}H$

The relationship between $\Delta_{sol}H$, $\Delta_{hyd}H$ and $\Delta_{lattice}H$ is best shown by an energy level diagram similar to that of a Born–Haber cycle as shown in **fig B**.

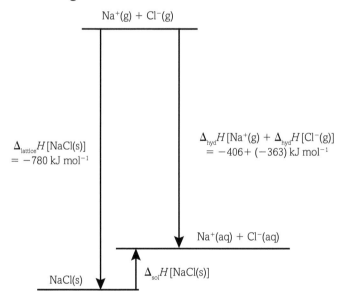

fig B Energy level diagram for the dissolving of sodium chloride.

Applying Hess's Law:

$\Delta_{lattice}H[\text{NaCl(s)}] + \Delta_{sol}H[\text{NaCl(s)}] = \Delta_{hyd}H[\text{Na}^+(g)] + \Delta_{hyd}H[\text{Cl}^-(g)]$

$\Delta_{sol}H[\text{NaCl(s)}] = -406 + (-363) \text{ kJ mol}^{-1} - (-780) \text{ kJ mol}^{-1}$

$= +11 \text{ kJ mol}^{-1}$

The relationship can also be shown in the form of a Hess cycle (shown in question 3).

Questions

1. When potassium fluoride dissolves in water, the lattice breaks up and the potassium and fluoride ions become hydrated.
   (a) Draw diagrams to represent (i) a hydrated potassium ion and (ii) a hydrated fluoride ion.
   (b) Name the type of interaction that occurs between the water molecules and the ion for both the hydrated potassium ion and the fluoride ion. Describe how each interaction occurs.

2. These data refer to some of the energy changes involved when magnesium chloride dissolves in water.

   $\text{Mg}^{2+}(\text{Cl}^-)_2(s) \xrightarrow{aq} \text{Mg}^{2+}(aq) + 2\text{Cl}^-(aq)$
   $\Delta_{sol}H[\text{Mg}^{2+}(\text{Cl}^-)_2(s)] = -155 \text{ kJ mol}^{-1}$

   $\text{Mg}^{2+}(g) + 2\text{Cl}^-(g) \rightarrow \text{Mg}^{2+}(\text{Cl}^-)_2(s)$
   $\Delta_{lattice}H[\text{Mg}^{2+}(\text{Cl}^-)_2(s)] = -2526 \text{ kJ mol}^{-1}$

   $\text{Mg}^{2+}(g) \xrightarrow{aq} \text{Mg}^{2+}(aq)$
   $\Delta_{hyd}H[\text{Mg}^{2+}(g)] = -1920 \text{ kJ mol}^{-1}$

   Use this information to calculate the enthalpy change of hydration of the chloride ion, $\Delta_{hyd}H[\text{Cl}^-(g)]$.

3. The diagram below is a Hess cycle for the dissolving of lithium fluoride in water.

   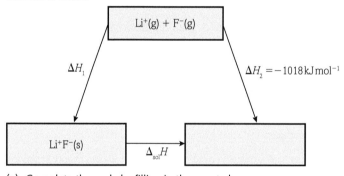

   (a) Complete the cycle by filling in the empty box.
   (b) State the name of the energy change represented by ΔH_1.
   (c) Apply Hess's Law to obtain an expression for $\Delta_{sol}H$ in terms of ΔH_1 and ΔH_2.
   (d) Calculate $\Delta_{sol}H[\text{Li}^+\text{F}^-(s)]$, given that $\Delta H_1 = -1031 \text{ kJ mol}^{-1}$.

4. The standard enthalpy change of solution of sodium fluoride is +0.3 kJ mol^{-1}.
   A sample of sodium fluoride of mass 1 g is added to 250 cm^3 of water in a beaker and stirred with a thermometer graduated in intervals of 1 °C.
   Explain what is likely to happen to the reading on the thermometer as the sodium fluoride dissolves. No calculation is necessary.

Key definitions

The **enthalpy change of solution**, $\Delta_{sol}H$, is the enthalpy change when one mole of an ionic solid dissolves in water to form an infinitely dilute solution.

The **enthalpy change of hydration**, $\Delta_{hyd}H$, is the enthalpy change when one mole of an ion in its gaseous state is completely hydrated by water.

13.2 1 Introduction to entropy

By the end of this section, you should be able to...

- understand that, since some endothermic reactions can occur without the input of heat, enthalpy changes alone do not control whether reactions occur
- recognise that entropy is a measure of the disorder of a system and that the natural direction of change is increasing total entropy (positive total entropy change)
- understand why entropy changes occur during changes of state

What makes a reaction occur?

Perhaps one of the most important questions to ask in chemistry is 'Will a reaction occur?'

We know that, once started, some chemical reactions simply 'go' with no further, continuous help from us.

For example, ammonia gas and hydrogen chloride gas react together at room temperature to form the white solid, ammonium chloride:

$$NH_3(g) + HCl(g) \rightarrow NH_4Cl(s)$$

Magnesium, once ignited, will burn in oxygen to form magnesium oxide:

$$Mg(s) + \tfrac{1}{2}O_2(g) \rightarrow MgO(s)$$

In other reactions, rather than the reactants changing completely into the products, a position of equilibrium is reached, with the final mixture containing a measurable amount of both reactants and products.

For example, ethanoic acid dissociates in water. In a 0.1 mol dm^{-3} solution of ethanoic acid, only about 1% of the ethanoic acid molecules are present as ions:

$$CH_3COOH(aq) \rightleftharpoons CH_3COO^-(aq) + H^+(aq)$$

Another example is the dimerisation of nitrogen dioxide in the gas phase:

$$2NO_2(g) \rightleftharpoons N_2O_4(g)$$

At a temperature of 298 K and a pressure of 100 kPa, the equilibrium mixture contains about 70% N_2O_4.

There are other types of reactions that simply do not occur at all, at least not without some help. For example, ammonium chloride does not spontaneously decompose into ammonia and hydrogen chloride. Magnesium oxide does not break apart to form magnesium and oxygen without some continuous intervention from us in the form of heating.

If we consider all of these types of reaction together, we see that the real difference between them is the *position* of equilibrium that is established. For some reactions, the position of equilibrium is so far over to the products side that, to all intents and purposes, the reaction has gone to completion. For some other reactions significant amounts of both reactants and products are present at equilibrium. For other reactions the equilibrium lies so far to the left that they appear not to take place at all.

So, perhaps, the better to question to ask is *not* 'Will a chemical reaction occur?', but 'What will be the position of equilibrium?' This is the question we hope to answer in this topic.

Exothermic and endothermic reactions

The reaction between magnesium and oxygen is exothermic and can be represented by the enthalpy level diagram shown in **fig A**.

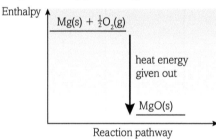

fig A Enthalpy level diagram for the reaction between magnesium and oxygen.

As the products have less energy than the reactants, we often say that the products are more *energetically stable* than the reactants.

It is very tempting to conclude that this reaction occurs *because* the magnesium oxide is energetically more stable than its elements, magnesium and oxygen. However tempting this argument may be, we must discard it, because experience tells us that many endothermic reactions occur at room temperature.

Let us consider the dimerisation of $NO_2(g)$ at 298 K:

$$2NO_2(g) \rightleftharpoons N_2O_4(g)$$

The enthalpy level diagram for this reaction is shown in **fig B**.

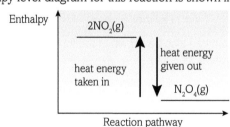

fig B Enthalpy level diagram for the dimerisation of NO_2.

62

This reaction is readily reversible and reaches a position of equilibrium at 298 K. The forward reaction leading to the formation of $N_2O_4(g)$ is exothermic. However, the backwards reaction is endothermic.

As we discovered in **Book 1 Section 10.1.1**, the position of equilibrium can be reached from either direction. This means, that if we place some $N_2O_4(g)$ in a sealed container at room temperature, some of it will decompose to form $NO_2(g)$. So, an endothermic reaction is taking place at room temperature without any continuous intervention from us. We say that the reaction is *spontaneous* (see later in this section).

Clearly, it is not only exothermic reactions that can take place spontaneously. The driving force for spontaneous endothermic reactions cannot be the formation of more energetically favourable, i.e. lower-energy, products but must involve another factor we have yet to consider.

That factor is a quantity known as **entropy**, and is governed by the *Second Law of Thermodynamics*.

But first, we need to appreciate what is meant by the term **spontaneous process**.

Spontaneous processes

A spontaneous process is one that takes place without continuous intervention from us.

A good example of a spontaneous process is the freezing of water to form ice. If water is placed in an environment at −20 °C it will turn into ice. However, the reverse will never happen: at −20 °C ice will never melt to form water.

Another example of a spontaneous process is the mixing of gases. If two unreactive gases are present in a container, they will mix completely in a process known as diffusion.

We can bring about the reverse of spontaneous processes by intervening. For example, we can melt ice by heating it, and we can separate a mixture of gases by liquefying followed by distillation.

The key point, however, is that the reverse of a spontaneous process never happens on its own. A mixture of gases will *never* separate of their own accord.

At the beginning of this section we stated that, in chemistry, we are concerned with whether or not chemical reactions will occur, and we suggested that perhaps the most important question to ask in chemistry is 'Why does a reaction occur?'.

From the point of view of thermodynamics, all reactions are reversible so we suggested that an even better question to ask is:

> When a reaction reaches a point of equilibrium, what determines whether the equilibrium will favour the reactants or the products, and to what extent?

This question is answered using the Second Law of Thermodynamics, in which the concept of entropy is introduced.

Entropy

The Second Law of Thermodynamics allows us to predict whether a process is likely to occur. It is, therefore, the key to understanding what drives chemical reactions and what determines the position of equilibrium. The Second Law of Thermodynamics introduces a term called entropy, which is a property of matter, just like density or energy. The simplest way to describe entropy is to say that it is a quantity associated with randomness or disorder. The natural direction of change is toward increasing total entropy (positive total entropy change).

The greater the degree of disorder, the greater the entropy. A gas has a greater entropy than its solid because, in a gas, the molecules are free to move around randomly. The entropy of a liquid is higher than that of its solid, but lower than that of its gas. This is because the molecules in a liquid are freer to move than those in a solid, but less free to move than those in a gas. So, as ice changes first of all into liquid water, and then into water vapour, the entropy increases at each stage.

The Second Law of Thermodynamics is sometimes misquoted. You may have come across the statement that in a spontaneous change, entropy always increases. This cannot be true: if it were, water could never freeze because this change involves a decrease in entropy.

Another example is the reaction between ammonia gas and hydrogen chloride gas:

$$NH_3(g) + HCl(g) \rightarrow NH_4Cl(s)$$

Clearly there is a reduction in the entropy of the system, since two gases are forming a solid, yet this reaction occurs spontaneously at room temperature.

The statement *entropy always increases* is almost correct. It needs to be expressed more carefully to give the Second Law of Thermodynamics:

> In a spontaneous process, the *total* entropy increases.

In **Section 13.2.2**, we explain what is meant by the term 'total entropy'.

Questions

1 For each of the following predict whether the change is accompanied by an increase or a decrease in total entropy.
 (a) $H_2O(g) \rightarrow H_2O(l)$
 (b) $I_2(s) \rightarrow I_2(g)$
 (c) $Na(l) \rightarrow Na(s)$

Key definitions

Entropy is a property of matter that is associated with the degree of disorder, or degree of randomness, of the particles.

A **spontaneous process** is one that takes place without continuous intervention from us.

13.2 ❷ Total entropy

By the end of this section, you should be able to...

- understand that the total entropy change in any reaction is the entropy change in the system added to the entropy change in the surroundings, as shown by the expression:

  $\Delta S_{total} = \Delta S_{system} + \Delta S_{surroundings}$

- calculate the entropy change of the system, ΔS_{system}, in a reaction, given the entropies of the reactants and products

- calculate the entropy change of the surroundings, and hence ΔS_{total}, using the expression:

  $\Delta S_{surroundings} = -\dfrac{\Delta H}{T}$

Total entropy change, ΔS_{total}

The total entropy change of a process is comprised of two components:

- the entropy change of the system, ΔS_{system}, and
- the entropy change of the surroundings, $\Delta S_{surroundings}$.

In a chemical reaction, the system is the species that are taking part in the reaction.

The surroundings is everything else. In practice this usually means the reaction vessel, e.g. test tube or beaker, and the air in the laboratory.

The total entropy change is defined as the sum of the entropy change of the system and the entropy change of the surroundings. That is:

$\Delta S_{total} = \Delta S_{system} + \Delta S_{surroundings}$

For a reaction to be spontaneous, ΔS_{total} must be positive. This is another way of expressing the Second Law of Thermodynamics.

Entropy change of the system

ΔS_{system} is calculated using the expression:

$\Delta S_{system} = \Sigma S \text{(products)} - \Sigma S \text{(reactants)}$

where S represents entropy and Σ represents 'the sum of'.

The standard entropy values of some substances are shown in **table A**.

[Standard refers to conditions of 100 kPa and 298 K.]

> **Did you know?**
>
> The official definition of the Second law of Thermodynamics does not use the term 'total entropy change'. Instead it refers to the 'entropy change of the universe'. These two terms have the same meaning. We will use the term 'total entropy change' throughout the book.

Gas	Entropy $S^\ominus$/J K^{-1} mol^{-1}	Liquid	Entropy $S^\ominus$/J K^{-1} mol^{-1}	Solid	Entropy $S^\ominus$/J K^{-1} mol^{-1}
$H_2O(g)$	188.7	$H_2O(l)$	69.9	$H_2O(s)$	47.9
H_2	130.6	CH_3OH	239.7	C(diamond)	2.4
O_2	205.0	CH_3CH_2OH	160.7	C(graphite)	5.7
N_2	191.6	C_6H_6	172.8	CaO	39.7
Cl_2	165.0			$CaCO_3$	92.9
CO_2	213.6				
NH_3	192.3				

table A

Enthalpy 13.2

Learning tip

Note that entropy values are usually quoted in J K^{-1} mol^{-1}, whereas enthalpy changes are usually quoted in kJ mol^{-1}. So, in any equation linking ΔS and ΔH, one of the energy terms will need a conversion of units.

WORKED EXAMPLE 1

Use the values in **table A** to calculate $\Delta S^\ominus_{system}$ for the following reaction:

$$CaCO_3(s) \rightarrow CaO(s) + CO_2(g)$$

$\Delta S^\ominus_{system} = S^\ominus[CaO(s)] + S^\ominus[CO_2(g)] - S^\ominus[CaCO_3(s)]$

$= 39.7 + 213.6 - 92.9$

$= +160.4$ J K^{-1} mol^{-1}

WORKED EXAMPLE 2

Use the values in **table A** to calculate $\Delta S^\ominus_{system}$ for the following reaction:

$$N_2(g) + 3H_2(g) \rightarrow 2NH_3(g)$$

$\Delta S^\ominus_{system} = 2 \times S^\ominus[NH_3(g)] - S^\ominus[N_2(g)] - 3 \times S^\ominus[H_2(s)]$

$= (2 \times 192.3) - 191.6 - (3 \times 130.6)$

$= -198.8$ J K^{-1} mol^{-1}

Entropy change of the surroundings

The entropy change of the surroundings, $\Delta S_{surroundings}$, is related to the enthalpy change of the reaction, ΔH, by the expression:

$$\Delta S_{surroundings} = -\frac{\Delta H}{T}$$

where T is the temperature in Kelvin.

For an *exothermic* reaction, where ΔH is negative, $\Delta S_{surroundings}$ will always be positive, so the entropy of the surroundings *increases*.

Conversely, for an *endothermic* reaction, $\Delta S_{surroundings}$ will always be negative, so the entropy of the surroundings *decreases*.

WORKED EXAMPLE 3

Calculate $\Delta S_{surroundings}$ at 298 K when one mole of hydrogen gas is burned in oxygen.

$$H_2(g) + \tfrac{1}{2}O_2(g) \rightarrow H_2O(l) \quad \Delta H = -286 \text{ kJ mol}^{-1}$$

$\Delta S_{surroundings} = -\dfrac{-286}{298}$

$= +0.960$ kJ K^{-1} mol^{-1} or $+960$ J K^{-1} mol^{-1}

Calculating the total entropy change, ΔS_{total}

Now that we know how to calculate both ΔS_{system} and $\Delta S_{surroundings}$, we are in a position to calculate the total entropy change, ΔS_{total}, for a reaction.

WORKED EXAMPLE 4

Using the information in **table A**, calculate the total entropy change at 298 K for the following reaction:

$$H_2(g) + \tfrac{1}{2}O_2(g) \rightarrow H_2O(l) \quad \Delta H = -286 \text{ kJ mol}^{-1}$$

$\Delta S_{system} = S[H_2O(l)] - S[H_2(g)] - \tfrac{1}{2} \times S[O_2(g)]$

$= 69.9 - 130.6 - (\tfrac{1}{2} \times 205)$

$= -163.2$ J K^{-1} mol^{-1}

$\Delta S_{total} = -163.2$ J K^{-1} mol^{-1} + 960 J K^{-1} mol^{-1}

$= +796.8$ J K^{-1} mol^{-1}

Summary

ΔS_{total} will be positive if:

- both $\Delta S_{surroundings}$ and ΔS_{system} are positive
- $\Delta S_{surroundings}$ is positive and ΔS_{system} is negative, but the magnitude of $\Delta S_{surroundings}$ > the magnitude of ΔS_{system}
- $\Delta S_{surroundings}$ is negative and ΔS_{system} is positive, but the magnitude of $\Delta S_{surroundings}$ < the magnitude of ΔS_{system}.

The role of temperature

The increase in entropy obtained by supplying a certain amount of heat energy to an object depends on the temperature of the system.

If an object is very cold the molecules are not moving around very much. Supplying some heat energy to the object will make the molecules move around more, so the entropy increases.

If we supply the same amount of heat energy to a much hotter object, the entropy will still increase, but not by as much as with the cold object. This is because in the hot object the molecules are already moving around vigorously and the increased degree of movement is less for the hot object.

Why does water freeze?

fig A Why does water freeze?

We will now apply these principles to explain why, under certain conditions, water will freeze.

Ice has lower entropy than liquid water, so ΔS_{system} is *negative*.

The process is exothermic, so $\Delta S_{surrondings}$ is *positive*.

If the magnitude of $\Delta S_{surroundings}$ > the magnitude of ΔS_{system}, then ΔS_{total} is *positive* and the water will freeze.

We will now calculate ΔS_{total} for the change of water into ice at +5 °C and −5 °C using the following data.

$$H_2O(l) \rightarrow H_2O(s) \quad \Delta H = -6010 \text{ J mol}^{-1}$$

- $S^\ominus(\text{water}) = 69.9 \text{ J K}^{-1} \text{ mol}^{-1}$
- $S^\ominus(\text{ice}) = 47.9 \text{ J K}^{-1} \text{ mol}^{-1}$

At +5 °C (278 K):

$$\Delta S_{system} = (47.9 - 69.9) = -22.0 \text{ J K}^{-1} \text{ mol}^{-1}$$

$$\Delta S_{surroundings} = -\frac{-6010}{278}$$

$$= +21.6 \text{ J K}^{-1} \text{ mol}^{-1}$$

$$\Delta S_{total} = -0.4 \text{ J K}^{-1} \text{ mol}^{-1}$$

The total entropy change is negative, meaning that the change is not thermodynamically spontaneous. The water will not freeze.

At −5 °C (268 K):

$$\Delta S_{system} = (47.9 - 69.9) = -22.0 \text{ J K}^{-1} \text{ mol}^{-1}$$

$$\Delta S_{surroundings} = -\frac{-6010}{268}$$

$$= +22.4 \text{ J K}^{-1} \text{ mol}^{-1}$$

$$\Delta S_{total} = +0.4 \text{ J K}^{-1} \text{ mol}^{-1}$$

The total entropy change is positive, meaning that the change is thermodynamically spontaneous. The water will freeze.

It is interesting to note that what has changed between +5 °C and −5 °C is the entropy of the surroundings. The reason that $\Delta S_{surroundings}$ has changed in magnitude is simply because of the temperature change. The same amount of heat energy has been transferred to the surroundings but, because the entropy of the surroundings has a higher temperature when at +5 °C than when at −5 °C, the change in entropy is smaller, as explained earlier.

So, the reason we put water into a freezer when we want to make ice is because the entropy change of the surroundings (i.e. essentially the air inside the freezer) is large enough to compensate for the decrease in entropy when the water freezes.

Questions

1. Hydrated cobalt(II) chloride dehydrates on heating according to the following equation:
$$CoCl_2.6H_2O(s) \rightarrow CoCl_2(s) + 6H_2O(l) \quad \Delta H^\ominus_{298 K} = +88.1 \text{ kJ mol}^{-1}$$
   (a) Calculate the standard entropy change of the system, $\Delta S^\ominus_{system}$.
   $S^\ominus[CoCl_2.6H_2O(s)] = 343.0 \text{ J K}^{-1} \text{ mol}^{-1}$
   $S^\ominus[CoCl_2(s)] = 109.2 \text{ J K}^{-1} \text{ mol}^{-1}$
   $S^\ominus[H_2O(l)] = 69.9 \text{ J K}^{-1} \text{ mol}^{-1}$
   (b) Calculate the standard entropy change of the surroundings, $\Delta S^\ominus_{surroundings}$, at 298 K.
   (c) Calculate the standard total entropy change, $\Delta S^\ominus_{total}$, at 298 K for this reaction.
   (d) Explain whether hydrated cobalt(II) chloride can be stored at 298 K without it dehydrating?

2. Use the data provided to calculate the total standard entropy change, $\Delta S^\ominus_{total}$, at 298 K, for the following reaction:
$$2Fe(s) + 1\tfrac{1}{2}O_2(g) \rightarrow Fe_2O_3(s)$$
   $\Delta H^\ominus = -822 \text{ kJ mol}^{-1}$
   $S^\ominus[Fe(s)] = 27.2 \text{ J K}^{-1} \text{ mol}^{-1}$
   $S^\ominus[O_2(g)] = 205.0 \text{ J K}^{-1} \text{ mol}^{-1}$
   $S^\ominus[Fe_2O_3(s)] = 90.0 \text{ J K}^{-1} \text{ mol}^{-1}$

13.2 ③ Understanding entropy changes

By the end of this section, you should be able to...

- understand why entropy changes to the system occur during:
  (i) reactions in which there is a change of state
  (ii) reactions in which there is a change in number of moles from reactants and products
  (iii) the dissolving of ionic solids in water

Reactions involving a change of state

We have already mentioned that, in general, entropy increases in the order:

solid < liquid < gas

So, we would expect an increase in the entropy of the system if a gas is produced from a reaction involving a solid and/or a liquid.

Example 1

When solid ammonium carbonate is added to pure ethanoic acid, bubbles of gas are rapidly produced. Despite its violent appearance, this is an endothermic reaction as can be shown by placing a thermometer in the acid before the ammonium carbonate is added. The temperature falls considerably as the reaction takes place.

$$2CH_3COOH(l) + (NH_4)_2CO_3(s)$$
$$\rightarrow 2CH_3COONH_4(aq) + H_2O(l) + CO_2(g)$$

Since the reaction is endothermic, $\Delta S_{surroundings}$ will be negative. However, there is a large increase in the entropy of the system, ΔS_{system}, because a gas is produced from a liquid and a solid. The magnitude of ΔS_{system} is greater than that of $\Delta S_{surroundings}$, and this makes ΔS_{total} positive, so the reaction is thermodynamically spontaneous.

Example 2

Hydrated barium hydroxide reacts with solid ammonium chloride in a rapid endothermic reaction at room temperature:

$$Ba(OH)_2.8H_2O(s) + 2NH_4Cl(s)$$
$$\rightarrow BaCl_2(s) + 10H_2O(l) + 2NH_3(g)$$

As in the previous example, the driving force of the reaction is ΔS_{system}, which overcomes the negative value of $\Delta S_{surroundings}$ caused by the endothermic nature of the reaction. The reactants are solids and the products are a solid, a liquid and a gas.

Example 3

Magnesium burns in oxygen to form solid magnesium oxide:

$$Mg(s) + \tfrac{1}{2}O_2(g) \rightarrow MgO(s)$$

The reaction is highly exothermic, so $\Delta S_{surroundings}$ will be positive. ΔS_{system} is negative since a solid and a gas are changing into a solid. However, the magnitude of the value of $\Delta S_{surroundings}$ is greater than that of ΔS_{system}, making ΔS_{total} positive. So, the reaction is thermodynamically spontaneous.

Reactions involving a change in number of moles from reactants to products

If you increase the number of moles that you have, you automatically increase the number of particles (i.e. atoms, molecules or ions) present. This will result in an increase in the number of ways that the particles can be arranged, and this increases the entropy of the system, making ΔS_{system} positive.

We will now look again at the reactions shown above, but this time considering the change in the number of moles from reactants to products.

Example 1

$$2CH_3COOH(l) + (NH_4)_2CO_3(s)$$
$$\rightarrow 2CH_3COONH_4(s) + H_2O(l) + CO_2(g)$$

Number of moles of reactants = 3
Number of moles of products = 4
So, ΔS_{system} is positive.

Example 2

$$Ba(OH)_2.8H_2O(s) + 2NH_4Cl(s)$$
$$\rightarrow BaCl_2(s) + 10H_2O(l) + 2NH_3(g)$$

Number of moles of reactants = 3
Number of moles of products = 13
So, ΔS_{system} is positive.

Example 3

$$Mg(s) + \tfrac{1}{2}O_2 \rightarrow MgO(s)$$

Number of moles of reactants = 1.5
Number of moles of products = 1
So, ΔS_{system} is negative.

Dissolving ionic solids in water

When an ionic solid dissolves in water, two changes take place:

- the lattice structure is broken down, and
- the ions become hydrated.

The breaking down of the lattice structure is an endothermic process, equivalent to the reverse of the lattice energy. It also results in an increased number of moles of particles present, so increases the entropy of the salt.

The hydration of the ions is an exothermic process, but results in the water molecules becoming more ordered as they arrange themselves in an orderly manner around the positive and negative ions. The increase in ordering of the water molecules produces a decrease in entropy of the water. This ordering of the water molecules is particularly significant when dissolving anhydrous solids in water.

To explain why some ionic solids are soluble in water, whilst others are insoluble, we need to look at both the enthalpy and entropy changes involved.

The solubility of an ionic solid is determined by the total entropy change for the solid.

$$\Delta S_{total} = \Delta S_{system} + \Delta S_{surroundings}$$

Since $\Delta S_{surroundings} = -\dfrac{\Delta_{sol}H}{T}$

This expression becomes:

$$\Delta S_{total} = \Delta S_{system} - \dfrac{\Delta_{sol}H}{T}$$

The value of ΔS_{total}, and hence the solubility of the solid, depends on the values of three factors:

- The entropy change of the system, ΔS_{system}.
- The enthalpy change of solution, $\Delta_{sol}H$.
- The temperature, in Kelvin, of the water, T.

Let us have a close look at the dissolving of ammonium nitrate crystals in water at a temperature of 298 K.

$$NH_4NO_3(s) \xrightarrow{aq} NH_4^+(aq) + NO_3^-(aq)$$

$\Delta_{sol}H = +25.8 \text{ kJ mol}^{-1}$

$\Delta S_{system} = S[NH_4^+(aq)] + S[NO_3^-(aq)] - S[NH_4NO_3(s)]$

$\qquad = +113.4 + 146.4 - 151.1 \text{ J K}^{-1} \text{ mol}^{-1}$

$\qquad = +108.7 \text{ J K}^{-1} \text{ mol}^{-1}$

$\Delta S_{surroundings} = -\dfrac{\Delta_{sol}H}{T} = -\dfrac{+25800}{298} = -86.6 \text{ J K}^{-1} \text{ mol}^{-1}$

$\Delta S_{total} = (+108.7 - 86.6) = +22.1 \text{ J K}^{-1} \text{ mol}^{-1}$

Since ΔS_{total} is positive, the dissolving of ammonium nitrate in water at 298 K is thermodynamically spontaneous. Since the activation energy for this process is very low, ammonium chloride is soluble in water at 298 K.

We are now in a position to consider the solubility in water of some other ionic solids.

Table A shows the relevant thermodynamic data for some solids at a temperature of 298 K. The values have been quoted to the nearest whole number.

Ionic solid	$\Delta_{sol}H$/kJ mol^{-1}	$\Delta S_{surroundings}$/J mol^{-1}	ΔS_{system}/J mol^{-1}	ΔS_{total}/J mol^{-1}	Solubility
NaCl	+4	−13	+43	+56	soluble
NH$_4$Cl	+15	−50	+167	+117	soluble
AgCl	+66	−221	+33	−188	insoluble
MgSO$_4$	−91	+305	−213	+92	soluble
CuSO$_4$	−73	+245	−192	+53	soluble
CaSO$_4$	−18	+60	−145	−85	insoluble

table A

Did you know?

When we state that the solids AgCl and CaSO$_4$ are insoluble in water, this is a simplification. Both solids have a measurable solubility in water, but the solubility is very small. What we are really saying is that the position of the equilibrium in both cases is very far over to the solid side of the equation. We will come back to this point when we discuss Gibbs energy, ΔG, in **Sections 13.3.1** and **13.3.2**.

Entropy 13.2

Learning tip

At this level it is assumed that ΔS_{system} does not change with temperature, since the entropies of both the reactants and the products change by similar amounts.

The magnitude of $\Delta S_{surroundings}$ will always decrease with increasing temperature. If $\Delta S_{surroundings}$ is negative, it will become less negative. If $\Delta S_{surroundings}$ is positive it will become less positive.

Evidence from thermodynamics gives no indication of the rate at which a reaction will occur. A reaction that is thermodynamically spontaneous at a given temperature may have a high activation energy. This could result in the reaction not taking place at all.

Questions

1. Predict whether there is likely to be an increase, a decrease or no change in the entropy of the system in each of the following reactions. In each case give your reasons.
   (a) $CuSO_4.5H_2O(s) \rightarrow CuSO_4(s) + 5H_2O(l)$
   (b) $HCl(g) + NH_3(g) \rightarrow NH_4Cl(s)$
   (c) $SO_2(g) + \frac{1}{2}O_2(g) \rightarrow SO_3(g)$
   (d) $Co(H_2O)_6^{2+}(aq) + EDTA^{2-}(aq) \rightarrow Co(EDTA)(aq) + 6H_2O(l)$

2. This question is about the following reaction:

   $H_2(g) + I_2(g) \rightarrow 2HI(g)$

   (a) State why you might predict that the entropy change in the system for this reaction is zero.
   (b) The actual ΔS_{system} for this reaction is +22 J K^{-1} mol^{-1}. Carry out some research into an alternative view of entropy that would explain why the value is not zero.
   [Note: this alternative view is beyond the scope of an A level course.]

13.3 1 The Second Law and Gibbs energy

By the end of this section, you should be able to…

- understand that the balance between the entropy change and the enthalpy change determines the feasibility of a reaction and is represented by the equation $\Delta G = \Delta H - T\Delta S_{system}$
- use the equation $\Delta G = \Delta H - T\Delta S_{system}$ to:
  (i) predict whether a reaction is thermodynamically feasible
  (ii) determine the temperature at which a reaction is thermodynamically feasible

Gibbs energy, ΔG

It is not always convenient to have to perform calculations involving ΔS_{system} and $\Delta S_{surroundings}$ in order to apply the Second Law of Thermodynamics. A much more convenient approach is to use a quantity called the *Gibbs energy*. Gibbs energy can be calculated from the properties of the system alone.

Gibbs energy is related to the entropy change of the system by the expression:

$$\Delta G = \Delta H - T\Delta S_{system}$$

where T is the temperature of the system in Kelvin.

A process or a reaction is thermodynamically feasible (i.e. thermodynamically spontaneous) if ΔG is negative, but is not thermodynamically feasible if ΔG is positive. If $\Delta G = 0$, then the reaction is in equilibrium.

Let us look at making ice once again, but this time considering Gibbs energy.

$$H_2O(l) \rightarrow H_2O(s) \qquad \Delta H = -6010 \text{ J mol}^{-1} \text{ and } \Delta S_{system} = -22.0 \text{ J K}^{-1} \text{ mol}^{-1}$$

At +5 °C (278 K):

$$\Delta G = -6010 - (278 \times -22.0) = +106 \text{ J mol}^{-1}$$

ΔG is positive so the process is not thermodynamically feasible. Water will not freeze at +5 °C (278 K).

At −5 °C (268 K):

$$\Delta G = -6010 - (268 \times -22.0) = -114 \text{ J mol}^{-1}$$

ΔG is negative so the process is thermodynamically feasible. Water will freeze at −5 °C (268 K).

You will notice that the outcomes are identical to those obtained by the calculation of ΔS_{total}. However, the calculations are much easier because we have to think only of the system and not the surroundings.

Summary

ΔG will be negative when:

- $\Delta H < 0$ and $\Delta S > 0$
- $\Delta H < 0$, $\Delta S < 0$ but the magnitude of $\Delta H >$ the magnitude of $T\Delta S$
- $\Delta H > 0$, $\Delta S > 0$ but the magnitude of $\Delta H <$ the magnitude of $T\Delta S$.

ΔG will always be positive when:

- $\Delta H > 0$ and $\Delta S < 0$.
- $\Delta H > 0$, $\Delta S > 0$ but the magnitude of $\Delta H >$ the magnitude of $T\Delta S$
- $\Delta H < 0$, $\Delta S < 0$ but the magnitude of $\Delta H <$ the magnitude of $T\Delta S$

Thermodynamic feasibility of chemical reactions

Let us now apply this concept to a chemical reaction; the thermal decomposition of calcium carbonate.

$$CaCO_3(s) \rightarrow CaO(s) + CO_2(g) \quad \Delta H = +178 \text{ kJ mol}^{-1}$$

The entropies of $CaCO_3(s)$, $CaO(s)$ and $CO_2(g)$ are 89, 40 and 214 J K^{-1} mol^{-1} respectively.

Calculate ΔG for the reaction at 25 °C and comment on the value obtained.

$$\Delta S_{system} = 40 + 214 - (+89) = +165 \text{ J K}^{-1} \text{ mol}^{-1}$$
$$\Delta G = (178 \times 1000) - (298 \times 165) = +128\,830 \text{ J mol}^{-1}$$

Since ΔG is positive, the reaction is not thermodynamically feasible. So calcium carbonate is thermodynamically stable at 25 °C.

We can also determine the minimum temperature at which the decomposition becomes thermodynamically feasible because at this temperature ΔG is zero.

When $\Delta G = 0$, $\Delta H = T \Delta S_{system}$. Rearranging this expression gives:

$$T = \frac{\Delta H}{\Delta S_{system}} = \frac{178\,000}{165} = 1079 \text{ K}$$

So, calcium carbonate starts to decompose when heated to a temperature of 1079 K (806 °C).

It is interesting to note that the value of ΔH does not vary much with temperature, but the value of ΔG changes significantly. The change in ΔG with temperature for the thermal decomposition of calcium carbonate is shown in **fig A**.

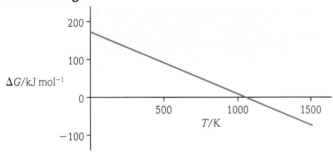

fig A Change in ΔG with temperature for the decomposition of $CaCO_3$.

Questions

1. In the Haber process for the manufacture of ammonia, a mixture of nitrogen and hydrogen is heated to a temperature of 823 K.

$$N_2(g) + 3H_2(g) \rightleftharpoons 2NH_3(g) \quad \Delta H^\ominus = -92.0 \text{ kJ mol}^{-1}$$

The standard entropies of nitrogen, hydrogen and ammonia are:
$S^\ominus[N_2(g)] = 191$ J K^{-1} mol^{-1}; $S^\ominus[H_2(g)] = 132$ J K^{-1} mol^{-1}; $S^\ominus[NH_3(g)] = 192$ J K^{-1} mol^{-1}.

(a) Show that the formation of ammonia from nitrogen and hydrogen gas is thermodynamically feasible at 298 K.

(b) Explain why this reaction is not thermodynamically feasible at high temperatures.

(c) Suggest why a temperature of 823 K is used despite the reaction being feasible at 298 K.

2. (a) The standard enthalpy changes for the melting and boiling of water are:

$$H_2O(s) \rightarrow H_2O(l) \quad \Delta H^\ominus = +6.01 \text{ kJ mol}^{-1}$$
$$H_2O(l) \rightarrow H_2O(g) \quad \Delta H^\ominus = +40.7 \text{ kJ mol}^{-1}$$

Explain why $\Delta H^\ominus$ for each of the changes is positive.

(b) The standard entropies of water in its three states are:
$S^\ominus[H_2O(g)] = 189$ J K^{-1} mol^{-1}; $S^\ominus[H_2O(l)] = 69.9$ J K^{-1} mol^{-1}; $S^\ominus[H_2O(s)] = 48.0$ J K^{-1} mol^{-1}

(i) Explain why there is an increase in entropy when water melts.

(ii) Explain why the increase in entropy for the change $H_2O(l) \rightarrow H_2O(g)$ is greater than that for the change $H_2O(s) \rightarrow H_2O(l)$.

(c) Use the data from parts (a) and (b) to calculate the minimum temperature at which water will freeze at standard pressure.

13.3 2 Gibbs energy and equilibrium

By the end of this section, you should be able to...

- use the equation $\Delta G = -RT\ln K$ to show that reactions that are thermodynamically feasible have large values for the equilibrium constant and vice versa
- understand why a reaction for which the ΔG value is negative may not occur in practice
- understand that reactions that are thermodynamically feasible may be inhibited by kinetic factors

Relationship between the equilibrium constant, K, and the Gibbs energy change, ΔG

The equilibrium constant and the Gibbs energy change are related by the equation:

$$\Delta G = -RT\ln K$$

where R is the gas constant (8.31 J mol^{-1} K^{-1})

This equation can be rearranged to give K as:

$$K = e^{\left(-\frac{\Delta G}{RT}\right)}$$

Learning tip

You may not have previously come across the concept of e. It is an important mathematical constant that is approximately equal to 2.718. The 'natural logarithim', ln, is the logarithim to base e. Using this information, you could have a go at rearranging the equation yourself.

If ΔG is negative, the exponent (the expression in brackets) is positive. The equilibrium constant will be greater than 1, which means that the products are favoured.

However, if ΔG is positive, the exponent will be negative giving an equilibrium constant of less than 1. In this case the reactants are favoured.

Notice that we are suggesting that if ΔG is negative, the reaction will reach a position of equilibrium that favours the products and if ΔG is positive, the reaction will reach a position of equilibrium that favours the reactants. We are not suggesting that if a reaction has a positive ΔG, then no reaction takes place. It is perfectly possible for *some* of the reactants to be converted into products, but impossible for them *all* to be converted into products.

This situation can easily be understood by observing the change in Gibbs energy with the percentage of B present for the equilibrium:

$$A \rightleftharpoons B$$

for which ΔG is positive.

The change in Gibbs energy for such a reaction is shown in **fig A**.

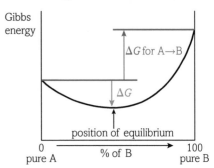

fig A Change in Gibbs energy against the percentage of B in the equilibrium mixture for the reaction A $\rightleftharpoons$ B.

Going from pure A to pure B would involve an increase in Gibbs energy and so is not allowed.

However, going from pure A to the equilibrium mixture involves a decrease in the Gibbs energy (ΔG is negative) and is allowed. A reaction with a positive ΔG is not 'forbidden' but simply comes to a position of equilibrium that favours the reactants.

Learning tip

In the equation $\Delta G = -RT\ln K$, K has no units. This value of the equilibrium constant is known as the thermodynamic equilibrium constant, as opposed to the experimental value of K which does have units (see **Section 11.1.1**).

Calculating K for a reaction

For the following reaction at 298 K:

$$SO_2(g) + \tfrac{1}{2}O_2(g) \rightarrow SO_3(g)$$

$$\Delta G = -71 \text{ kJ mol}^{-1}$$

$$K = e^{\left(-\frac{\Delta G}{RT}\right)} = e^{\left(-\frac{-71 \times 1000}{8.31 \times 298}\right)} = 2.83 \times 10^{12}$$

The equilibrium constant is very large, suggesting that the equilibrium position will be almost totally in favour of the products in this reaction.

Both the sign and the magnitude of ΔG tell us something about the position of equilibrium of a reaction. Because of the exponential relationship between ΔG and K, once ΔG becomes more negative than a certain value, K becomes so large that the reaction has effectively gone to completion. That is, all the reactants have been converted into products.

Similarly, once ΔG becomes more positive than a certain value K becomes so small that the equilibrium lies entirely to the reactants side. That is, the reaction does not proceed to a significant extent and hardly any reactants, if any at all, have been converted into products.

Fig B shows the relationship between K and ΔG.

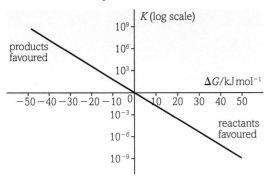

fig B Relationship between the equilibrium constant, K, and the Gibbs energy change, ΔG.

Once ΔG is more negative than about $-40\,\text{kJ}\,\text{mol}^{-1}$, K is so large that all of the reactants have effectively changed into products.

Once ΔG is greater than $+40\,\text{kJ}\,\text{mol}^{-1}$, K is so small that essentially none of the reactants have changed into products.

Relationship between the equilibrium constant and temperature

The two equations:

$$\Delta G = -RT\ln K$$

and:

$$\Delta G = \Delta H - T\Delta S_{system}$$

both involve temperature. If we combine the two we obtain the expression:

$$-RT\ln K = \Delta H - T\Delta S_{system}$$

Dividing both sides by $-RT$ gives:

$$\ln K = -\frac{\Delta H}{RT} + \frac{\Delta S_{system}}{R}$$

This equation demonstrates that the way the equilibrium constant varies with temperature depends upon the sign of ΔH.

If the forward reaction is exothermic, ΔH is negative, and the term $-\dfrac{\Delta H}{RT}$ is positive. If the temperature is increased, $-\dfrac{\Delta H}{RT}$ will become less positive and so $\ln K$, and hence K, decreases. Thus, increasing the temperature for an exothermic reaction results in the equilibrium shifting to the left, that is in the endothermic direction. If the forward reaction is endothermic, ΔH is positive, and the term $-\dfrac{\Delta H}{RT}$ is negative. If the temperature is increased, $-\dfrac{\Delta H}{RT}$ will become less negative and so $\ln K$, and hence K, increases.

Both of these conclusions are in agreement with the qualitative statements we made in **Book 1 Section 10.1.2**.

Special cases

There are two factors to consider when applying the concept of Gibbs energy change to predict the thermodynamic feasibility of a chemical reaction.

1 Kinetic stability

A negative value for ΔG indicates that the reaction is thermodynamically feasible, but it does not necessarily mean that the reaction will take place.

An example is the reaction between hydrogen and oxygen at 298 K.

$$H_2(g) + \tfrac{1}{2}O_2(g) \rightarrow H_2O(l) \qquad \Delta G^\ominus = -229\,\text{kJ}\,\text{mol}^{-1}$$

This reaction has a large negative ΔG value, but a mixture of hydrogen gas and oxygen gas does not spontaneously react to form water at 298 K. This is because the activation energy for the reaction is very high, so very few, if any, collisions result in reaction. We say that the reactants are *kinetically stable*.

2 Non-standard conditions

Standard Gibbs energy changes, $\Delta G^\ominus$, are calculated for a specific set of standard conditions, usually 298 K and 100 kPa pressure. Any solutions involved would be at a concentration of $1\,\text{mol}\,\text{dm}^{-3}$.

The following reaction between hydrochloric acid and manganese(IV) oxide is not thermodynamically feasible under standard conditions because the value for $\Delta G^\ominus$ is positive.

$$4HCl(aq) + MnO_2(s) \rightarrow MnCl_2(aq) + H_2O(l) + Cl_2(g)$$

However, this reaction is commonly used to generate chlorine in the laboratory. If concentrated hydrochloric acid ($\sim 10\,\text{mol}\,\text{dm}^{-3}$) is used and the reaction mixture is heated, the reaction becomes thermodynamically feasible because ΔG is negative under these conditions.

Questions

1. The equation for the disproportionation of carbon monoxide into carbon and carbon dioxide is:

    $$2CO(g) \rightleftharpoons C(s) + CO_2(g)$$

    Some thermodynamic data for this reaction is given below:

    $\Delta S^\ominus(298\,\text{K}) = -175.0\,\text{J}\,\text{K}^{-1}\,\text{mol}^{-1}$
    $\Delta H^\ominus(298\,\text{K}) = -172.5\,\text{kJ}\,\text{mol}^{-1}$
    $\Delta G^\ominus(298\,\text{K}) = -120.1\,\text{kJ}\,\text{mol}^{-1}$

    (a) Calculate the equilibrium constant, K, for this reaction.
    (b) Calculate the temperature above which the disproportionation of carbon monoxide ceases to be thermodynamically feasible.
    (c) Suggest why carbon monoxide does not spontaneously disproportionate at 298 K.

2. Consider the reaction:

    $$C(s) + H_2O(g) \rightleftharpoons CO(g) + H_2(g)$$

    Some thermodynamic data for this reaction are given below:

    $\Delta S^\ominus(298\,\text{K}) = -142.9\,\text{J}\,\text{K}^{-1}\,\text{mol}^{-1}$
    $\Delta H^\ominus(298\,\text{K}) = -135.0\,\text{kJ}\,\text{mol}^{-1}$

    (a) Calculate the minimum temperature at which this reaction becomes thermodynamically feasible. What assumptions have you made in your calculation?
    (b) The Gibbs energy change for this reaction at 500 K is $+633.0\,\text{kJ}\,\text{mol}^{-1}$. Calculate the equilibrium constant, K, for this reaction at 500 K.

13.3 Some applications of Gibbs energy: further understanding

By the end of this section, you should be able to...

- understand why some ionic salts are soluble in water whilst others are insoluble
- understand why hydrofluoric acid is classified as a weak acid
- understand the trend in acid strength of the chloroethanoic acids

This section will take you beyond the scope of the current A level course, but has been included in order to show you some useful applications of the concept of entropy.

Solubility of salts

Soluble salts

In **Section 13.2.3** we saw that by considering entropy changes of both the system and the surroundings we can explain why a salt such as ammonium nitrate dissolves in water despite the fact that the process is endothermic. We are now going to revisit the question of why some salts are soluble and others are not, but this time view it by considering the Gibbs energy change of the system.

When trying to predict the extent to which a salt, for example M^+X^-, is soluble, we simply need to look at the Gibbs energy of solution, $\Delta_{sol}G^\ominus$, for the process:

$$M^+X^-(s) \xrightarrow{aq} M^+(aq) + X^-(aq)$$

If $\Delta_{sol}G^\ominus$ is negative, then the products are favoured at equilibrium and the salt is soluble.

If $\Delta_{sol}G^\ominus$ is positive, then the solid salt is favoured at equilibrium and so the salt is insoluble (or more accurately *sparingly soluble*, as we have already identified that a process with a positive Gibbs energy change can take place to a certain extent as long as the magnitude of ΔG is not too large).

The equilibrium constant for the reaction is given by the expression:

$$K = [M^+(aq)][X^-(aq)]$$

[You will remember that in a heterogeneous equilibrium the solid does not contribute to the expression.]

This equilibrium constant is called the solubility product and is usually given the symbol K_{sp}.

The relationship between $\Delta_{sol}G^\ominus$ and K_{sp} is given by the expression:

$$\Delta_{sol}G^\ominus = -RT\ln K_{sp}$$

The relationship between $\Delta_{sol}G^\ominus$, $\Delta_{sol}H^\ominus$ and ΔS_{system} is given by:

$$\Delta_{sol}G^\ominus = \Delta_{sol}H^\ominus - T\Delta S_{system}$$

Let us first of all consider three soluble salts, calcium nitrate, magnesium sulfate and sodium nitrate. The relevant data for each is given in **table A**. [T = 298 K]

Salt	$\Delta_{sol}H^\ominus$/J mol^{-1}	ΔS_{system}/J K^{-1} mol^{-1}	$T\Delta S_{system}$/J mol^{-1}	$\Delta_{sol}G^\ominus$/J mol^{-1}	K_{sp}
Ca(NO$_3$)$_2$	−19 000	+45	+13 000	−32 000	410 000
MgSO$_4$	−91 000	−210	−63 000	−28 000	82 000
NaNO$_3$	+21 000	+90	+27 000	−6000	11

table A

Each of these salts is soluble at 298 K, as is shown by the negative $\Delta_{sol}G^\ominus$ and the large values of K_{sp}.

Let us now consider why each of these salts is soluble in water.

Calcium nitrate

$\Delta_{sol}H^\ominus$ is negative so $\Delta_{hyd}H$ of the ions must be greater than $\Delta_{lattice}H$ of $Ca(NO_3)_2$.

ΔS_{system} is positive, so the increase in entropy produced by the breaking up of the lattice must be greater than the decrease in entropy of the water produced by the hydration of the ions.

Since $\Delta_{sol}H^\ominus < 0$ and $\Delta S_{system} > 0$, $\Delta_{sol}G^\ominus$ is negative (see **Section 13.2.1**). Calcium nitrate is soluble in water as both the enthalpy and the entropy terms are favourable.

Magnesium sulfate

Again, $\Delta_{sol}H^\ominus$ is negative. However, this time ΔS_{system} is also negative, since the ordering of the water molecules around the hydrated ions is much greater because both ions have a double positive charge and the charge density of the magnesium ion is that much greater than that of the calcium ion.

However, $\Delta_{sol}G^\ominus$ is still negative because the magnitude of $\Delta_{sol}H^\ominus$ is greater than the magnitude of $T\Delta S_{system}$ (see **Section 13.2.1**).

Magnesium sulfate is soluble because the favourable enthalpy term outweighs the unfavourable entropy term.

Sodium nitrate

$\Delta_{sol}H^\ominus$ and ΔS_{system} are both positive. However, the magnitude of $\Delta_{sol}H^\ominus$ is less than the magnitude of $T\Delta S_{system}$, so once again $\Delta_{sol}G^\ominus$ is negative (see **Section 13.2.1**).

Sodium nitrate is soluble because the unfavourable enthalpy term is outweighed by the favourable entropy term.

Sparingly soluble salts

Salt	$\Delta_{sol}H^\ominus$/J mol^{-1}	ΔS_{system}/J K^{-1} mol^{-1}	$T\Delta S_{system}$/J mol^{-1}	$\Delta_{sol}G^\ominus$/J mol^{-1}	K_{sp}
AgCl	+66 000	+34	+10 000	+56 000	1.5 × 10^{-10}
BaSO$_4$	+19 000	−105	−31 000	+51 000	1.1 × 10^{-9}
CaCO$_3$	−12 000	−200	−59 000	+47 000	5.7 × 10^{-9}

table B

Each of these salts is only sparingly soluble at 298 K, as is shown by the positive $\Delta_{sol}G^\ominus$ and the very small values of K_{sp}.

Let us now consider why each of these salts is only sparingly soluble in water.

Silver chloride

$\Delta_{sol}H^\ominus$ and ΔS_{system} are both positive. However, the magnitude of $\Delta_{sol}H^\ominus$ is greater than the magnitude of $T\Delta S_{system}$, so $\Delta_{sol}G^\ominus$ is positive (**Section 13.2.1**).

Silver chloride is only sparingly soluble because the unfavourable enthalpy term outweighs the favourable entropy term.

Barium sulfate

$\Delta_{sol}H^\ominus$ is positive and ΔS_{system} is negative. However, the magnitude of $T\Delta S_{system}$ is greater than that of $\Delta_{sol}H^\ominus$, which leads to a positive value of $\Delta_{sol}G^\ominus$.

Barium sulfate is sparingly soluble because both the enthalpy and the entropy terms are unfavourable.

Calcium carbonate

$\Delta_{sol}H^\ominus$ and ΔS_{system} are both negative, but the magnitude of $T\Delta S_{system}$ is greater than the magnitude of $\Delta_{sol}H^\ominus$. Hence, $\Delta_{sol}G^\ominus$ is positive (**Section 13.2.1**).

Calcium carbonate is sparingly soluble because the favourable enthalpy term is outweighed by the unfavourable entropy term.

Strength of acids

Hydrofluoric acid, HF(aq)

Why is hydrofluoric acid a weak acid, whereas hydrochloric acid is a strong acid?

It is often stated that hydrofluoric acid is a weak acid because of the high strength of the H—F bond; i.e. that the incomplete dissociation of HF molecules in water is enthalpy driven. The following data should convince you that this is not correct, and that entropy plays the predominant role.

Let us first of all look at the case of hydrochloric acid, which we know to be a strong acid in dilute aqueous solution. The two important energy terms for the dissociation of HCl in water are:

Enthalpy change:

$$\text{HCl(aq)} \rightarrow \text{H}^+(\text{aq}) + \text{Cl}^-(\text{aq}) \qquad \Delta H^\ominus = -58 \text{ kJ mol}^{-1}$$

Entropy change:

$$\text{HCl(aq)} \rightarrow \text{H}^+(\text{aq}) + \text{Cl}^-(\text{aq}) \qquad \Delta S^\ominus = -55 \text{ J K}^{-1} \text{ mol}^{-1}$$

The dissociation in water of HCl has a favourable (i.e. negative) $\Delta H^\ominus$, but an unfavourable (i.e. negative) $\Delta S^\ominus$. From these values we can calculate $\Delta G^\ominus$ at 298 K as follows.

$$\begin{aligned}\Delta G^\ominus &= \Delta H^\ominus - T\Delta S^\ominus \\ &= -58 - (298 \times -55 \times 10^{-3}) \\ &= -58 + 16.4 \\ &= -41.6 \text{ kJ mol}^{-1}\end{aligned}$$

This gives a value for the equilibrium constant, K, of 1.98×10^7, which is consistent with the description of HCl as a strong acid. The favourable enthalpy term outweighs the unfavourable entropy term.

Let us now turn our attention to hydrofluoric acid. The two important energy terms this time are:

Enthalpy change:

$$\text{HF(aq)} \rightarrow \text{H}^+(\text{aq}) + \text{F}^-(\text{aq}) \qquad \Delta H^\ominus = -9 \text{ kJ mol}^{-1}$$

Entropy change:

$$\text{HF(aq)} \rightarrow \text{H}^+(\text{aq}) + \text{F}^-(\text{aq}) \qquad \Delta S^\ominus = -71 \text{ J K}^{-1} \text{ mol}^{-1}$$

Calculating the value for $\Delta G^\ominus$ at 298 K, we obtain:

$$\begin{aligned}\Delta G^\ominus &= \Delta H^\ominus - T\Delta S^\ominus \\ &= -9 - (298 \times -71 \times 10^{-3}) \\ &= -9 + 21.2 \\ &= +12.2 \text{ kJ mol}^{-1}\end{aligned}$$

This gives a value for K of 7.73×10^{-3}, which is consistent with the description of HF as a weak acid.

On this occasion the entropy term, $T\Delta S^\ominus$, dominates making $\Delta G^\ominus$ positive. Hence, it is the entropy term that is the deciding factor in making HF a weak acid. Since the H—F bond strength is incorporated into the enthalpy term, we can discount this as the major contributing factor.

Another factor that contributes to the acid strength of HF is hydrogen bonding. The hydrogen bonding in HF is strong and this has an effect on the dissociation of the HF molecules in aqueous solution.

The chloroethanoic acids

Ethanoic acid, CH_3COOH, is a weak acid. However, as chlorine atoms are substituted for the hydrogen atoms in the methyl group the acid strength increases, as shown by the pK_a vales in **table C**.

Name of acid	Formula of acid	pK_a
ethanoic acid	CH_3COOH	4.76
chloroethanoic acid	$CH_2ClCOOH$	2.86
dichloroethanoic acid	$CHCl_2COOH$	1.29
trichloroethanoic acid	CCl_3COOH	0.65

table C

A simple explanation for what is happening is that the electron-withdrawing chlorine atoms are polarising the O—H bond in the carboxyl group, making it easier to ionise. It can also be argued that the electronegative chlorine atoms are helping to stabilise the anion formed upon ionisation.

Did you know?

Here we can use enthalpy and entropy to explain the pK_a trend observed for the chloroethanoic acids. Consider the equilibrium:

$$CH_{3-n}Cl_nCOOH(aq) \rightleftharpoons CH_{3-n}Cl_nCOO^-(aq) + H^+(aq)$$

where n = 0,1, 2 or 3.

	CH_3COOH	$CH_2ClCOOH$	$CHCl_2COOH$	CCl_3COOH
$\Delta H^\ominus$/kJ mol^{-1}	−0.08	−4.6	−0.7	+1.2
$\Delta S^\ominus$/J K^{-1} mol^{-1}	−91.6	−70.2	−27.0	−5.8
$T\Delta S^\ominus$/kJ mol^{-1}	−27.3	−20.9	−8.1	−1.7
$\Delta G^\ominus$/kJ mol^{-1}	+27.2	+16.3	+7.3	+2.9

table D

The important observation to make is that the enthalpy terms, $\Delta H^\ominus$, for all but CCl_3COOH are negligible compared to the entropy terms, $T\Delta S^\ominus$. The variation in pK_a values, and hence strength of acid, is therefore entirely an entropy effect and has nothing to do with the strength or polarity of the O—H bond in the carboxyl group.

The explanation for the increasing acidity must centre on the $\Delta S^\ominus$ values, which are becoming less negative from CH_3COOH to CCl_3COOH. The electron-withdrawing chlorine atoms are producing a spreading out of the negative charge across the whole of the anion. The charge is therefore less concentrated and so the water molecules are less tightly held, leading to a smaller reduction in entropy.

The case of CCl_3COOH is a little different from the others because the $\Delta H^\ominus$ and $T\Delta S^\ominus$ values are comparable. However, $\Delta H^\ominus$ for this acid is the only one that is positive and therefore unfavourable. Despite this CCl_3COOH is the most acidic. The less unfavourable entropy term is predominant over the enthalpy term.

THINKING BIGGER

HYDROGEN REVOLUTION

Like many car manufacturers, Toyota is looking to build a lead in the next generation of hydrogen-powered cars. On sale since 2015 is the new Toyota Mirai: '… *the nearest thing yet to an ultimate eco-car*'. The following extract from the car manufacturer's blog explains how they work.

HOW DOES TOYOTA'S FUEL CELL VEHICLE WORK?

fig A The Toyota Mirai fuel cell car came second in the What Car? 2015 Reader Award.

The revolution starts here… The countdown to the launch of Toyota's advanced new fuel cell vehicle has finally begun.

On sale in 2015, the Toyota Mirai has been described as the nearest thing yet to the ultimate eco-car, and a key step in finding a solution to energy demands and emissions issues associated with traditional petrol and diesel engines.

But Toyota's new fuel cell vehicle is much more than the realisation of cutting-edge science theory.

How do fuel cell vehicles actually work?

A fuel cell converts fuel into electricity by forcing it to react with oxygen.

Hydrogen is the most common fuel used in today's fuel cells, but almost any hydrocarbon, including gas and alcohol, can be used. Fuel cells require a constant supply of fuel and oxygen to sustain the electricity generating reaction.

It's worth pointing out that the idea of fuel cells is nothing new; in fact, the first examples were designed in the mid-1800s. However, it took more than 100 years for the idea to get off the ground – literally, as NASA refined their use for the Apollo Moon project.

This environmentally friendly and highly energy-efficient process of generating electricity in a fuel cell produces no tailpipe emissions, but lots of pure water – great news, if you are running one inside a spaceship. Back on Earth the same qualities make fuel cell vehicles ideal for achieving sustainable mobility, which is why Toyota has been striving to make this technology widely available as soon as possible.

That understood, it is now necessary to explain the functions of the two primary components used in a fuel cell vehicle.

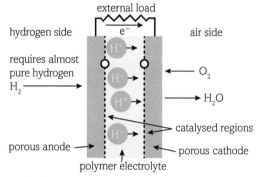

fig B How a fuel cell makes electricity.

Fuel cell

The fuel cell generates electricity through a chemical reaction between hydrogen and oxygen. This is achieved by supplying hydrogen to the negative anode of the fuel cell and ambient air to the positive cathode.

A fuel cell consists of individual cells within a membrane electrode assembly (MEA) sandwiched between separators. The MEA consists of a polymer electrolyte membrane with positive and negative catalyst layers on either side. Each cell yields less than one volt of electricity, so several hundred cells are connected in series to increase the voltage. This combined body of cells is called a stack, which is commonly referred to as a fuel cell unit.

Though it is possible to use other elements in a fuel cell, the advantage of hydrogen is its high energy efficiency. Since electricity can be produced directly from hydrogen without combustion, it is possible to convert 83% of the energy within a hydrogen molecule into electricity. This is more than double the energy efficiency of a petrol-powered engine.

High-pressure hydrogen tank

Hydrogen is stored in two high-pressure (70 MPa) tanks. The innermost layer features a polyamide resin liner that has high strength and superb resistance to hydrogen permeation. This is necessary because the diameter of hydrogen molecules is the smallest known to science and tend to escape through inferior materials.

Further use of optimal materials has increased tank capacity (historically, most FCVs have needed four separate tanks to improve capacity and therefore cruising range) and reduced weight. This can be seen in the winding angle, tension, volume and wall thickness of the carbon fibre used in the outer shell.

Where else will I encounter these themes?

| Book 1 | 11 | 12 | 13 |

Thinking Bigger

Let us start by considering the nature of the writing in the article.

1. Write a list of any advantages and disadvantages of fuel cell vehicles that are mentioned in the blog post. Do you think that the post provides a balanced viewpoint of the technology? Justify your answer.

> Always remember to consider the source when you read scientific writing. Does the author have an interest in presenting a particular viewpoint?

Now we will look at the chemistry in detail. Some of these questions will link to topics earlier in this book or **Book 1**, so you may need to combine concepts from different areas of chemistry to work out the answers.

2. a. Write a balanced chemical equation for the reaction between hydrogen and oxygen to produce water.
   b. Explain why the volumes of hydrogen to oxygen must be in the ratio of 2:1 to ensure complete combustion.
   c. The enthalpy of combustion of one mole of hydrogen to produce water is -286 kJ mol^{-1} under standard conditions. The extract says that '83% of this energy can be converted into electricity' in a fuel cell. Calculate the energy converted to electricity by 50 dm^3 of hydrogen (measured at standard temperature and pressure) using the enthalpy of combustion value given.

> If the word 'explain' is used in a question, your answer needs to include a justification. This could involve reasoning or a mathematical explanation.

3. Use the data in the table below to answer this question.

Half-equation	Standard electrode potential/eV
$H^+ + e \rightleftharpoons \frac{1}{2} H_2$	0.00
$\frac{1}{2} O_2 + 2H^+ + 2e \rightleftharpoons H_2O$	+1.23

   a. Write down the balanced equation for the reaction of hydrogen with oxygen in a fuel cell and calculate the cell emf.
   The ΔG for this reaction can be calculated from the expression:
   $$\Delta G = -nFE_{cell}$$
   where F is the Faraday constant (96 500 C mol^{-1}) and n = number of electrons transferred.
   b. What does ΔG represent and what is its significance for a reaction?
   c. Calculate a value for ΔG from the information above giving units for your answer.
   d. How do the values calculated in questions **2c** and **3c** compare? Why would you expect them to be: (i) similar, (ii) different?

4. State two challenges that manufacturers of fuel cell vehicles may face.

Activity

The concept of Gibbs energy ΔG is a central one to chemistry in that it links a number of areas of the syllabus:
- $\Delta G = -RT\ln K$
- $\Delta G = -nFE$
- $\Delta G = \Delta H - T\Delta S_{system}$

Choose one of the three equations above and give a short (5-8 slides) presentation on the application of the equation to an aspect of the chemistry you have covered to date. You should also point out the limitations of using the equation in making predictions about chemical reactions.

• From *Toyota The Official Blog of Toyota UK* by Joe Clifford blog.toyota.co.uk/how-does-toyotas-fuel-cell-vehicle-work

Did you know?

Although predicted as early as 1935, it was only in 2011 that the creation of the metallic state of hydrogen was finally reported. At a pressure of 260 billion Pa, scientists at the Max Planck institute in Germany reported that this form of hydrogen had finally been made. Metallic hydrogen helps explain, among other things, the strong magnetic field generated by the 'gas giant' Jupiter, which unlike Earth has no iron core.

13 Exam-style questions

1 (a) Identify which equation defines the lattice energy of the compound X^+Y^-. [1]

   A $X(g) + Y(g) \rightarrow X^+Y^-(s)$
   B $X(s) + Y(s) \rightarrow X^+Y^-(s)$
   C $X^+(g) + Y^-(g) \rightarrow X^+Y^-(s)$
   D $X^+(s) + Y^-(s) \rightarrow X^+Y^-(s)$

   (b) Identify which of the following compounds has the largest (i.e. the most exothermic) lattice energy. [1]

   A caesium iodide
   B lithium fluoride
   C rubidium chloride
   D sodium bromide

   (c) The Born-Haber cycle shown can be used to calculate the lattice energy for magnesium oxide.

   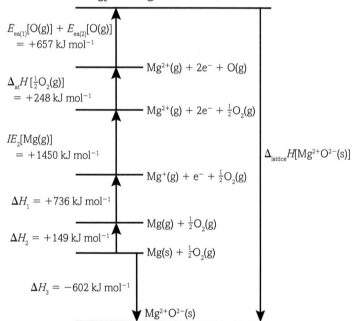

   (i) Give the name of each of the enthalpy changes labelled ΔH_1, ΔH_2 and ΔH_3. [3]
   (ii) Write the missing formulae from the top of the Born-Haber cycle. Include state symbols. [2]
   (iii) The equations for the first and second electron affinities of oxygen are:

   $O(g) + e^- \rightleftharpoons O^-(g)$ $\quad E_{ea(1)} = -141\,\text{kJ mol}^{-1}$
   $O^-(g) + e^- \rightleftharpoons O^{2-}(g)$ $\quad E_{ea(2)} = +798\,\text{kJ mol}^{-1}$

   Suggest why the second electron affinity of oxygen is endothermic. [2]
   (iv) Use the data in the Born-Haber cycle to calculate the lattice energy of magnesium oxide. [2]

   [Total: 11]

2 Lattice energies can be found *either* via experimentally determined energy changes, using the Born–Haber cycle, *or* by theoretical calculation using the ionic model for the crystalline compound.

   (a) State the energy changes that need to be determined in order to find the lattice energy of sodium fluoride, using the Born-Haber cycle. [4]

   (b) The table lists the experimentally determined and theoretically calculated lattice energies for some compounds.

Compound	NaF	NaI	AgF	AgI
Experimental value/kJ mol^{-1}	−918	−705	−958	−889
Theoretical value/kJ mol^{-1}	−912	−687	−920	−778

   Comment on the discrepancies between the experimental and theoretical values of the lattice energies. [4]

   (c) The enthalpy changes of hydration, in kJ mol^{-1}, of the ions in the compounds listed in the table in part (b) are:

   Na^+ −406 $\quad Ag^+$ −464 $\quad F^-$ −506 $\quad I^-$ −293

   (i) Using the experimental lattice energy values, show by calculation which of the four compounds is the most soluble. [2]
   (ii) Explain what other information is necessary in order to make a valid comparison of the solubility of the four compounds. [2]

   [Total: 12]

3 Entropy changes are important in determining the feasibility of a process.

   (a) For each of the four processes, predict the likely sign of the entropy change of the system, ΔS_{system}. In each case justify your answer. [4]

   (i) $N_2(g) + 3H_2(g) \rightarrow 2NH_3(g)$
   (ii) $H_2O(s) \rightarrow H_2O(l)$
   (iii) $2Na(s) + O_2(g) \rightarrow Na_2O_2(s)$
   (iv) $S(s) + O_2(g) \rightarrow 2SO_2(g)$

(b) Magnesium oxide is used as a refractory lining for furnaces. It can be made by heating magnesium carbonate.

$$MgCO_3(s) \rightarrow MgO(s) + CO_2(g) \quad \Delta H^\ominus = +117 \text{ kJ mol}^{-1}$$

The table shows the standard entropies at 298 K of $MgCO_3(s)$, $MgO(s)$ and $CO_2(g)$.

Substance	$MgCO_3(s)$	$MgO(s)$	$CO_2(g)$
$S^\ominus$/J K^{-1} mol^{-1}	65.7	27.0	214.0

(i) Use the data to show that the magnesium carbonate is stable at 298 K. [4]

(ii) Calculate the minimum temperature at which magnesium carbonate will decompose. [2]

[Total: 10]

4 (a) The overall equation for the reaction between two chemicals, J and M, is

$$J + 2M \rightarrow Q + R$$

This reaction occurs spontaneously at room temperature. Identify which of the following must be true? [1]

A $\Delta H^\ominus$ is negative
B $\Delta H^\ominus$ is positive
C $\Delta S^\ominus_{total}$ is negative
D $\Delta S^\ominus_{total}$ is positive

(b) Barium carbonate decomposes at high temperature to form barium oxide and carbon dioxide:

$$BaCO_3(s) \rightarrow BaO(s) + CO_2(g)$$

Barium carbonate is thermodynamically stable at 298 K because for this reaction:

A the activation energy is high.
B the standard enthalpy change of reaction, $\Delta H^\ominus$, is positive.
C the standard entropy change of the system ($\Delta S^\ominus_{system}$) is positive.
D the standard entropy change of the system ($\Delta S^\ominus_{system}$) is negative. [1]

(c) The equation for the combustion of hydrogen is:

$$H_2(g) + \tfrac{1}{2}O_2(g) \rightarrow H_2O(l) \quad \Delta H^\ominus = -285.5 \text{ kJ mol}^{-1}$$

The table gives the standard entropies at 298 K for $H_2(g)$, $O_2(g)$ and $H_2O(l)$.

Substance	$H_2(g)$	$O_2(g)$	$H_2O(l)$
$S^\ominus$/J K^{-1} mol^{-1}	131.0	205.0	69.9

(i) Calculate the standard entropy change of the system at 298 K, $\Delta S^\ominus_{system}$, for the combustion of hydrogen. [2]

(ii) Calculate the standard entropy change of the surroundings at 298 K, $\Delta S^\ominus_{surroundings}$, for the combustion of hydrogen. [2]

(iii) Calculate the total standard entropy change at 298 K, $\Delta S^\ominus_{total}$, for the combustion of one mole of hydrogen. Give your answer to an appropriate number of significant figures. [3]

(d) Explain why hydrogen does not react with oxygen unless the mixture is ignited. [1]

[Total: 10]

5 Ammonia can be made by reacting together nitrogen and hydrogen. The reaction is reversible.

$$N_2(g) + 3H_2(g) \rightleftharpoons 2NH_3(g) \quad \Delta_r H^\ominus (700 \text{ K}) = -110.2 \text{ kJ mol}^{-1}$$

$$\Delta S^\ominus_{total} (700 \text{ K}) = -78.7 \text{ J K}^{-1} \text{ mol}^{-1}$$

(a) The $^\ominus$ symbol represents standard conditions of a specified temperature and standard pressure. In this instance, the temperature specified is 700 K. State the standard pressure for thermodynamic measurements. [1]

(b) Calculate $\Delta S^\ominus_{surroundings}$ (700 K). [2]

(c) Calculate $\Delta S^\ominus_{system}$ (700 K). [2]

(d) If the reaction mixture is left for long enough, in contact with a suitable catalyst, it is possible to establish an equilibrium between the nitrogen, hydrogen and the ammonia.

Comment on what the value of $\Delta S^\ominus_{total}$ (700 K) indicates about the relative proportions of nitrogen, hydrogen and ammonia in the equilibrium mixture at 700 K. [1]

(e) (i) Deduce how the value of $\Delta S^\ominus_{surroundings}$ changes as the temperature increases above 700 K. [1]

(ii) Explain how this change affects the value of $\Delta S^\ominus_{total}$ and the equilibrium constant, K_p, as the temperature increases. [3]

(iii) Assuming that the reaction reaches equilibrium, use your answer to (e)(ii) to give a possible disadvantage of using a temperature higher than 700 K. [1]

(iv) State one advantage of using a temperature higher than 700 K. [1]

[Total: 12]

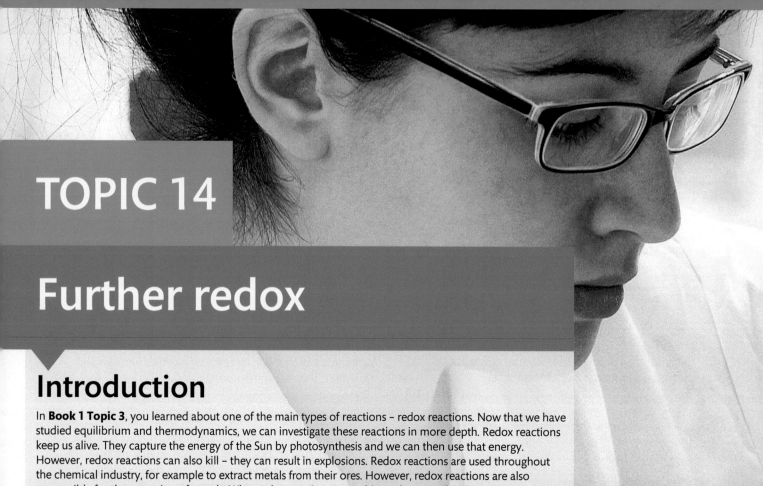

TOPIC 14

Further redox

Introduction

In **Book 1 Topic 3**, you learned about one of the main types of reactions – redox reactions. Now that we have studied equilibrium and thermodynamics, we can investigate these reactions in more depth. Redox reactions keep us alive. They capture the energy of the Sun by photosynthesis and we can then use that energy. However, redox reactions can also kill – they can result in explosions. Redox reactions are used throughout the chemical industry, for example to extract metals from their ores. However, redox reactions are also responsible for the corrosion of metals. What redox reactions can achieve, they can also destroy.

In **Topic 13**, we saw that some chemical reactions take place spontaneously. In this topic we shall look at the branch of chemistry know as electrochemistry, which deals with the use of spontaneous reactions to produce electricity.

This topic also has a link to **Topic 12**. In that topic we saw that a great deal of chemistry can be discussed in terms of the transfer of protons. In this topic we shall see that another large area of chemistry can be understood in terms of the transfer of another fundamental particle – the electron. Proton transfer is the basis of acid–base reactions; electron transfer is the basis of redox reactions.

All the maths you need

- Recognise and make use of appropriate units in calculations
- Recognise and use expressions in decimal and ordinary form
- Use an appropriate number of significant figures
- Change the subject of an equation
- Solve algebraic equations

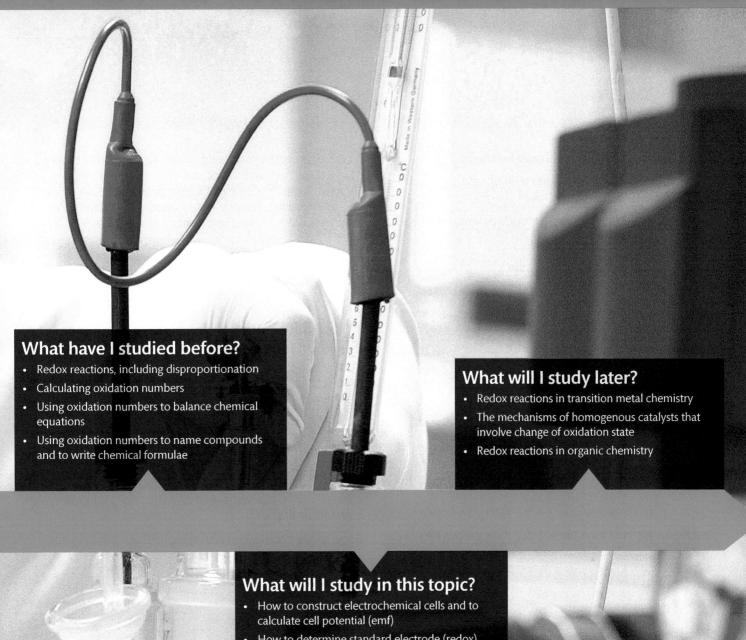

What have I studied before?
- Redox reactions, including disproportionation
- Calculating oxidation numbers
- Using oxidation numbers to balance chemical equations
- Using oxidation numbers to name compounds and to write chemical formulae

What will I study later?
- Redox reactions in transition metal chemistry
- The mechanisms of homogenous catalysts that involve change of oxidation state
- Redox reactions in organic chemistry

What will I study in this topic?
- How to construct electrochemical cells and to calculate cell potential (emf)
- How to determine standard electrode (redox) potentials
- Using standard electrode (redox) potentials to predict feasibility of chemical reactions
- Storage cells
- Redox titrations

14.1 1 Standard electrode (redox) potentials

By the end of this section, you should be able to...

- understand what is meant by the term 'standard electrode (redox) potential', $E^{\ominus}$
- understand that the standard electrode potential, $E^{\ominus}$, refers to conditions of:
  (i) 298 K temperature
  (ii) 100 kPa (1 bar) pressure of gases
  (iii) 1.00 mol dm^{-3} concentration of ions
- understand the features of a standard hydrogen electrode and understand why a reference electrode is necessary
- understand that different methods are used to measure the standard electrode potentials of:
  (i) metals or non-metals in contact with their ions in aqueous solution
  (ii) ions of the same element with different oxidation numbers
- understand that standard electrode potentials can be listed as an electrochemical series

Revision of oxidation and reduction

You should remember from **Book 1** that there are two very useful definitions of oxidation and reduction. One is in terms of loss or gain of electrons. The other is in terms of changes to the oxidation number of an element.

In terms of electrons:
- Oxidation is the loss of electrons.
- Reduction is the gain of electrons.

In terms of changes in oxidation number:
- An element in a species is oxidised when its oxidation number increases.
- An element in a species is reduced when its oxidation number decreases.

These definitions apply to all elements whether they are in the s-, p- or d-block of the Periodic Table.

How well you will understand the topic of electrode potentials depends very heavily on your appreciation of redox reactions. If you are at all unsure of your understanding, then we suggest that you revisit the topic in **Book 1**.

Standard electrode (redox) potential

The two terms 'standard electrode potential' and 'standard redox potential' mean the same. The reactions involved in the measurement of a standard electrode potential are redox reactions. Hence the term redox potential is also commonly used. In line with the Edexcel specification, we will use the term 'standard electrode potential' throughout the book.

There are many ways of teaching this topic, but very often a fundamental mistake is made at the outset. It is sometimes forgotten to emphasise that the reactions involved are *equilibria*. Too often you will find that the equations are written as if they were one-way reactions, rather than reversible. This very simple mistake can make the whole topic much more difficult to understand than is necessary.

Throughout this topic we shall avoid this problem by always talking in terms of equilibria.

Background: the different tendencies of metals to release electrons to form positive ions

When a metal such as magnesium or copper is placed in water, there is a very small tendency for the metal atoms to lose electrons and go into solution as positive ions:

$$M(s) \rightarrow M^{n+}(aq) + ne^- \text{ (where M = Mg or Cu)}$$

The electrons will remain on the surface of the metal.

In a very short time there will be a build-up of electrons on the surface of the metal, and the resulting negative charge attracts positive ions. In this way, a layer of positive ions is formed surrounding the metal.

Some of the positive ions will regain their electrons from the surface of the metal and return to form part of the metal surface:

$$M^{n+}(aq) + ne^- \rightarrow M(s)$$

Eventually, a dynamic equilibrium will be established in which the rate at which ions are leaving the surface of the metal to go into solution is the same as the rate at which they are joining it from solution. This equilibrium is represented by the equation:

$$M^{n+}(aq) + ne^- \rightleftharpoons M(s)$$

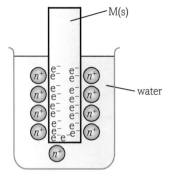

fig A Metal M immersed in water. n^+ represents metal ions.

The equations for magnesium and copper are:

$$Mg^{2+}(aq) + 2e^- \rightleftharpoons Mg(s)$$
$$Cu^{2+}(aq) + 2e^- \rightleftharpoons Cu(s)$$

The difference between magnesium and copper is that the equilibrium position will be further to the left-hand side for magnesium than for copper, because the tendency of magnesium to release electrons is greater than that of copper. So, with magnesium there will be a greater negative charge on the metal, and more positive ions in solution.

Learning tip

Note how the two equilibria are written. By convention they are written with the electrons on the *left*-hand side. Make sure you stick to this convention at all times.

In the case of both magnesium and copper, there is a potential difference between the metal and the solution. However, the potential difference is greater with magnesium than with copper. In each case, we would like to measure the potential difference between the metal and the solution. This is called the *absolute* potential difference.

However, it is not possible to measure the absolute potential difference between a metal electrode and its solution. The reason is that although it is a simple matter to connect the metal electrode to one terminal of a voltmeter, the other terminal would have to be connected to the solution, and the only way of doing this is to dip another piece of metal into the solution. This second piece of metal would create its own potential difference. So, you would be measuring the potential difference between the two pieces of metal, and not between the original metal and the solution.

The way to solve this problem is to create a reference electrode and then measure the difference in potential between this reference electrode and the metal electrode.

The standard hydrogen electrode

The reference electrode of choice is the standard hydrogen electrode. This electrode consists of hydrogen gas at a pressure of 100 kPa (1 bar) bubbling over a piece of platinum foil dipped into a solution of hydrochloric acid (or sulfuric acid) with a hydrogen ion concentration of 1 mol dm^{-3}, at a temperature of 298 K.

The surface of the platinum foil is covered in porous platinum. Porous platinum has a large surface area and allows an equilibrium between hydrogen ions in solution and hydrogen gas to be established quickly.

$$H^+(aq) + e^- \rightleftharpoons \tfrac{1}{2}H_2(g)$$

It is this equilibrium that we are going compare with all others.

Fig B shows a cross-section of a typical standard hydrogen electrode.

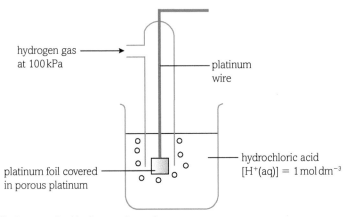

fig B A standard hydrogen electrode.

The importance of using standard conditions

You will remember that the position of an equilibrium can be changed by altering the conditions. If we are going to make fair comparisons, it is therefore necessary to standardise the conditions used. The standard conditions are:

- gas pressure, 100 kPa (1 bar)
- temperature, 298 K
- concentration of ions in solution, 1 mol dm^{-3}.

These standard conditions apply to all equilibria.

Measuring a standard electrode potential

We are now in a position to appreciate how to measure the standard electrode potential of a metal ion | metal system, such as that between magnesium ions and magnesium (Mg^{2+} | Mg).

To do this, we connect the standard hydrogen electrode to the magnesium electrode via a circuit containing a high resistance voltmeter. **Fig C** shows the arrangement of the apparatus.

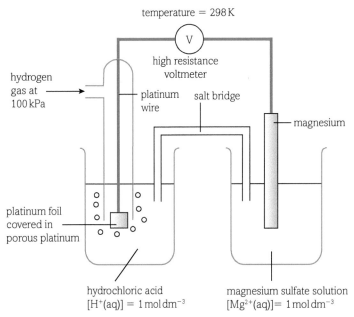

fig C Using a standard hydrogen electrode to measure a standard electrode potential.

The two components of the apparatus are known as half-cells. The two half-cells combine make a complete cell.

The salt bridge is needed to complete the electrical circuit. It usually contains a concentrated solution of potassium nitrate in the form of liquid or a gel. It functions by allowing the movement of ions. Theoretically, the salt bridge can contain any ionic salt, but neither of the ions present should interfere with the components of the half-cells. For example, if the right-hand half-cell contains a solution of silver nitrate, then potassium chloride would be unsuitable for the salt bridge because the chloride ions would interact with the silver ions to form a precipitate of silver chloride.

There is a reason why a high resistance voltmeter is used. Ideally, the voltmeter should have infinite resistance so that there is

Did you know?

The best instrument to use to measure the potential difference is a potentiometer. No electrons flow when this instrument is used, so both half-cell reactions are in equilibrium.

However, the best practical way of measuring the potential difference is to use a high resistance voltmeter.

---

no flow of electrons (i.e. no current flowing) around the external circuit. If this were possible, the reading on the voltmeter would represent the difference in potential between the two half-cells when *both reactions are in equilibrium*. The two reactions concerned are:

$$H^+(aq) + e^- \rightleftharpoons \tfrac{1}{2}H_2(g)$$
$$Mg^{2+}(aq) + 2e^- \rightleftharpoons Mg(s)$$

With the magnesium half-cell connected to the standard hydrogen electrode, the reading on the voltmeter is 2.37 V.

If the magnesium half-cell is replaced by a copper half-cell (i.e. copper metal dipped into a solution of copper sulfate with $[Cu^{2+}(aq)] = 1$ mol dm^{-3}), then the reading on the voltmeter would be 0.34 V.

There is, however, a further issue to be considered. In the case of magnesium, if the voltmeter was taken out of the circuit, electrons would flow from the magnesium electrode to the hydrogen electrode. This means that magnesium is the negative electrode of the cell.

With copper, electrons would flow from the hydrogen electrode to the copper electrode, making copper the positive electrode of the cell.

This difference in behaviour is recognised by using a sign convention. A negative sign (−) indicates that the metal electrode is negative with respect to the hydrogen electrode, whilst a positive sign indicates the reverse.

The difference in potential measured, together with the appropriate sign, is called the *standard electrode potential* of the metal. Standard electrode potential is given the symbol $E^\ominus$.

Standard electrode potentials are quoted together with the relevant half-cell reaction. For example, for magnesium and copper they are:

$$Mg^{2+}(aq) + 2e^- \rightleftharpoons Mg(s) \quad E^\ominus = -2.37 \text{ V}$$
$$Cu^{2+}(aq) + 2e^- \rightleftharpoons Cu(s) \quad E^\ominus = +0.34 \text{ V}$$

By convention, the standard electrode potential of the standard hydrogen electrode is zero.

$$H^+(aq) + e^- \rightleftharpoons \tfrac{1}{2}H_2(g) \quad E^\ominus = 0.00 \text{ V}$$

Some other $E^\ominus$ values, together with the relevant half-cell reactions, are shown in **table A**.

Half-cell reaction	$E^\ominus$/V
$Ag^+(aq) + e^- \rightleftharpoons Ag(s)$	+0.80
$Al^{3+}(aq) + 3e^- \rightleftharpoons Al(s)$	−1.66
$Cr^{3+}(aq) + 3e^- \rightleftharpoons Cr(s)$	−0.74
$Fe^{2+}(aq) + 2e^- \rightleftharpoons Fe(s)$	−0.44
$Ni^{2+}(aq) + 2e^- \rightleftharpoons Ni(s)$	−0.25
$Zn^{2+}(aq) + 2e^- \rightleftharpoons Zn(s)$	−0.76

table A

Learning tip

The sign of a standard electrode potential indicates the polarity of the electrode relative to the hydrogen electrode. This sign is fixed and does not change if the equation for the half-cell reaction is reversed, as sometimes stated elsewhere. Remember, standard electrode potentials are measured when *no electrons are flowing* and hence both half-cell reactions are *in equilibrium*.

So, if the half-cell reaction for magnesium is represented as:
$Mg(s) \rightleftharpoons Mg^{2+}(aq) + 2e^-$, the $E^\ominus$ value is still −2.37 V. It does not change to +2.37 V.

Standard electrode potential is described as being a *'sign invariant quantity'*. This means that however the reaction is represented, the sign *remains the same*.

What do $E^\ominus$ values tell us?

It is important to remember that the standard electrode potential of a metal ion | metal half-cell is the potential difference measured when the half-cell is connected to a standard hydrogen electrode.

The potential difference provides a *comparison* between the position of the metal ion | metal equilibrium and the position of equilibrium in the hydrogen electrode.

To illustrate this point, let us consider the following two standard electrode potentials:

$$Mg^{2+}(aq) + 2e^- \rightleftharpoons Mg(s) \quad E^\ominus = -2.37\,V$$
$$H^+(aq) + 2e^- \rightleftharpoons \tfrac{1}{2}H_2(g) \quad E^\ominus = 0.00\,V$$
$$Cu^{2+}(aq) + 2e^- \rightleftharpoons Cu(s) \quad E^\ominus = +0.34\,V$$

The negative sign for $Mg^{2+}(aq)$ | $Mg(s)$ indicates that the equilibrium position of this reaction is further to the left than the equilibrium position of the reaction in the hydrogen electrode.

The positive sign for $Cu^{2+}(aq)$ | $Cu(s)$ indicates that the equilibrium position of this reaction is further to the right than the equilibrium position of the reaction in the hydrogen electrode.

This is another way of saying that magnesium releases electrons more readily than hydrogen (and also than copper), but that copper releases electrons less readily than hydrogen.

Magnesium is, therefore, a better reducing agent than both hydrogen and copper. In addition, hydrogen is a better reducing agent than copper. This point will be developed later in the topic.

Learning tip

A reducing agent is a species that reduces another species by adding one or more electrons to it. When a reducing agent reacts it loses electrons and is, therefore, oxidised.

An oxidising agent is a species that oxidises another species by removing one or more electrons from it. When an oxidising agent reacts it gains electrons and is, therefore, reduced.

Learning tip

Remember, a negative $E^\ominus$ does not suggest that the equilibrium of the half-cell reaction lies to the left. It merely means that it lies *further to the left* than the equilibrium of the standard hydrogen electrode.

In the same way, a positive $E^\ominus$ means that the equilibrium of the half-cell reaction lies *further to the right* than the equilibrium of the standard hydrogen electrode.

Summary

$E^\ominus$ values provide a method of comparing the positions of equilibria when metal atoms lose electrons to form ions in solution.

- The more negative the $E^\ominus$ value, the further the equilibrium lies to the left, i.e. the more readily the metal loses electrons to form ions.
- The more positive (or less negative) the $E^\ominus$ value, the further the equilibrium lies to the right, i.e. the less readily the metal loses electrons to form ions.

Later in this topic we will extend this to include both molecules and ions losing electrons.

Electromotive force

As already mentioned, when we are determining a standard electrode potential, the potential difference is measured when *no electrons are flowing* (i.e. no current is flowing) through the external circuit. The potential difference measured under these conditions is called the **electromotive force (emf)** of the cell. The standard emf of a cell is given the symbol $E^\ominus_{cell}$ and is sometimes called the standard cell potential.

Emf values can be positive or negative, depending on the reference point against which the potential difference is measured. When measuring standard electrode potentials, the reference point is the potential of the standard hydrogen electrode, which is set at zero. The emf of a cell in which the hydrogen electrode is the positive electrode will have a negative value. Conversely, the emf of a cell in which the hydrogen electrode is the negative electrode will have a positive value.

This now allows us to define standard electrode potential as follows:

The standard electrode potential of a half-cell is the emf of a cell containing the half-cell connected to the standard hydrogen electrode. Standard conditions of 298 K, 100 kPa pressure of gases and solution concentrations of 1 mol dm^{-3} apply.

Measuring standard electrode potentials of more complicated redox systems

Systems involving gases

Suppose we wanted to measure the standard electrode potential of the following redox system:

$$\tfrac{1}{2}Cl_2(g) + e^- \rightleftharpoons Cl^-(aq)$$

We can do this by setting up a half-cell in which chlorine gas is bubbled into a solution containing chloride ions. However, in order to establish an equilibrium between the chlorine molecules and the chloride ions, and also to provide an electrical connection to the external circuit, we would need to place a piece of platinum into the solution. **Fig D** shows the set up.

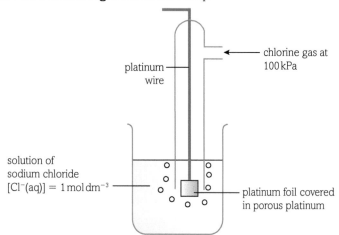

fig D $\tfrac{1}{2}Cl_2(g)|Cl^-(aq)$ half-cell.

This half-cell is then connected to a standard hydrogen electrode and the emf is measured in the usual way. The measured emf is +1.36 V. Hence:

$$\tfrac{1}{2}Cl_2(g) + e^- \rightleftharpoons Cl^-(aq) \quad E^\ominus = +1.36 \text{ V}$$

Other systems

1. Non-metal elements and their ions in solution

Examples: $\tfrac{1}{2}Br_2(aq) | Br^-(aq)$ and $\tfrac{1}{2}I_2(aq) | I^-(aq)$.

To determine the $E^\ominus$ of the redox system $\tfrac{1}{2}Br_2 | Br^-$, a half-cell containing a solution of Br_2 and Br^- ions, each of concentration 1 mol dm^{-3} is connected to a standard hydrogen electrode. A similar set up is used to determine the $E^\ominus$ of the redox system $\tfrac{1}{2}I_2 | I^-$.

$$\tfrac{1}{2}Br_2(aq) + e^- \rightleftharpoons Br^-(aq) \quad E^\ominus = +1.09 \text{ V}$$
$$\tfrac{1}{2}I_2(aq) + e^- \rightleftharpoons I^-(aq) \quad E^\ominus = +0.54 \text{ V}$$

2. Ions of the same element with different oxidation numbers

Example: $Fe^{3+}(aq) | Fe^{2+}(aq)$

The half-cell is arranged as shown in **fig E**.

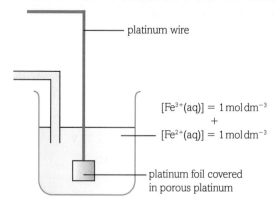

fig E $Fe^{3+}(aq)|Fe^{2+}(aq)$ half-cell.

Note that each ion has a concentration of 1 mol dm^{-3}.

$$Fe^{3+}(aq) + e^- \rightleftharpoons Fe^{2+}(aq) \quad E^\ominus = +0.77 \text{ V}$$

The electrochemical series

Arranging redox equilibria in order of their $E^\ominus$ values

The electrochemical series is built up by arranging various redox equilibria in order of their standard electrode (redox) potentials. The most negative $E^\ominus$ values are placed at the top of the series, and the most positive at the bottom.

Table B shows part of the electrochemical series.

Half-cell reaction	$E^\ominus$/V
$Li^+(aq) + e^- \rightleftharpoons Li(s)$	−3.03
$K^+(aq) + e^- \rightleftharpoons K(s)$	−2.92
$Ca^{2+}(aq) + 2e^- \rightleftharpoons Ca(s)$	−2.87
$Na^+(aq) + e^- \rightleftharpoons Na(s)$	−2.71
$Al^{3+}(aq) + 3e^- \rightleftharpoons Al(s)$	−1.66
$Zn^{2+}(aq) + 2e^- \rightleftharpoons Zn(s)$	−0.76
$Cr^{3+}(aq) + 3e^- \rightleftharpoons Cr(s)$	−0.74
$Fe^{2+}(aq) + 2e^- \rightleftharpoons Fe(s)$	−0.44
$H^+(aq) + e^- \rightleftharpoons \tfrac{1}{2}H_2(g)$	0.00
$Cu^{2+}(aq) + 2e^- \rightleftharpoons Cu(s)$	+0.34
$\tfrac{1}{2}I_2(aq) + e^- \rightleftharpoons I^-(aq)$	+0.54
$Ag^+(aq) + e^- \rightleftharpoons Ag(s)$	+0.80
$\tfrac{1}{2}Br_2(aq) + e^- \rightleftharpoons Br^-(aq)$	+1.09
$\tfrac{1}{2}Cl_2(aq) + e^- \rightleftharpoons Cl^-(aq)$	+1.36
$\tfrac{1}{2}F_2(aq) + e^- \rightleftharpoons F^-(aq)$	+2.87

table B

Reducing agents and oxidising agents

The species on the right-hand side of the half-cell reactions are all capable of behaving as reducing agents because they can lose electrons.

The most powerful reducing agent in **table B** is lithium because its redox system has the most negative $E^\ominus$ value; the equilibrium position of its half-cell reaction is furthest to the left.

The least powerful reducing agent in the list is the fluoride ion, F⁻(aq) because its redox system has the least negative (most positive) $E^\ominus$ value. The position of equilibrium of its half-cell reaction is furthest to the right.

The species on the left-hand side of the half-cell reactions are all capable of acting as oxidising agents because they can gain electrons.

The most powerful oxidising agent is fluorine and the least powerful is the lithium ion, Li^+(aq).

Table C summarises these statements.

least powerful oxidising agent ↓	Oxidised form	Reduced form	most powerful reducing agent ↑	$E^\ominus$/V
	Li^+(aq)	Li(s)		−3.04
	K^+(aq)	K(s)		−2.92
	Ca^{2+}(aq)	Ca(s)		−2.87
	Na^+(aq)	Na(s)		−2.71
	Al^{3+}(aq)	Al(s)		−1.66
	Zn^{2+}(aq)	Zn(s)		−0.76
	Cr^{3+}(aq)	Cr(s)		−0.74
	Fe^{2+}(aq)	Fe(s)		−0.44
	H^+(aq)	$\frac{1}{2}H_2$(g)		0.00
	Cu^{2+}(aq)	Cu(s)		+0.34
	$\frac{1}{2}I_2$(aq)	I^-(aq)		+0.54
	Ag^+(aq)	Ag(s)		+0.80
	$\frac{1}{2}Br_2$(aq)	Br^-(aq)		+1.09
most powerful oxidising agent	$\frac{1}{2}Cl_2$(aq)	Cl^-(aq)	least powerful reducing agent	+1.36
	$\frac{1}{2}F_2$(aq)	F^-(aq)		+2.87

table C

Did you know?

The standard electrode potentials of lithium, sodium, calcium and potassium cannot be determined experimentally because each of these metals reacts to completion with water; i.e. they do not establish an equilibrium. The $E^\ominus$ values for these metals are calculated from thermodynamic data.

The same is true for fluorine, which also reacts with water.

Key definition

The **electromotive force (emf)** is the standard electrode potential of a half-cell (measured under standard conditions of 298 K, 100 kPa pressure and concentrations of 1 mol dm⁻³) connected to a standard hydrogen electrode.

Summary

The more negative the $E^\ominus$ value, the more the equilibrium position lies toward the left – and the more readily the species on the right loses electrons. The more negative the $E^\ominus$ value, the more powerful the reducing agent.

The more positive the $E^\ominus$ value, the more the equilibrium position lies toward the right – and the more readily the species of the left loses electrons. The more positive the value, the more powerful the oxidising agent.

Questions

1. Draw a labelled diagram of apparatus that could be used to measure the standard electrode potential of a Cu^{2+}(aq)|Cu(s) half-cell.

2. The standard electrode potential of the $\frac{1}{2}Br_2$(aq)|Br^-(aq) half-cell is +1.09 V. Explain what is meant by this statement and state the significance of the positive sign.

3. What is meant by the term 'electromotive force (emf)'. How is the electromotive force of a cell measured?

14.1 2 Electrochemical cells

By the end of this section, you should be able to...

- write cell diagrams using the conventional representation of half-cells
- calculate a standard emf, $E^\ominus_{cell}$, by combining two standard electrode potentials

Electrochemical cells

An electrochemical cell is a device for producing an electric current from chemical reactions. It is constructed from two half-cells. **Fig A** shows the apparatus used to construct an electrochemical cell from a Zn^{2+} | Zn half-cell and a Cu^{2+} | Cu half-cell under standard conditions.

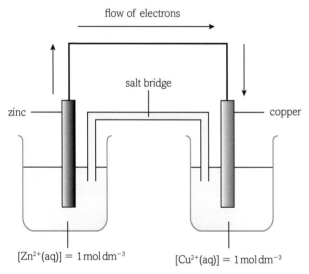

fig A An electrochemical cell consisting of Zn^{2+}|Zn and Cu^{2+}|Cu half-cells.

The relevant standard electrode potentials for the redox systems involved are:

$Zn^{2+}(aq) + 2e^- \rightleftharpoons Zn(s)$ $E^\ominus = -0.76$ V

$Cu^{2+}(aq) + 2e^- \rightleftharpoons Cu(s)$ $E^\ominus = +0.34$ V

The $E^\ominus$ value for the Zn^{2+} | Zn half-cell is the more negative, so the zinc electrode will be the negative electrode of the cell. When the cell is in operation, i.e. it is generating an electric current, electrons will flow through the external circuit from the zinc electrode to the copper electrode.

The reactions taking place under these conditions are shown below.

At the negative electrode: $Zn(s) \rightarrow Zn^{2+}(aq) + 2e^-$

At the positive electrode: $Cu^{2+}(aq) + 2e^- \rightarrow Cu(s)$

The overall cell reaction is:

$Zn(s) + Cu^{2+}(aq) \rightarrow Zn^{2+}(aq) + Cu(s)$

Cell diagrams

It is not always convenient to draw a diagram of the full apparatus for a cell. To simplify matters, chemists use a shorthand notation to represent half-cells. The half-cell made from zinc ions and zinc metal is written as:

$Zn^{2+}(aq)$ | $Zn(s)$ $E^\ominus = -0.76$ V

The solid vertical line indicates a phase boundary, in this case between an aqueous phase, $Zn^{2+}(aq)$, and a solid phase, $Zn(s)$.

Some other examples are:

$Mg^{2+}(aq)$ | $Mg(s)$ $E^\ominus = -2.37$ V

$Cu^{2+}(aq)$ | $Cu(s)$ $E^\ominus = +0.34$ V

$Al^{3+}(aq)$ | $Al(s)$ $E^\ominus = -1.66$ V

$Ag^+(aq)$ | $Ag(s)$ $E^\ominus = +0.80$ V

When the electrode, i.e. the electrical connection between the solution and the external circuit, is a piece of platinum foil, the following convention is used:

$Pt(s)$ | $Fe^{3+}(aq), Fe^{2+}(aq)$ $E^\ominus = +0.77$ V

Notice that because there is no phase boundary between $Fe^{3+}(aq)$ and $Fe^{2+}(aq)$, a comma is used to separate them, not a solid vertical line.

The standard hydrogen electrode is represented as follows:

$H^+(aq)$ | $\frac{1}{2}H_2(g)$ | $Pt(s)$

These shorthand notations can now be used to represent a cell comprising of two half-cells. Convention dictates two things:

1. The two reduced forms of the species are shown on the outside of the cell diagram.
2. The positive electrode is shown on the right-hand side of the cell diagram.

Applying these conventions produces the following cell diagram for a cell formed by combining the $Zn^{2+}(aq)$ | $Zn(s)$ and the $Cu^{2+}(aq)$ | $Cu(s)$ half-cells:

$Zn(s)$ | $Zn^{2+}(aq)$ ⁞⁞ $Cu^{2+}(aq)$ | $Cu(s)$

The double vertical lines (⁞⁞) represent the salt bridge.

The emf, $E^\ominus_{cell}$, of this cell is simply the *difference* between two standard electrode potentials of the two half-cells.

$Zn(s)$ | $Zn^{2+}(aq)$ ⁞⁞ $Cu^{2+}(aq)$ | $Cu(s)$

$E^\ominus = -0.76$ V $E^\ominus = +0.34$ V

The difference between these two numbers is 1.10, so the emf of the cell is 1.10 V. To indicate that the right-hand electrode, i.e. the copper, is the positive electrode of the cell, the emf is given a positive (+) sign.

So, the complete cell diagram is:

$$Zn(s) \mid Zn^{2+}(aq) \mathbin{\|} Cu^{2+}(aq) \mid Cu(s) \quad E^\ominus_{cell} = +1.10 \text{ V}$$

Learning tip

Note that the sign for the $E^\ominus$ of the $Zn^{2+}|Zn$ half-cell is still given as negative, despite the half-cell being written in reverse. As mentioned in **Section 14.1.1**, the sign of a standard electrode potential remains the same no matter how the half-cell is represented.

Did you know?

It is also possible to calculate the emf of the cell from a cell diagram by subtracting the $E^\ominus$ value of the left-hand half-cell from that of the right-hand half-cell:

$$E^\ominus_{cell} = E^\ominus_{right} - E^\ominus_{left}$$

It is important to remember, however, that you must not change the sign of the $E^\ominus$ value of the left-hand half-cell, even though the reaction is written as an oxidation and not a reduction.

What is meant by the 'difference' between two standard electrode potentials?

The easiest way to explain this is to represent the two values on a number scale, such as the one shown in **fig B**.

```
+ 1.00

      + 0.34  ┐
 0.00         │ difference = 1.10
      − 0.76  ┘
− 1.00
```

fig B Example of a number scale to calculate an emf.

To move from one number to the other on the number scale involves a change of 1.10 units.

Breaking the cell convention

Chemists always reserve the right to break a convention if it suits their purpose. This is exactly what we do when we draw cell diagrams to represent the measurement of a standard electrode potential. In this case, the standard hydrogen electrode is always written on the left-hand side.

The cell diagrams for the set-up used when measuring the standard electrode potential of the $Zn^{2+}(aq) \mid Zn(s)$ and the $Cu^{2+}(aq) \mid Cu(s)$ half-cells are therefore:

$$Pt(s) \mid \tfrac{1}{2}H_2(g) \mid H^+(aq) \mathbin{\|} Zn^{2+}(aq) \mid Zn(s) \quad E^\ominus_{cell} = -0.76 \text{ V}$$

$$Pt(s) \mid \tfrac{1}{2}H_2(g) \mid H^+(aq) \mathbin{\|} Cu^{2+}(aq) \mid Cu(s) \quad E^\ominus_{cell} = +0.34 \text{ V}$$

As before, the sign of $E^\ominus_{cell}$ indicates the polarity of the right-hand electrode. Zinc is the negative electrode of the cell formed in combination with the standard hydrogen electrode. Copper is the positive electrode of the cell formed in combination with the standard hydrogen electrode.

Questions

1. (a) Draw a labelled diagram of apparatus that can be used to construct a cell, under standard conditions, from a $Zn^{2+}|Zn$ half-cell and a $Fe^{3+}, Fe^{2+}|Pt$ half-cell.
   (b) The standard electrode potentials for the half-cells are:
   $$Zn^{2+}|Zn \quad E^\ominus = -0.76 \text{ V}$$
   $$Fe^{3+}, Fe^{2+}|Pt \quad E^\ominus = +0.77 \text{ V}$$
   Explain the direction of electron flow in the external circuit when this cell is in use.
   (c) Give the cell diagram for this cell and calculate the emf ($E^\ominus_{cell}$) of the cell.

2. Chlorine may be prepared in the laboratory by reacting dilute hydrochloric acid with potassium manganate(VII). The standard electrode potentials that relate to this reaction are:
   $$\tfrac{1}{2}Cl_2(g) + e^- \rightleftharpoons Cl^-(aq) \quad E^\ominus = +1.36 \text{ V}$$
   $$MnO_4^-(aq) + 8H^+(aq) + 5e^- \rightleftharpoons Mn^{2+}(aq) + 4H_2O(l) \quad E^\ominus = +1.51 \text{ V}$$
   (a) Calculate the emf, $E^\ominus_{cell}$, of a cell constructed from these two redox systems.
   (b) Explain the direction of electron flow that would take place in the external circuit of this cell when in use.
   (c) Give the cell diagram for this cell.

14.1 3 Standard electrode potentials and thermodynamic feasibility

By the end of this section, you should be able to...

- predict the thermodynamic feasibility of a reaction using standard electrode potentials
- understand how disproportionation reactions relate to standard electrode potentials
- understand the limitations of predictions made using standard electrode potentials
- understand the importance of conditions when measuring the electrode potential, E
- understand that $E^\ominus_{cell}$ is directly proportional to the total entropy change, $\Delta S^\ominus_{total}$, and to $\ln K$ for a cell reaction

Making predictions using standard electrode potentials

Using standard electrode potentials is one way of measuring how easily a species loses electrons. This method provides information on how far to the left an equilibrium is relative to the equilibrium in the standard hydrogen electrode. For example, for the following two equilibria, the equilibrium of the top reaction lies further to the left than that of the hydrogen electrode, and the equilibrium for the bottom reaction lies further to the right than that of the hydrogen electrode.

$$Zn^{2+}(aq) + 2e^- \rightleftharpoons Zn(s) \quad E^\ominus = -0.76\,V$$
$$Cu^{2+}(aq) + 2e^- \rightleftharpoons Cu(s) \quad E^\ominus = +0.34\,V$$

It follows, therefore, that the position of equilibrium of the $Zn^{2+}(aq) \mid Zn(s)$ reaction lies further to the left than that of the $Cu^{2+}(aq) \mid Cu(s)$ reaction.

If these two equilibria are linked by combining the two half-cells to make an electrochemical cell then electrons will flow from the zinc electrode to the copper electrode. Therefore, the equilibria will be disturbed and the following reactions will take place in each half-cell:

$$Zn(s) \rightarrow Zn^{2+}(aq) + 2e^-$$
$$Cu^{2+}(aq) + 2e^- \rightarrow Cu(s)$$

Electrons flow from the half-cell with the more negative $E^\ominus$ value to the half-cell with the less negative (i.e. more positive) $E^\ominus$ value.

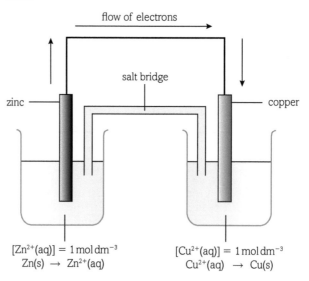

fig A Electrochemical cell showing direction of flow of electrons.

Standard electrode potential 14.1

Linking this to a test tube reaction

Is the following reaction thermodynamically feasible?

Zn(s) + Cu²⁺(aq) → Zn²⁺(aq) + Cu(s)

That is, will zinc displace copper from an aqueous solution containing copper(II) ions?

To answer this question, we will look at the $E^\ominus$ values for the relevant half-cell reactions:

Zn²⁺(aq) + 2e⁻ ⇌ Zn(s) $E^\ominus = -0.76$ V   equilibrium 1
Cu²⁺(aq) + 2e⁻ ⇌ Cu(s) $E^\ominus = +0.34$ V   equilibrium 2

The $E^\ominus$ value for equilibrium 1 is more negative than that for equilibrium 2. So, the position of equilibrium 1 will shift to the left, releasing electrons, and the position of equilibrium 2 will shift to the right, accepting electrons. This means that the reaction between zinc and copper(II) ions is thermodynamically feasible.

Fig B illustrates this process.

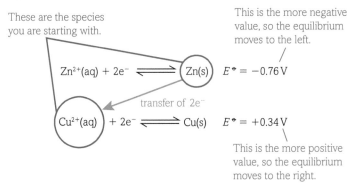

fig B Reaction of zinc with copper(II) ions in aqueous solution.

Will zinc react with dilute sulfuric acid?

To answer this question, we need to consider the following reaction:

Zn(s) + 2H⁺(aq) → Zn²⁺(aq) + H₂(g)

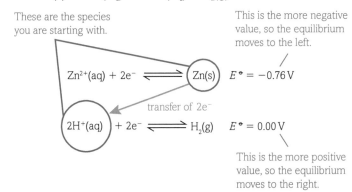

fig C Reaction of zinc with dilute sulfuric acid.

As $E^\ominus$[Zn²⁺(aq) | Zn(s)] is more negative than $E^\ominus$[2H⁺(aq) | H₂(g)], the reaction is thermodynamically feasible, as zinc will release electrons to the hydrogen ions.

Will copper react with dilute sulfuric acid?

To answer this question, we will consider the following reaction:

Cu(s) + 2H⁺(aq) → Cu²⁺(aq) + H₂(g)

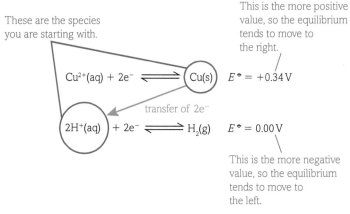

fig D Copper does not react with dilute sulfuric acid.

This time, $E^\ominus$[Cu²⁺(aq) | Cu(s)] is more positive than $E^\ominus$[2H⁺(aq) | H₂(g)], so the reaction is not thermodynamically feasible because copper will not release electrons to hydrogen ions. The two equilibria tend to move in the opposite directions to those required for reaction.

Interestingly, the reverse reaction:

Cu²⁺(aq) + H₂(g) → Cu(s) + 2H⁺(aq)

is thermodynamically feasible. However, no reaction takes place when hydrogen is bubbled into an aqueous solution containing copper(II) ions because the activation energy for the reaction is very large. The reactants are therefore *kinetically stable*.

The reaction between manganese(IV) oxide and hydrochloric acid

One method of preparing chlorine in the laboratory is to react manganese(IV) oxide with hydrochloric acid:

MnO₂(s) + 4HCl(aq) → Mn²⁺(aq) + 2Cl⁻(aq) + 2H₂O(l) + Cl₂(g)

However, the standard electrode potentials suggest that this reaction is not feasible.

MnO₂(s) + 4H⁺(aq) + 2e⁻ ⇌ Mn²⁺(aq) + 2H₂O(l)
$E^\ominus = +1.23$ V   equilibrium 1

Cl₂(g) + 2e⁻ ⇌ 2Cl⁻(aq)
$E^\ominus = +1.36$ V   equilibrium 2

The $E^\ominus$ value for equilibrium 2 is more positive (i.e. less negative) than that for equilibrium 1. So, chloride ions cannot release electrons to MnO₂, and this is required for the reaction shown above to take place. The reaction is not thermodynamically feasible under standard conditions, in which the concentration of the hydrochloric acid, and therefore the concentration of both the hydrogen ions and chloride ions, is 1 mol dm⁻³.

The key to making the reaction take place is to increase the concentration of hydrochloric acid.

If concentrated hydrochloric acid (~10 mol dm⁻³) is used, then the concentration of both the hydrogen ions and the chloride ions increases. The effect is to shift the position of equilibrium 1 to the right, and the position of equilibrium 2 to the left.

$$\xrightarrow{\text{equilibrium shifts to the right}}$$
$$MnO_2(s) + 4H^+(aq) + 2e^- \rightleftharpoons Mn^{2+}(aq) + 2H_2O(l)$$
$$Cl_2(g) + 2e^- \rightleftharpoons 2Cl^-(aq)$$
$$\xleftarrow{\text{equilibrium shifts to the left}}$$

fig E Equilibria for the reaction of manganese(IV) oxide with hydrochloric acid.

As a result, the electrode potential, E, for equilibrium 1 becomes more positive because the redox system is now a better electron acceptor.

The electrode potential for equilibrium 2 becomes less positive because the redox system is now a better electron releaser.

The net effect is that the electrode potential of equilibrium 2 becomes less positive (more negative) than that of equilibrium 1, so the chloride ions can now release electrons to manganese(IV) oxide. The reaction has become thermodynamically feasible under these non-standard conditions.

Summary

- The thermodynamic feasibility of a chemical reaction can be predicted using standard electrode potentials.
- Although the standard electrode potentials indicate that a reaction is thermodynamically feasible, it may not take place for two reasons:
  - the reactants may be kinetically stable because the activation energy for the reaction is very large, and
  - the reaction may not be taking place under standard conditions.
- A reaction that is not thermodynamically feasible under standard conditions may become feasible when the conditions are altered.
- Changing the conditions may alter the electrode potential, E, of a half-cell because the position of equilibrium of the half-cell reaction may change.

Disproportionation reactions

In **Book 1 Topic 3**, we learnt that a disproportionation reaction is one in which an element in a species is simultaneously oxidised and reduced in the same reaction.

A typical disproportionation reaction is the conversion of copper(I) ions in aqueous solution into copper(II) ions and copper atoms.
$$Cu^+(aq) + Cu^+(aq) \rightarrow Cu^{2+}(aq) + Cu(s)$$

This reaction can be explained in terms of standard electrode potentials.

$$Cu^{2+}(aq) + e^- \rightleftharpoons Cu^+(aq) \quad E^\ominus = +0.15\,V \quad \text{equilibrium 1}$$
$$Cu^+(aq) + e^- \rightleftharpoons Cu(s) \quad E^\ominus = +0.52\,V \quad \text{equilibrium 2}$$

The $E^\ominus$ value for equilibrium 1 is more negative than that for equilibrium 2, so the position of equilibrium 1 will move to the left, and the equilibrium position of 2 will move to the right. Therefore, Cu^+ ions will both release and accept electrons to one another to form Cu^{2+} ions and Cu atoms. That is, Cu^+ ions disproportionate into Cu^{2+} ions and Cu atoms.

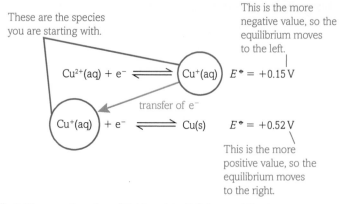

fig F Disproportionation of Cu^+ ions into Cu^{2+} ions and Cu atoms.

Relationship between total entropy and $E^\ominus_{cell}$

The standard Gibbs energy change, $\Delta G^\ominus$, taking place in a cell and the emf of the cell, $E^\ominus_{cell}$, are related by the following expression:
$$\Delta G^\ominus = -nFE^\ominus_{cell}$$

where n is the number of moles of electrons involved in the cell reaction and F is the Faraday constant.

$$\Delta G = \Delta H - T\Delta S_{system}$$

but, $\Delta S_{surroundings} = -\dfrac{\Delta H}{T}$

therefore, $\Delta H = -T\Delta S_{surroundings}$

and hence:
$$\Delta G = -T\Delta S_{surroundings} - T\Delta S_{system} = -T(\Delta S_{surroundings} + \Delta S_{system})$$
$$= -T\Delta S_{total}$$

As $\Delta G^\ominus = -T\Delta S^\ominus_{total}$, the expression can be written as:
$$T\Delta S^\ominus_{total} = nFE^\ominus_{cell}$$

As n is a constant for a given cell reaction, and F is also a constant, it follows that, at a given temperature, the total entropy change of the cell reaction is proportional to the emf of the cell.

$$\Delta S^\ominus_{total} \propto E^\ominus_{cell}$$

If $E^\ominus_{cell}$ is positive, the reaction as written from left to right in the cell diagram is thermodynamically feasible because $\Delta S^\ominus_{total}$ will be positive.

For example, for the following cell diagram:
$$Zn(s) \mid Zn^{2+}(aq) \;\|\; Cu^{2+}(aq) \mid Cu(s) \quad E^\ominus_{cell} = +1.10\,V$$

the half-cell reactions that will take place in the cell are:
$$Zn(s) \rightarrow Zn^{2+}(aq) + 2e^- \quad \text{and} \quad Cu^{2+}(aq) + 2e^- \rightarrow Cu(s)$$

giving an overall reaction of:
$$Zn(s) + Cu^{2+}(aq) \rightarrow Zn^{2+}(aq) + Cu(s)$$

In the following cell diagram:
$$Pt(s) \mid \tfrac{1}{2}H_2(g) \mid H^+(aq) \;\|\; Zn^{2+}(aq) \mid Zn(s) \quad E^\ominus_{cell} = -0.76\,V$$

the half-cell reactions as written from left to right will not occur, as $E^\ominus_{cell}$ is negative. Instead the half-cell reactions will occur from right to left:
$$Zn(s) \rightarrow Zn^{2+}(aq) + 2e^- \quad \text{and} \quad H^+(aq) + e^- \rightarrow \tfrac{1}{2}H_2(g)$$

giving an overall reaction of:
$$Zn(s) + 2H^+(aq) \rightarrow Zn^{2+}(aq) + H_2(g)$$

This means that $E^\ominus_{cell}$ can be used to predict the direction of a cell reaction.

Relationship between the equilibrium constant and $E^\ominus_{cell}$

The standard Gibbs energy change for a reaction and the equilibrium constant, K, are related by the following expression:
$$\Delta G^\ominus = -RT\ln K$$

However:
$$\Delta G^\ominus = -nFE^\ominus_{cell}$$

Therefore:
$$RT\ln K = nFE^\ominus_{cell}$$

or:
$$\ln K = \frac{nFE^\ominus_{cell}}{RT}$$

As n, F and R are constants, it follows that at a given temperature, T, $\ln K$ is proportional to $E^\ominus_{cell}$:
$$\ln K \propto E^\ominus_{cell}$$

$E^\ominus_{cell}$ can therefore be used to calculate the thermodynamic equilibrium constant, K, for a cell reaction.

Questions

1 (a) Use the following data to explain why, under standard conditions, VO^{2+} can be used to reduce MnO_4^- in acidic solution.

$VO_2^+(aq) + 2H^+(aq) + e^- \rightleftharpoons VO^{2+}(aq) + H_2O(l)$ $E^\ominus = +1.00$ V
$MnO_4^-(aq) + 8H^+(aq) + 5e^- \rightleftharpoons Mn^{2+}(aq) + 4H_2O(l)$ $E^\ominus = +1.51$ V

(b) Write an ionic equation for the reaction that takes place.

2 Use the following data to explain to which oxidation state vanadium(V), in the form of VO_2^+, can be reduced by zinc.

$VO_2^+(aq) + 2H^+(aq) + e^- \rightleftharpoons VO^{2+}(aq) + H_2O(l)$ $E^\ominus = +1.00$ V
$VO^{2+}(aq) + 2H^+(aq) + e^- \rightleftharpoons V^{3+}(aq) + H_2O(l)$ $E^\ominus = +0.34$ V
$V^{3+}(aq) + e^- \rightleftharpoons V^{2+}(aq)$ $E^\ominus = -0.26$ V
$V^{2+}(aq) + 2e^- \rightleftharpoons V(s)$ $E^\ominus = -1.18$ V
$Zn^{2+}(aq) + 2e^- \rightleftharpoons Zn(s)$ $E^\ominus = -0.76$ V

3 (a) Use the following data to explain why hydrogen peroxide, H_2O_2, decomposes into water and oxygen in aqueous solution.

$H_2O_2(aq) + 2H^+(aq) + 2e^- \rightleftharpoons 2H_2O(l)$ $E^\ominus = +1.77$ V
$O_2(g) + 2H^+(aq) + 2e^- \rightleftharpoons H_2O_2(l)$ $E^\ominus = +0.68$ V

(b) Write an equation for this decomposition.
(c) Explain, using oxidation numbers, why this reaction is classified as disproportionation.

4 $Fe^{3+}(aq)$ ions can be reduced by $I^-(aq)$ ions. The equation for the reaction is:
$$Fe^{3+}(aq) + I^-(aq) \rightarrow Fe^{2+}(aq) + \tfrac{1}{2}I_2(aq)$$
The standard electrode potentials for $Fe^{3+}(aq)$, $Fe^{2+}(aq)$ and $\tfrac{1}{2}I_2(aq)$, $I^-(aq)$ are +0.77 V and +0.54 V respectively.

[The gas constant, R, is 8.31 J mol^{-1} K^{-1}. The Faraday constant, F, = 96 500 C mol^{-1}.]
(a) Calculate the emf of a cell made by combining these two half-cells.
(b) Calculate the standard Gibbs energy change, $\Delta G^\ominus$, for the cell reaction.
(c) Calculate the value of the equilibrium constant, K, for the cell reaction.

14.2 1 Storage cells and fuel cells

By the end of this section, you should be able to...
- understand the application of standard electrode potentials to storage cells
- understand that the energy released on the reaction of a fuel with oxygen is utilised in a fuel cell to generate a voltage
- understand the electrode reactions that occur in a hydrogen-oxygen fuel cell

Storage cells

A storage cell, or secondary cell, is a cell that can be recharged by passing a current through it in the opposite direction to the flow of current generated by the cell.

Storage cells, often commonly called batteries, are used in appliances such as torches, battery operated radios, electronic calculators, cordless house phones and mobile phones to generate an electric current.

One of the first storage cells to be invented was the rechargeable nickel-cadmium (NiCd) cell, which was invented in 1899. It uses a nickel compound, NiO(OH), for the positive electrode, cadmium for the negative electrode and aqueous potassium hydroxide as the electrolyte.

The two half-cell reactions are:

$$Cd(OH)_2(s) + 2e^- \rightleftharpoons Cd(s) + 2OH^-(aq)$$
$$E^\ominus = -0.88 \text{ V}$$

$$NiO(OH)(s) + H_2O(l) + e^- \rightleftharpoons Ni(OH)_2(s) + OH^-(aq)$$
$$E^\ominus = +0.52 \text{ V}$$

The $E^\ominus$ value of the upper equilibrium is the more negative of the two. Hence, the equilibrium position of this reaction shifts to the left when the cell is generating a flow of electrons (i.e. an electric current).

So, when the cell is generating a flow of electrons the two reactions that take place are:

$$Cd(s) + 2OH^-(aq) \rightarrow Cd(OH)_2(s) + 2e^-$$
$$NiO(OH)(s) + H_2O(l) + e^- \rightarrow Ni(OH)_2(s) + OH^-(aq)$$

giving an overall cell reaction of:

$$Cd(s) + 2NiO(OH)(s) + 2H_2O(l) \rightarrow Cd(OH)_2(s) + Ni(OH)_2(s)$$

The equations must be multiplied up so the number of electrons in each half reaction is the same.

Under standard conditions, the emf is 1.40 V. Under the conditions present in a commercial cell, the potential difference produced is about 1.2 V.

To recharge the cell, an external potential difference is applied that reverses the reactions shown above.

The major disadvantage of these cells is that cadmium is very toxic. Therefore, great care needs to be taken when disposing of them.

Many other storage cells are now produced, including the nickel metal hydride (NiMH) and silver-zinc (Ag-Zn) cells.

Perhaps the most widely known storage cell is the lead–acid accumulator, which is commonly used for car batteries.

fig A A car battery.

Fuel cells

A fuel cell produces a voltage from the chemical reaction of a fuel with oxygen. Several fuel cells have been developed, with the most common using hydrogen as the fuel (the hydrogen-oxygen fuel cell). In this fuel cell, hydrogen is supplied externally as a gas and the cell can operate as long as the fuel supply is maintained. Other fuel cells use methanol/ethanol as the fuels.

The hydrogen-oxygen fuel cell

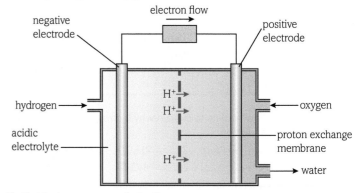

fig B A hydrogen-oxygen fuel cell.

In the hydrogen-oxygen fuel cell, both metal electrodes are coated with platinum, which catalyses the reactions that take place at the electrodes. The two reactions that take place in the presence of an acidic electrolyte are shown below.

At the negative electrode: $H_2(g) \rightarrow 2H^+(aq) + 2e^-$

At the positive electrode: $\frac{1}{2}O_2(g) + 2H^+(aq) + 2e^- \rightarrow H_2O(l)$

The hydrogen ions (protons) pass through the proton exchange membrane, which allows them to enter the compartment containing the positive electrode, where they can react with oxygen.

The overall cell reaction is:
$$H_2(g) + \tfrac{1}{2}O_2(g) \rightarrow H_2O(l)$$

There is also a hydrogen-oxygen fuel cell that has an alkaline electrolyte. The electrode reactions for this cell are shown below.

At the negative electrode: $H_2(g) + 2OH^-(aq) \rightarrow 2H_2O(l) + 2e^-$

At the positive electrode: $\tfrac{1}{2}O_2(g) + 2H^+(aq) + 2e^- \rightarrow H_2O(l)$

The overall cell reaction is the same as above:
$$H_2(g) + \tfrac{1}{2}O_2(g) \rightarrow H_2O(l)$$

Advantages of hydrogen-oxygen fuel cells
- They offer an alternative to the direct use of fossil fuels such as petrol and diesel.
- Unlike fossil fuels, their products do not include pollutants, such as carbon monoxide, carbon dioxide and oxides of nitrogen.
- They are lighter and more efficient than engines that use fossil fuels.

Problems associated with hydrogen-oxygen fuel cells

Hydrogen is explosive, so great care has to be taken when transporting it. We will consider the following options in turn.
- compressing the gas
- adsorbing it onto the surface of a suitable solid material.
- absorbing it into a suitable material.

Compressing the gas has disadvantages in terms of safety. Transporting hydrogen under pressure is particularly hazardous.

Many metals adsorb hydrogen, but so far no firm conclusion has been reached as to the best one to use. One of the most promising areas of research is the use of carbon nanotubes to adsorb hydrogen.

Many metal hydrides absorb hydrogen, which can be released under the correct conditions. The problem is that high temperatures are required to release the hydrogen.

Another problem is the supply of hydrogen. At the moment most of the hydrogen is produced from methane, which is a finite source. It can be produced using renewable sources, but it is an expensive process. Integrated wind-to-hydrogen (power-to-gas) plants, using electrolysis of water, are exploring technologies to deliver costs low enough, and quantities great enough, to compete with traditional energy sources.

Fig C shows a prototype car that uses fuel cell technology.

fig C A car that uses a hydrogen-oxygen fuel cell.

Questions

1. Nickel-cadmium (NiCd) cells are used in many electronic devices because they are very reliable, economical and easy to use.
   The standard electrode potentials for the redox systems in NiCd cells are:

   $Cd(OH)_2(s) + 2e^- \rightleftharpoons Cd(s) + 2OH^-(aq)$ $E^\ominus = -0.88\,V$
   $NiO(OH)(s) + H_2O(l) + e^- \rightleftharpoons Ni(OH)_2(s) + OH^-(aq)$ $E^\ominus = +0.52\,V$

   (a) What is the standard cell potential, $E^\ominus_{cell}$, of a NiCd cell?
   (b) Construct the overall reaction that takes place when a NiCd cell is producing an electric current.
   (c) Use oxidation numbers to show which element is oxidised and which is reduced during the discharge of a NiCd cell.

2. Fuel cells are being developed as an alternative to the use of petrol and diesel in cars. The most common fuel used in these cells is hydrogen, but cells using other fuels are being developed.
   In a methanol fuel cell, the fuel is supplied at the negative electrode, with oxygen at the positive electrode.
   The ionic half-equation for the reaction of oxygen at the positive electrode is:
   $$\tfrac{1}{2}O_2(g) + 2H^+(aq) + 2e^- \rightarrow H_2O(l)$$
   (a) Write a chemical equation for the complete combustion of methanol in oxygen.
   (b) Use the equation in (a), and the ionic half-equation for the reaction of oxygen at the positive electrode, to deduce an ionic half-equation for the reaction that takes place at the negative electrode.
   (c) Give two advantages of using fuel cells compared with the combustion of petrol or diesel.
   (d) Give an advantage of using methanol, rather than hydrogen, in a fuel cell for use in cars.

3. (a) What is the essential difference between a fuel cell and a conventional electrochemical cell?
   (b) State two ways that hydrogen might be stored as a fuel for use in cars.
   (c) Suggest why some scientists consider that using hydrogen as a fuel for cars uses up more energy than using petrol or diesel as the fuel.

14.2 2 Redox titrations

By the end of this section, you should be able to...

- carry out structured and non-structured titration calculations including Fe^{2+}/MnO_4^- and $I_2/S_2O_3^{2-}$
- understand the methods used in redox titrations

Redox titrations with potassium manganate(VII), $KMnO_4$

Potassium manganate(VII) is a powerful oxidising agent. It is used for the quantitative estimation of many reducing agents, especially compounds of iron(II) and for ethanedioic acid and its salts.

In acidic solution, the half-equation for the reduction of the manganate(VII) ion is:

$$MnO_4^-(aq) + 8H^+(aq) + 5e^- \rightarrow Mn^{2+}(aq) + 4H_2O(l)$$

In alkaline solution, potassium manganate(VII) yields manganese(IV) oxide as a brown precipitate, and this interferes with the end point colour. For this reason, therefore, potassium manganate(VII) is almost always used in solutions that are sufficiently acidic to prevent the formation of manganese(IV) oxide.

As the titration proceeds, manganese(II) ions, Mn^{2+}, accumulate but, at the dilution used, they give a colourless solution. As soon as the potassium manganate(VII) is in excess, the solution becomes pink. It therefore acts as its own indicator – the end point is the first permanent pink colour.

Titration of potassium manganate(VII) with iron(II) ions

In this reaction, the iron(II) ions are oxidised and the manganate(VII) ions are reduced.

The ionic half-equations involved are:

$$MnO_4^-(aq) + 8H^+(aq) + 5e^- \rightarrow Mn^{2+}(aq) + 4H_2O(l)$$
$$Fe^{2+}(aq) \rightarrow Fe^{3+}(aq) + e^-$$

The overall equation for the reaction is:

$$MnO_4^-(aq) + 8H^+(aq) + 5Fe^{2+}(aq)$$
$$\rightarrow Mn^{2+}(aq) + 4H_2O(l) + 5Fe^{3+}(aq)$$

An aqueous solution containing MnO_4^- ions is purple. In reasonably concentrated solutions containing Fe^{2+}, Fe^{3+} and Mn^{2+}, the ions have the following colours:

Fe^{2+}(aq) – pale green
Fe^{3+}(aq) – yellow
Mn^{2+}(aq) – pale pink

However, the solutions used in titrations are usually so dilute that they appear colourless.

The method of performing the titration is:

- Pipette an accurately measured volume, usually 25.0 cm³, of the iron(II) solution into a conical flask and then add a small volume of dilute sulfuric acid.
- Slowly add potassium manganate(VII) solution of accurately known concentration from a burette and swirl the mixture.
- The potassium manganate(VII) solution will turn colourless until all of the iron(II) ions have been oxidised.
- The addition of one more drop of potassium manganate(VII) solution will turn the mixture pale pink.

fig A End point colour in a potassium manganate(VII) titration.

WORKED EXAMPLE:

25.0 cm³ of iron(II) sulfate solution reacts with 22.40 cm³ of 0.0200 mol dm⁻³ potassium manganate(VII) solution. Calculate the concentration of the iron(II) sulfate.

$$5Fe^{2+}(aq) \equiv MnO_4^-(aq)$$

25.0 cm³ of x mol dm⁻³ Fe^{2+} ≡ 22.40 cm³ of 0.0200 mol dm⁻³ MnO_4^-

$$\frac{25.0x}{22.40 \times 0.0200} = \frac{5}{1}$$

$$x = \frac{5 \times 22.40 \times 0.200}{25.0} = 0.896$$

$$[FeSO_4(aq)] = 0.896 \text{ mol dm}^{-3}$$

Titration of potassium manganate(VII) with ethanedioic acid

In this reaction, the ethanedioic acid is reduced to carbon dioxide. The ionic half-equation for this reduction is:

$$H_2C_2O_4(aq) \rightarrow 2H^+(aq) + 2CO_2(g) + 2e^-$$

The overall equation for the reaction is:

$$2MnO_4^-(aq) + 6H^+(aq) + 5H_2C_2O_4(aq)$$
$$\rightarrow 2Mn^{2+}(aq) + 8H_2O(l) + 10CO_2(g)$$

Aqueous ethanedioic acid is pipetted into the conical flask (Care: ethanedioic acid is extremely poisonous, so a pipette filler *must* be used) and is then acidified with dilute sulfuric acid. Aqueous potassium manganate(VII) is added from the burette. However, the reaction is very slow so the ethanedioic acid is heated to around 60 °C before starting the titration. Even then, the reaction is slow to begin with and the pink colour does not immediately disappear when the first sample of potassium manganate(VII) is added. This may lead you to erroneously think that the titration is complete. After the initial sample is added, the Mn^{2+} ions produced act as a catalyst and the reaction speeds up, allowing you to titrate normally and obtain an accurate end point.

Potassium manganate(VII) can also be used to estimate the amount of ethanedioate ion present in a solution. The equation for the reaction is:

$$2MnO_4^-(aq) + 16H^+(aq) + 5C_2O_4^{2-}(aq) \rightarrow 2Mn^{2+}(aq) + 8H_2O(l) + 10CO_2(g)$$

Did you know?
A reaction in which a product acts as a catalyst is said to be autocatalysed. Autocatalysis is discussed in **Section 16.1.5**.

Investigation
A redox titration – the determination of the number of moles of water of crystallisation per mole of hydrated iron(II) sulfate crystals

Procedure
1. Bring to the boil 100 cm³ of dilute sulfuric acid and then allow it to cool to room temperature.
2. Weigh accurately a weighing bottle containing approximately 7 g of hydrated iron(II) sulfate crystals.
3. Dissolve the crystals in the cool acid and reweigh the bottle. Transfer the solution to a 250 cm³ volumetric flask. Add several washings from the beaker and then make up to the mark. Shake well.
4. Titrate a 25.0 cm³ sample of the iron(II) solution, as before, with the standard potassium manganate(VII) solution.
5. Perform repeat titrations until you obtain concordant results.

Learning tip
The dilute sulfuric acid is boiled to remove dissolved oxygen that might otherwise oxidise some of the iron(II) ions to iron(III).

Calculations
1. Calculate the mean titre from your concordant results.
2. Calculate the concentration of the iron(II) sulfate solution.
3. Calculate the mass of iron(II) sulfate in the 250 cm³ of solution.
4. Calculate the mass of water in the mass of crystals taken and hence calculate the number of moles of water per mole of salt.

Sample results
Concentration of potassium manganate(VII) solution = 0.0198 mol dm⁻³
Mass of hydrated iron(II) sulfate = 6.84 g
Mean titre = 24.95 cm³
[Molar masses: $FeSO_4$ = 151.9 g mol⁻¹; H_2O = 18.0 g mol⁻¹]

$$5FeSO_4 \equiv KMnO_4$$

25.0 cm³ of x mol dm⁻³ $FeSO_4 \equiv$ 22.40 cm³ of 0.0198 mol dm⁻³ $KMnO_4$

$$\frac{25.0x}{24.95 \times 0.0198} = \frac{5}{1}$$

$[FeSO_4(aq)]$ = 0.0988 mol dm⁻³

Mass of $FeSO_4$ in 250 cm³

$$x = \frac{5 \times 24.95 \times 0.0198}{25.0} = 0.0988$$

Mass of water = (6.84 − 3.75) = 3.09 g

If the formula of the hydrated salt is $FeSO_4.yH_2O$, then

Mass of $FeSO_4$ in 250 cm³ = 0.0988 × 151.9 × $\frac{250}{1000}$ = 3.75 g

$$\frac{18.0y}{151.9} = \frac{3.09}{3.75}$$

$$y = \frac{3.09 \times 151.9}{18.0 \times 3.75}$$

$$= 6.95$$

∴ The formula of hydrated iron(II) sulfate is $FeSO_4.7H_2O$

Redox titrations with iodine and sodium thiosulfate

Thiosulfate ions reduce iodine to iodide ions. The ionic half-equations involved are:

$$2S_2O_3^{2-}(aq) \rightarrow S_4O_6^{2-}(aq) + 2e^-$$
$$I_2(aq) + 2e^- \rightarrow 2I^-(aq)$$

The overall equation for the reaction is:

$$2S_2O_3^{2-}(aq) + I_2(aq) \rightarrow S_4O_6^{2-}(aq) + 2I^-(aq)$$

The reaction can be used for the direct estimation of iodine, or for the estimation of a substance that can take part in a reaction that produces iodine.

The indicator used in titrating iodine with sodium thiosulfate is starch solution. With free iodine it produces a deep blue-black colour. The blue-black colour disappears as soon as sufficient sodium thiosulfate has been added to react with all of the iodine.

The procedure is as follows.

- Add sodium thiosulfate solution from the burette to the iodine solution until the original brown colour of the iodine changes to pale yellow.
- Add a few drops of starch solution to produce a blue coloration.
- Add the sodium thiosulfate solution drop by drop until the blue-black solution turns colourless.

If the starch is added too early, it adsorbs some of the iodine and reduces the accuracy of the titration.

If starch were not added, it would be very difficult to detect the end point. The colour of the iodine solution becomes very faint towards the end of the reaction and this makes it difficult to accurately assess when the end point has been obtained.

Investigation

A redox titration – determination of the percentage of copper(II) ions in a sample of hydrated copper(II) sulfate

The copper(II) sulfate solution must be free from anything but a trace of mineral acid, otherwise the end point is not accurate. This solution produces iodine when potassium iodide is added. The ionic equation for this reaction is:

$$2Cu^{2+}(aq) + 4I^-(aq) \rightarrow 2CuI(s) + I_2(aq)$$

The iodine produced is then titrated with standard sodium thiosulfate solution.

Procedure

1. Weigh accurately a weighing bottle containing approximately 6 g of hydrated copper(II) sulfate crystals.
2. Transfer the crystals to a 250 cm³ volumetric flask and reweigh the bottle.
3. Add sodium carbonate solution until a *slight* permanent bluish precipitate of copper(II) carbonate is formed.
4. Acidify the mixture with a little dilute ethanoic acid until a clear blue solution is formed. Make up the solution to 250 cm³ with deionised (or distilled) water and shake well.
5. Pipette 25.0 cm³ of the solution into a conical flask and add about 1.5 g of solid potassium iodide.
6. Titrate the iodine produced with sodium thiosulfate solution, adding starch solution just as the colour of the mixture changes to pale yellow. The end point is achieved when you can see a white suspension of insoluble copper(I) iodide, CuI(s), in a colourless solution.
7. Perform repeat titrations until you obtain concordant results.

Calculation

Calculate the percentage of copper(II) ions in the sample of hydrated copper(II) sulphate crystals.

Sample results

Concentration of sodium thiosulfate solution = 0.0995 mol dm^{-3}

Mass of hydrated copper(II) sulfate crystals = 5.85 g

Mean titre = 23.45 cm^3

[Molar mass of Cu^{2+} = 63.5 g mol^{-1}]

$$2Cu^{2+} \equiv I_2 \equiv 2Na_2S_2O_3$$

∴ $\quad Cu^{2+} \equiv Na_2S_2O_3$

25.0 cm^3 of x mol dm^{-3} Cu^{2+} ≡ 23.45 cm^3 of 0.0995 mol dm^{-3} $Na_2S_2O_3$

∴ $\quad$ 25.0 x = 23.45 × 0.0995

∴ $\quad\quad x$ = 0.0933

Solution = 0.0933 mol^{-1} dm^{-3} × 63.5 g mol^{-1} × $\frac{250}{1000}$ = 1.48 g

Percentage of Cu^{2+} in the sample = $\frac{1.48}{5.85}$ × 100 = 25.3%

Questions

1. An iron nail of mass 1.50 g was reacted with dilute sulfuric acid. The resulting solution was transferred to a volumetric flask and made up to 250 cm^3 with deionised water.
   A 25.0 cm^3 sample of this solution reacted with exactly 24.40 cm^3 of 0.0218 mol dm^{-3} potassium manganate(VII) solution.
   The ionic equation for the reaction between the iron nail and the acid is:
   $$Fe(s) + 2H^+(aq) \rightarrow Fe^{2+}(aq) + H_2(g)$$
   Calculate the percentage by mass of iron in the nail.
   [Molar mass of iron = 55.8 g mol^{-1}]

2. A 50 cm^3 volume of a solution of hydrogen peroxide was diluted with water to 1.00 dm^3. A 25.0 cm^3 sample of this solution was acidified and then titrated against 0.0200 mol dm^{-3} potassium manganate(VII). The mean titre value was 23.90 cm^3. Calculate the concentration, in mol dm^{-3}, of the original solution of hydrogen peroxide.
   The ionic half-equation for the oxidation of hydrogen peroxide is:
   $$H_2O_2(aq) \rightarrow 2H^+(aq) + O_2(g) + 2e^-$$

3. Calculate the volume of a 0.0200 mol dm^{-3} solution of potassium manganate(VII) required to completely oxidise 1.00 g of iron(II) ethanedioate (FeC_2O_4).
   [Molar mass of FeC_2O_4 = 143.8 g mol^{-1}]

4. An aqueous solution contains a mixture of iron(II) ions and iron(III) ions. A 25.0 cm^3 volume of the solution required 21.60 cm^3 of 0.0210 mol dm^{-3} potassium manganate(VII) for oxidation.
   A separate 25.0 cm^3 sample of the solution was reacted with an excess of zinc amalgam in order to reduce the iron(III) ions to iron(II). This sample, after filtration to remove the excess zinc, required 44.40 cm^3 of 0.0210 mol dm^{-3} potassium manganate(VII) solution for complete oxidation.
   Calculate the concentrations, in mol dm^{-3}, of both the iron(II) and iron(III) ions in the aqueous solution.

5. When bleaching powder reacts with a dilute acid, chlorine is liberated. This chlorine is available for bleaching and is known as 'available' chlorine.
   A 2.50 g sample of bleaching powder is added to an excess of potassium iodide solution and the mixture is then acidified. The solution is placed in a volumetric flask and made up to 250 cm^3 with deionised water. A 25.0 cm^3 volume of this solution required 23.20 cm^3 of 0.105 mol dm^{-3} sodium thiosulfate solution to oxidise the iodine present.
   Calculate the percentage of 'available' chlorine in the 2.50 g sample of bleaching powder.
   $$Cl_2(aq) + 2I^-(aq) \rightarrow 2Cl^-(aq) + I_2(aq)$$
   [Molar mass of Cl_2 = 71.0 g mol^{-1}]

6. A 25.0 cm^3 sample of of 0.0210 mol dm^{-3} potassium peroxydisulfate, $K_2S_2O_8$, was treated with excess potassium iodide. The iodine liberated reacted with 21.00 cm^3 of 0.0500 mol dm^{-3} sodium thiosulfate solution.
   Suggest a likely ionic equation for the reaction between $S_2O_8^{2-}$ ions and I^- ions.

THINKING BIGGER

ELECTROLYTE EVOLUTION

One of the reasons that your laptops and smartphones are significantly lighter and slimmer than those of 10 years ago is not so much because of advances in computing power but because of advances in rechargeable battery technology.

BATTERY POWER

Primary cells

These are the disposable cells – they are discharged once and discarded – that for over a century powered small and portable equipment. Nowadays as we become more waste conscious we find that many of these cells are giving way to rechargeable (secondary) cells and batteries.

The most common primary cells are based on the zinc-manganese dioxide couple: either zinc-carbon cells or alkaline manganese cells. For a short period some manufacturers offered mercury cells, which were replaced by zinc-air batteries, and some companies now produce 3V lithium-MnO_2 cells.

Zinc-carbon

This cell became commercially available in the late 1800s and was a dry cell version of the original wet Leclanché cell. (The latter was made up of a conducting solution (electrolyte) of ammonium chloride with a negative terminal of zinc and a positive terminal of manganese dioxide.) Replacing the liquid electrolyte with a gel and then sealing the whole cell made it suitable for domestic applications where portability was a key feature. These cells were developed and marketed by Ever Ready for use in radios and torches and are still popular today, though Ever Ready has now changed to Energizer and many other makes have become available.

The dry cell zinc-carbon chemistry performs well in applications where there is intermittent use such as flashlights but performance is not so good in devices that put a heavy drain upon the cell. In such applications polarisation and loss of output results. However, the cell chemistry recovers when left idle for a while – the well-known case of the dead battery that comes back to life. Polarisation is caused by the products of the electrode reactions building up on the electrode surface and preventing new reactants arriving. On standing, diffusion occurs and new reactants reach the electrode as the products disperse.

Anode: Zn metal
Cathode: MnO_2 powder with graphite powder for electrical conduction
Electrolyte: $NH_4Cl(aq)$ and/or $ZnCl_2(aq)$

In earlier cells the zinc doubled as the outer can which tended to leak as the cell ran down and the zinc passed into solution.

$Zn(s) + 4NH_3(aq) + 2MnO_2(s) + 2H_3O^+(aq)$
$\rightarrow [Zn(NH_3)_4]^{2+}(aq) + 2Mn(OH)_3(s)$

fig A Alkaline-manganese batteries.

Alkaline-manganese

These batteries developed from the zinc-carbon cell and became available for domestic use in the 1960s and quickly gained in popularity because they were less prone to polarisation, had greater capacity and were less likely to leak.

Here the anode is powdered zinc, which provides a greater reactive surface area and thus more power. The electrolyte is an alkali in contrast to the previous cell which was acidic.

Anode: Zn metal powder

Cathode: MnO_2 powder with graphite powder for electrical conduction

Electrolyte: KOH(aq)

$Zn(s) + MnO_2(s) + 2H_2O(l)$ ($Mn(OH)_2(s) + Zn(OH)_2(s)$ E = +1.5 eV

Development of these batteries continues and recently Panasonic introduced a vacuum forming process to compact manganese dioxide and graphite powders, which the company claims gives improved capacity.

Where else will I encounter these themes?

Book 1 | 11 | 12 | 13

Thinking Bigger 14

Let us start by considering the nature of the writing in the article.

> 1. In the final sentence of the extract, the author tells us that Panasonic claim their vacuum forming process for cathodes of alkaline-manganese batteries gives improved capacity. What stages would the technology need to go through before Panasonic's claim was proven?

Think about the role of the scientific community in validating new knowledge and ensuring integrity.

Now we will look at the chemistry in detail. Some of these questions will link to topics earlier in this book or **Book 1**, so you may need to combine concepts from different areas of chemistry to work out the answers.

> 2. Write half-equations for the reactions taking place at the anode and the cathode of the zinc-carbon cell.
> 3. The electrolyte for this cell is $NH_4Cl(aq)$. Why would you expect this electrolyte to be acidic? You may use equations to help explain your answer.
> 4. Give details of the bonding in the $[Zn(NH_3)_4]^{2+}$ complex and explain why this species is colourless in solution.
> 5. In the alkaline-manganese cell, the standard electrode potential for Zn is $-0.76\,V$. Calculate the standard electrode potential for the reduction of manganese(IV) oxide to manganese(II) hydroxide given that the $E_{cell} = 1.50\,V$.

Equations, diagrams and calculations can all be useful tools to help justify an answer, particularly when a question uses the command words 'explain' or 'justify'.

Think about the electronic configuration of the Zn^{2+} ion. You may wish to revisit this question once you have completed Topic 15.

Activity

One of the major technological developments over the last few years has been the lithium ion battery. This has enabled mobile devices to be rechargeable and light-weight. Produce a presentation on the technology behind lithium ion batteries. Your presentation should be 5–8 slides in length and should include:

- the reactions involved in generating the voltage
- the advantages of lithium ion batteries over earlier secondary cells
- any disadvantages associated with lithium ion batteries.

Did you know?

In 2013, some five billion lithium ion batteries were sold to supply electronic devices such as phones, laptops, tablets, cameras, power tools and even electric cars.

● From *After Battery Power* by Tony Hargreaves. http://www.rsc.org/education/eic/issues/2008Mar/BatteryPower.asp

14 Exam-style questions

1. Brass is an alloy of copper and zinc. It is used to make decorative ornaments.

   (a) A sample of brass is analysed to find the percentage of copper it contains.
   This is the method used:
   - 0.500 g of brass was reacted to form a solution containing copper(II) ions:
   $Cu(s) \rightarrow Cu^{2+}(aq)$
   - The solution is reacted with an excess of potassium iodide to produce iodine:
   $2Cu^{2+}(aq) + 4I^-(aq) \rightarrow 2CuI(s) + I_2(aq)$
   - The iodine is titrated against sodium thiosulfate(VI), using starch indicator:
   $I_2(aq) + 2S_2O_3^{2-}(aq) \rightarrow 2I^-(aq) + S_4O_6^{2-}(aq)$
   - 22.3 cm³ of 0.200 mol dm⁻³ sodium thiosulfate(VI) is required.

   Use the data to calculate the percentage of copper in the sample of brass. [5]

   (b) A student carries out the titration but forgets to add the starch indicator.
   (i) Predict the colour change that the student observes at the endpoint **without** the starch indicator. [2]
   (ii) Comment on why the colour change at the endpoint is easier to observe if starch indicator is used. [1]

   [Total: 8]

2. The table shows the formulae of some vanadium ions that can exist in aqueous solution, together with the colour of each ion.

	VO_3^-(aq)	VO^{2+}(aq)	V^{3+}(aq)	V^{2+}(aq)
oxidation number of vanadium				
colour of ion	yellow	blue	green	violet

   (a) Complete the table by inserting the oxidation numbers. [1]

   (b) Iron powder is added to an acidified solution of ammonium vanadate(V), NH_4VO_3 and the mixture is shaken until no further change takes place.
   Use the following data to explain what you would expect to observe in the sequence of reactions that take place. Your answer should consider **all** of the electrode reactions listed. [5]

	Electrode reaction	$E^\ominus$/V
1	$V^{2+}(aq) + 2e^- \rightleftharpoons V(s)$	−1.18
2	$V^{3+}(aq) + e^- \rightleftharpoons V^{2+}(aq)$	−0.26
3	$VO^{2+}(aq) + 2H^+(aq) + e^- \rightleftharpoons V^{3+}(aq)$	+0.34
4	$VO_3^-(aq) + 4H^+(aq) + e^- \rightleftharpoons VO^{2+}(aq) + 2H_2O(l)$	+1.00
5	$Fe^{2+}(aq) + 2e^- \rightleftharpoons Fe(s)$	−0.44

   [Total: 6]

3. An electrochemical cell is set up based on the following two half-cell reactions:

   $Mg^{2+}(aq) + 2e^- \rightleftharpoons Mg(s)$ $E^\ominus = -2.37$ V
   $Ni^{2+}(aq) + 2e^- \rightleftharpoons Ni(s)$ $E^\ominus = -0.25$ V

   (a) (i) Draw a labelled diagram of the cell operating under standard conditions. [4]
   (ii) Explain the direction of electron flow in the external circuit. [2]
   (iii) Write an ionic equation to represent the overall reaction in the cell when current is flowing. [1]

   (b) Calculate the standard emf, $E^\ominus_{cell}$, of this cell. [1]

   (c) Explain how the emf of the cell changes when water is added to the Mg^{2+}(aq) | Mg(s) half-cell. [3]

   [Total: 11]

4. (a) Draw a suitably labelled diagram to show how you would measure the standard electrode potential of the Cu^{2+}(aq) | Cu(s) half-cell. [6]

   (b) The table contains some standard electrode (redox) potentials involving copper and its ions.

	Electrode reaction	$E^\ominus$/V
1	$Cu^{2+}(aq) + e^- \rightleftharpoons Cu^+(aq)$	+0.15
2	$Cu^{2+}(aq) + 2e^- \rightleftharpoons Cu(s)$	+0.34
3	$Cu^+(aq) + e^- \rightleftharpoons Cu(s)$	+0.52

   (i) Use the data in the table to explain why the following reaction is likely to occur. [2]
   $2Cu^+(aq) \rightarrow Cu^{2+}(aq) + Cu(s)$
   (ii) Use oxidation numbers to explain why this reaction is classified as a disproportionation reaction. [3]

   [Total: 11]

104

5. Chlorine gas can be prepared in the fume cupboard of a laboratory by reacting hydrochloric acid with potassium manganate(VII). The two standard electrode (redox) potentials that relate to this reaction are:

$\frac{1}{2}Cl_2(g) + e^- \rightleftharpoons Cl^-(aq)$ $E^\ominus = +1.36\,V$

$MnO_4^-(aq) + 8H^+(aq) + 5e^- \rightleftharpoons Mn^{2+}(aq) + 4H_2O(l)$ $E^\ominus = +1.52\,V$

(a) State what is meant by the term **standard electrode potential**. [2]

(b) Calculate the standard emf, $E^\ominus_{cell}$, for an electrochemical cell constructed from these two redox systems. [1]

(c) (i) Construct an ionic equation for the reaction that takes place between hydrochloric acid and potassium manganate(VII). Include state symbols. [2]

(ii) Use oxidation numbers to explain which species is acting as the reducing agent in this reaction. [2]

(d) When very dilute solutions of potassium manganate(VII) and hydrochloric acid are added together, there is no visible change. Explain, using the changes in electrode potentials that occur, why no reaction takes place when the solutions are diluted. [3]

[Total: 10]

6. Vanadium(V), in the form of vanadate(V) ions, can be reduced in acidified solution by adding sulfur dioxide.

A student carries out an experiment to determine the new oxidation state of the vanadium after the reaction has taken place. He uses this method:

- He dissolves 2.24 g of ammonium vanadate(V), NH_4VO_3, (molar mass = 116.9 g mol^{-1}) in water and makes up the solution to 250 cm^3.
- In a fume cupboard, he bubbles sulfur dioxide through the solution until there is no further colour change.
- He then boils the solution for five minutes and allows the solution to cool.
- He labels this solution '**Solution Y**'.
- He then titrates 25.0 cm^3 of solution Y against 0.0200 mol dm^{-3} KMnO$_4$ solution. He finds that 38.40 cm^3 of KMnO$_4$ are required to oxidise the vanadium.

In this titration:
- The manganate(VII) ions are reduced from the +7 to the +2 oxidation state.
- The vanadium in solution Y is oxidised to the +5 oxidation state.

(a) Give a reason why the student carried out the experiment in a fume cupboard. [1]

(b) Use the data to show that the oxidation state of the vanadium in solution Y is +3. [5]

(c) The student repeats the experiment with the same mass of ammonium vanadate(V), but this time he boils the solution for only one minute. He finds that the volume of potassium manganate(VII) solution required in the titration has increased. Explain this observation. [2]

[Total: 8]

7. The diagram shows a hydrogen-oxygen fuel cell.

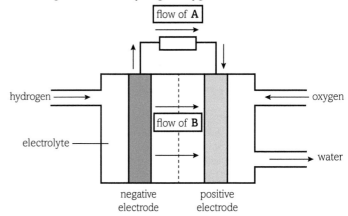

(a) Identify the particles represented by **A** and **B**. [2]

(b) The two electrodes are coated with a metal catalyst. Name a suitable metal to use as the catalyst. [1]

(c) State two advantages of fuel cells over the use of fossil fuels in cars. [2]

(d) Apart from cost, state two disadvantages of using fuels cells rather than fossil fuels in cars. [2]

(e) The zinc–silver oxide cell is used in button cells for watch batteries. It is based on the following half-cell reactions:

$Zn^{2+}(aq) + 2e^- \rightleftharpoons Zn(s)$ $E^\ominus = -0.76\,V$

$Ag_2O(s) + H_2O(l) + 2e^- \rightleftharpoons 2Ag(s) + 2OH^-(aq)$ $E^\ominus = +0.34\,V$

(i) Write the cell diagram for this cell. [2]

(ii) Write the equation for the overall reaction taking place in the cell when it is in use. Include state symbols. [2]

(iii) Calculate the standard emf, $E^\ominus_{cell}$, of the cell. [1]

(iv) Use your answer to part (iii) to calculate the standard Gibbs energy change, $\Delta G^\ominus$, for the reaction occurring in the cell. [F = 96 500 C mol^{-1}] [2]

(v) Use your answer to part (iv) to calculate the equilibrium constant, K, for the reaction at 298 K.

[R = 8.31 J mol^{-1} K^{-1}] [2]

[Total: 16]

TOPIC 15
Transition metals

Introduction

In **Book 1**, you learned about the Periodic Table, and in particular the elements of Groups 1, 2 and 7. These three groups show typical properties of s-block and p-block elements and clear trends in physical and chemical properties. For example, in Group 1, the first ionisation energy decreases down the group, and in Group 7 the reactivity decreases down the group. You learned to understand simple chemical reactions of these elements and their compounds and how to test for ions such as chloride and iodide.

In this topic, you will learn about transition metals, in particular those in the first row between the s-block element calcium and the p-block element gallium. Transition metals have some distinctive properties and uses not generally found in s-block and p-block elements, for example:

- the elements have important uses as industrial catalysts
- they help to remove pollutants from car exhausts
- they are important structural metals, such as iron and copper
- they are used in jewellery, especially platinum, gold and silver
- their compounds have bright colours, such as copper(II) sulfate
- they form ions and compounds with different oxidation numbers, such as iron(II) and iron(III) compounds
- they form complexes, some of which are vital to life, such as haemoglobin, and *cis*-platin, which is used in cancer treatment.

Many of these properties will be explained, especially those that involve complexes. The term 'complex' will be carefully explained, but don't think that 'complex' means 'complicated' – it doesn't!

All the maths you need

- Investigate the geometry of different transition metal complexes

What have I studied before?

- Writing electronic configurations for elements and ions
- Using oxidation numbers to consider whether species are oxidised or reduced
- How dative covalent bonds form
- How to predict the shapes of molecules and ions
- The meanings of *cis-* and *trans-* in stereoisomerism
- Predicting how changes in conditions affect the position of an equilibrium

What will I study later?

- The f-block elements, also known as the inner transition elements or rare earth elements
- Uses of f-block elements in modern technology, such as the use of neodymium in headphone speakers, hard drives, electric and hybrid cars and wind turbines

What will I study in this topic?

- Understand how the variety of oxidation numbers can be explained in terms of electronic configurations
- The meanings of some new terms, such as ligand, complex, monodentate and multidentate
- How carbon monoxide prevents the transport of oxygen through the bloodstream
- The two different ways in which transition metals and their compounds can act as catalysts
- How carbon monoxide and oxides of nitrogen are removed from vehicle exhausts by catalytic converters

15.1 1 Transition metal electronic configurations

By the end of this section, you should be able to...

- recognise that transition metals are d-block elements that form one or more stable ions with incompletely filled d-orbitals
- deduce the electronic configurations of atoms and ions of the d-block elements of Period 4 (Sc–Zn), given the atomic number and charge (if any)
- understand why transition metals show variable oxidation numbers

Which elements are the transition metals?

Where are they found in the Periodic Table?

You will be familiar with the Periodic Table, and the numbers of the different groups. Group 1 is on the far left, and on the far right are the noble gases, which are in a group numbered 0 (or sometimes 8). When you studied periodicity using the elements in Periods 2 and 3, you ignored the big gap between Groups 2 and 3. However, when you look at Period 4 you can see that there are elements between Groups 2 and 3 – some of the **transition metals** are there.

You will also be familiar with the idea of classifying elements as belonging to one of the blocks (s, p, d or f) of the Periodic Table. The transition metals are in the d-block of the Periodic Table. In this topic we will concentrate on the 10 elements in Period 4.

What is a transition metal?

It is easier to list the characteristics of transition metals than to explain fully what they are. To describe them as being in the d-block only indicates their location and not their characteristics. Most definitions of a transition metal refer to the electronic configurations of their atoms or ions (see the Key definition for a simple definition).

A fuller definition is quoted in the Edexcel specification:

'Transition metals are d-block elements that form one or more stable ions with incompletely filled d orbitals.'

fig A These are some useful objects containing transition metals and their alloys.

Characteristics of transition metals

Transition metals:

- are hard solids
- have high melting and boiling temperatures
- can act as catalysts
- form coloured ions and compounds
- form ions with different oxidation numbers
- form ions with incompletely-filled d-orbitals.

Of the 10 d-block elements in Period 4, the first and the last (scandium and zinc) do not have some of these characteristics, so although it is correct to describe them as d-block elements, they are not considered to be transition metals. In particular, each forms only one ion and their compounds are not coloured.

Electronic configurations

The elements

You learned in **Book 1 Topic 1**, how to write the electronic configurations of the elements from hydrogen to krypton, which includes the 10 d-block elements in Period 4. For reference, we show these in full in **table A** using the spdf notation, with electron-in-box diagrams for the electrons in the 4s and 3d orbitals. Copper and chromium do not follow the expected pattern. Where irregularities in the spdf notation exist, they are shown in red. In the electron-in-box diagrams, paired electrons are shown with a green background, and unpaired electrons with a blue background.

It is sometimes acceptable to use an abbreviated form of electronic configuration which represents the complete shells with reference to the previous noble gas. For example, the electronic configuration of titanium is shown in full as $1s^2 2s^2 2p^6 3s^2 3p^6 4s^2 3d^2$, but can be abbreviated as $[Ar]4s^2 3d^2$.

The ions

When these elements form ions, they lose one or more electrons. Scandium loses both of its 4s electrons and its only 3d electron, forming the Sc^{3+} ion. Zinc loses both of its 4s electrons and none of its 3d electrons, forming the Zn^{2+} ion. Because these two elements form only one ion each, and these ions have no incompletely-filled d-orbitals, they are not classified as transition metals.

Principles of transition metal chemistry | 15.1

Element	Z	Electronic configuration	Electrons-in-box diagram for 4s and 3d electrons
scandium	21	$1s^2\ 2s^2\ 2p^6\ 3s^2\ 3p^6\ 4s^2\ 3d^1$	↑↓ | ↑
titanium	22	$1s^2\ 2s^2\ 2p^6\ 3s^2\ 3p^6\ 4s^2\ 3d^2$	↑↓ | ↑ ↑
vanadium	23	$1s^2\ 2s^2\ 2p^6\ 3s^2\ 3p^6\ 4s^2\ 3d^3$	↑↓ | ↑ ↑ ↑
chromium	24	$1s^2\ 2s^2\ 2p^6\ 3s^2\ 3p^6\ 4s^1\ 3d^5$	↑ | ↑ ↑ ↑ ↑ ↑
manganese	25	$1s^2\ 2s^2\ 2p^6\ 3s^2\ 3p^6\ 4s^2\ 3d^5$	↑↓ | ↑ ↑ ↑ ↑ ↑
iron	26	$1s^2\ 2s^2\ 2p^6\ 3s^2\ 3p^6\ 4s^2\ 3d^6$	↑↓ | ↑↓ ↑ ↑ ↑ ↑
cobalt	27	$1s^2\ 2s^2\ 2p^6\ 3s^2\ 3p^6\ 4s^2\ 3d^7$	↑↓ | ↑↓ ↑↓ ↑ ↑ ↑
nickel	28	$1s^2\ 2s^2\ 2p^6\ 3s^2\ 3p^6\ 4s^2\ 3d^8$	↑↓ | ↑↓ ↑↓ ↑↓ ↑ ↑
copper	29	$1s^2\ 2s^2\ 2p^6\ 3s^2\ 3p^6\ 4s^1\ 3d^{10}$	↑ | ↑↓ ↑↓ ↑↓ ↑↓ ↑↓
zinc	30	$1s^2\ 2s^2\ 2p^6\ 3s^2\ 3p^6\ 4s^2\ 3d^{10}$	↑↓ | ↑↓ ↑↓ ↑↓ ↑↓ ↑↓

table A Electronic configurations of the elements scandium to zinc.

When the other eight elements form ions, they lose their 4s electrons before their 3d electrons. Each element can lose a variable number of electrons, and so forms ions with different oxidation numbers. The higher oxidation numbers are not found in simple ions – for example manganese forms an ion with oxidation number +2 (the Mn^{2+} ion), but the manganese ion with oxidation number +7 is MnO_4^-, not Mn^{7+}. Transition metal ions involving higher oxidation numbers usually contain an electronegative element, often oxygen. **Table B** shows the most common oxidation numbers and examples of compounds formed.

Element	Common oxidation numbers						Compounds		
Ti			+3	+4				Ti_2O_3	$TiCl_4$
V		+2	+3	+4	+5			VCl_3	V_2O_5
Cr			+3			+6		$CrCl_3$	K_2CrO_4
Mn		+2		+4			+7	$MnCl_2$ MnO_2	$KMnO_4$
Fe		+2	+3					$FeCl_2$	Fe_2O_3
Co		+2	+3					$CoSO_4$	$CoCl_3$
Ni		+2						$NiSO_4$	
Cu	+1	+2						Cu_2O	$CuSO_4$

table B The most common oxidation numbers and compounds for Period 4 transition elements.

You can see that the highest common oxidation number increases from Ti (+4) to Mn (+7) as all the 4s and 3d electrons become involved in bonding. From Fe to Cu, the increasing nuclear charge means that the electrons are attracted more strongly and are less likely to be involved in bonding. Therefore, ions with higher oxidation numbers are less common.

Questions

1 Write out the full and abbreviated electronic configurations for the element chromium.

2 Describe, with reference to its electrons, how an atom of manganese becomes an ion with an oxidation number of +4.

Key definition

A **transition metal** is an element that forms one or more stable ions with incompletely filled d-orbitals.

15.1 2 Ligands and complexes

By the end of this section, you should be able to...

- understand what is meant by the term 'ligand'
- understand that dative (coordinate) bonding is involved in the formation of complex ions
- understand that a complex ion is a central metal ion surrounded by ligands
- understand the meaning of the term 'coordination number'

Symbols and equations without ligands

An example of the symbol of a metal ion is Na^+, the sodium ion. You will also have used the symbol $Na^+(aq)$, which shows that the sodium ion is dissolved in water. You may have had the process of dissolving solid sodium chloride in water explained in terms of the slightly negative ($\delta-$) oxygen atoms in the water molecules being attracted to the positively charged sodium ions and keeping them in solution. You may have a mental picture of Na^+ being surrounded by an unspecified number of water molecules, with these water molecules leaving and being replaced by others.

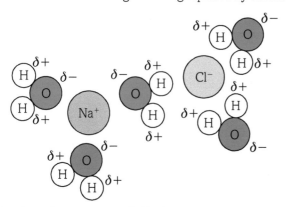

fig A Positive sodium ions surrounded by $\delta-$ oxygen atoms in water molecules, and negative chloride ions surrounded by $\delta+$ hydrogen atoms in water molecules.

When you write equations for the reactions of transition metal ions, you can sometimes use the same method. For example, the iron(II) ion can be shown as Fe^{2+}, or $Fe^{2+}(aq)$ to show that it is dissolved in water.

The equation for a displacement reaction involving a transition metal ion is:

$$Mg(s) + Fe^{2+}(aq) \rightarrow Mg^{2+}(aq) + Fe(s)$$

This equation clearly shows that the reaction is an example of redox, as magnesium transfers electrons to iron(II) ions.

However, it is often better to consider the reactions of transition metal ions in a different way in order to better understand the reactions. This method will be described next.

Symbols and equations with ligands

The ions of non-transition metals tend to have larger radii than those of transition metals in the same period of the Periodic Table. For example, the ionic radius of K^+ is 0.133 nm, and that of Fe^{2+} is 0.076 nm. The relatively small size of transition metal ions enables them to attract electron-rich species more strongly, including the water molecules present in aqueous solutions. These water molecules are attracted to the transition metal ions so strongly that they form a specific number of bonds, usually six, giving a structure that can be represented in diagrams such as these:

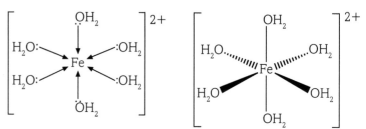

fig B Here are two ways to represent a complex ion. The one on the left shows one of the lone pairs of electrons on the oxygen atoms and the dative covalent bonds with arrows. The one on the right shows the 3D arrangement of bonds and water molecules around the metal ion.

In **fig B** (left):

- the bonds are shown with arrow heads, indicating that they are dative (coordinate) bonds – one of the lone pairs of electrons on each oxygen atom is used to form the bond
- the whole structure is shown inside square brackets, and the original charge of the Fe^{2+} ion is shown outside the bracket
- the water molecules are arranged in a regular pattern around the Fe – this arrangement can be explained in terms of the electron pair repulsion theory.

In **fig B** (right):

- solid wedges represent bonds coming out of the plane of the paper
- hatched wedges represent bonds going behind the plane of the paper.

Although it is often useful to think about diagrams like this, when writing equations it is more usual to abbreviate them. The example above would be shown as $[Fe(H_2O)_6]^{2+}$. Notice that the square brackets and the position of the charge are the same in this abbreviated form.

These water molecules, and other electron-rich species that can form dative bonds in the same way, are called **ligands**. The complete formulae are called **complexes**. As most complexes have charges, they are often called **complex ions**, but where there is no overall charge, it is better to use the term 'complex'. The total number of dative bonds around the metal ion is called the **coordination number**.

Principles of transition metal chemistry | 15.1

Examples of ligands

Many species can act as ligands, but those that we will meet most often are shown in **table A**, which also shows how the name of the ligand changes when used in the name of a complex.

Ligand	Formula	Charge	Name in complex
water	H_2O	0	aqua
hydroxide	^-OH	−1	hydroxo
ammonia	NH_3	0	ammine
chloride	Cl^-	−1	chloro

table A

Note these points about the names used in complexes:

- ligands with a negative charge end in -o
- ammine should not be confused with the organic term amine.

Naming complexes

Table B shows some examples to illustrate the naming system used.

Complex	Name	Notes
$[Fe(H_2O)_6]^{2+}$	hexaaquairon(II)	In order, there is: • the number of ligands • the name of the ligand • the name of the metal ion • the oxidation number of the metal ion.
$[FeCl_4]^-$	tetrachloroferrate(III)	The overall charge is 1− because the ion is formed from one Fe^{3+} ion and four Cl^- ions. '-ate' is added to the metal name to show that it is now part of a negatively charged complex ion. As it is negatively charged, a Latin name for the metal is used (in Latin, iron is ferrum).
$[Cu(NH_3)_4(H_2O)_2]^{2+}$	tetraamminediaquacopper(II)	With two different ligands, they appear in alphabetical order (so ammine before aqua), ignoring the tetra- and di- parts of the name.

table B

Questions

1. Explain why the methane molecule cannot act as a ligand.

2. An equation for a reaction involving a transition metal complex is:

    $[Cu(H_2O)_6]^{2+} + 2OH^- \rightarrow [Cu(H_2O)_4(OH)_2] + 2H_2O$

    Explain why the first product is a complex but not a complex ion.

Key definitions

A **ligand** is a species that uses a lone pair of electrons to form a dative bond with a metal ion.
A **complex** is a species containing a metal ion joined to ligands.
A **complex ion** is a complex with an overall positive or negative charge.
The **coordination number** is the number of dative bonds in the complex.

Learning tip

You can work out the charge on the metal ion in a complex by considering the overall charge on the complex and knowing the charges on the ligands. For example, $[FeCl_4]^{2-}$ has an overall charge of 2−, and there are four chloride ions each with a charge of 1−, so the charge on the metal ion must be 2+.

15.1 3 The origin of colour in complexes

By the end of this section, you should be able to...

- understand that transition metals form coloured ions in solution
- understand that the colour of aqueous ions, and other complex ions, results from the splitting of the energy levels of the d-orbitals by ligands
- understand why there is a lack of colour in some aqueous ions and other complex ions

A complex explanation!

Colour is very important in our world – think of traffic lights, paint charts, clothing and the colours of flowers. There are scientific explanations for colour in plants, paints and dyes, but these explanations are often specific to the type of chemical substance being considered.

As far as transition metal complexes are concerned, a full explanation includes three main concepts:

- the electromagnetic spectrum and the colour wheel
- the connection between colour, energy and wavelength
- the electronic configurations of transition metal ions.

The electromagnetic spectrum

We first need to find out about the **electromagnetic spectrum**. You are probably familiar with the idea that the visible part of the electromagnetic spectrum (sometimes described as 'white light') is made up of a mixture of colours (sometimes described as the colours of the rainbow).

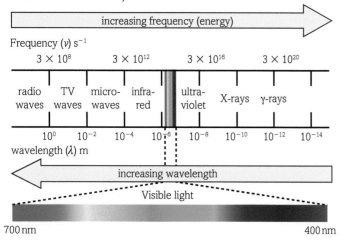

fig A The diagram shows how the visible part of the spectrum fits into the complete electromagnetic spectrum.

So, how many colours make up white light? A traditional answer is seven:

red   orange   yellow   green   blue   indigo   violet

However, seven is a purely arbitrary number, and this list of colours does not contain some that are now more familiar, such as cyan and magenta.

The real answer is that visible light is made up of an infinite number of colours! If we take red light to have a wavelength of 700 nm and violet light to have a wavelength of 400 nm, then each precise value within this range (such as 651.0 and 438.4) represents a different wavelength and therefore a different colour. There is clearly no limit to the number of values in this range.

The colour wheel

The colours in a colour wheel

Now consider the colours in the visible spectrum shown as a wheel, with the red and violet ends of the spectrum next to each other. There are many different colour wheels with varying numbers of colours (just try a web search), but here is one that contains seven colours corresponding to those listed above. Clockwise from the top, they are shown in order from red to violet.

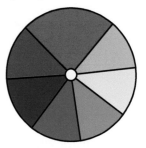

fig B This is one example of a colour wheel.

Complementary colours

In colour wheels, **complementary colours** are shown opposite to each other. In this one, red is opposite blue and green. When white light is passed through a solution containing a transition metal complex, some wavelengths of light are absorbed by the complex. The light emerging will therefore contain proportionately more of the complementary colour. So, if a complex absorbs red light, the light emerging will look blue or green.

fig C A range of transition metal solutions. From left to right: Ti^{2+}, V^{3+}, VO^{2+}, Cr^{3+}, $Cr_2O_7^{2-}$, Mn^{2+}, MnO_4^-, Fe^{3+}, Co^{2+}, Ni^{2+} and Cu^{2+}.

Colour depends on electrons in 3d energy levels

A complete explanation of colour, or absence of colour, in aqueous solutions of metal ions, is quite complicated, but the following explanation should help you to understand the basics. We will use an aqueous solution containing copper(II) ions as an example.

An aqueous solution of zinc sulfate is colourless but an aqueous solution of copper(II) sulfate is blue. As the sulfate ion is present in both solutions, we can assume that any difference in colour is caused by the zinc ions and copper(II) ions. Zinc and copper are next to each other in the Periodic Table, so why should copper, but not zinc, form coloured ions?

The electronic configurations of the Zn^{2+} and Cu^{2+} ions are shown below in the electrons-in-boxes format.

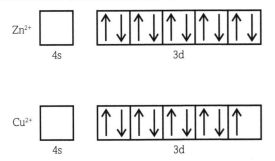

fig D The diagram shows the configurations of the 4s and 3d electrons in Zn^{2+} and Cu^{2+}.

Ions that have completely filled 3d energy levels (such as Zn^{2+}) and ions that have no electrons in their 3d energy levels (such as Sc^{3+}) are not coloured. However, the copper(II), or Cu^{2+}, ion has only nine electrons in the 3d energy level, so it is not completely filled. When water ligands are attached to the copper(II) ion, the energy level splits into two levels with slightly different energies. The lower energy level contains six electrons and the higher energy level contains three electrons.

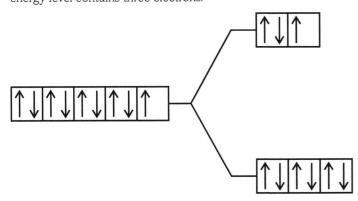

fig E In the Cu^{2+} ion, the water ligands split the 3d energy level into lower and higher energy levels.

If one of the electrons in the lower energy level absorbs energy from the visible spectrum, it can move to the higher energy level. Movement from a lower energy level to a higher energy level is called 'promotion' or 'excitation'. When an electron moves to a higher energy level, the amount of energy it absorbs depends on the difference in energy between the two levels – the bigger the energy difference, the more energy the electron absorbs. It is important to know that the amount of energy gained by the electron is directly proportional to the frequency of the absorbed light and inversely proportional to the wavelength of the light. In the case of the Cu^{2+} ion, the small difference in energy levels means that low frequency (or high wavelength) radiation is absorbed from the red end of the spectrum. Therefore, blue light is transmitted.

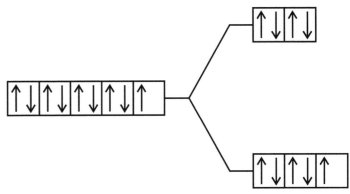

fig F Note that one electron has moved from the lower energy level to the higher energy level.

Learning tip

Try to find a way to remember that red light has a longer wavelength than violet light. The Red Sea (between Africa and Asia) is a long (and narrow) sea. Perhaps you can find a better way!

Questions

1. Solutions containing $Fe^{3+}(aq)$ ions are yellow. Explain which part of the electromagnetic spectrum is absorbed by these ions.

2. State why solutions containing $Al^{3+}(aq)$ ions are colourless.

Key definitions

The **electromagnetic spectrum** is the range of all wavelengths and frequencies of all the types of radiation.

Complementary colours are colours opposite each other on a colour wheel.

15.1 4 Common shapes of complexes

By the end of this section, you should be able to...

- understand why complexes with six-fold coordination have an octahedral shape, for example those formed by metal ions with H_2O, OH^- and NH_3 as ligands
- understand that transition metal ions may form tetrahedral complexes with relatively large ligands such as Cl^-

Predicting the shapes of complexes

Using electron pair repulsion theory

You are already familiar with the use of the electron pair repulsion theory to predict and explain the shapes of simple molecules and ions. It is quite easy to extend this idea to predicting and explaining the shapes of complexes. A ligand bonds to the central metal ion by donating a pair of electrons to form a dative bond. The only differences are that you should ignore the 3d electrons in the transition metal ion and the overall charge on the complex – just count the number of electrons donated by the ligands. **Table A** shows how this theory can be applied in exactly the same way to predict the shape of a transition metal complex and the shape of a simple molecule.

Number of ligands	Electrons donated	Shape	Bond angle	Example	Analogous example
6	12	octahedral	90°	$[Co(NH_3)_6]^{2+}$	SF_6
4	8	tetrahedral	109.5°	$[CuCl_4]^{2-}$	CH_4
2	4	linear	180°	$[Ag(NH_3)_2]^+$	$BeCl_2$

table A The electron pair repulsion theory can predict the shapes of simple molecules and complex ions.

Octahedral complexes

The most common ligands in most octahedral complexes are water, ammonia and the hydroxide ion. Although these ligands have different numbers of lone pairs of electrons (ammonia has one, water has two and the hydroxide ion has three), each ligand uses only one lone pair to form a coordinate bond with the transition metal ion. As they contain six ligands, the complexes are sometimes described as having **six-fold coordination**. Note that the electron pair donor in each of these ligands is an element in Period 2 of the Periodic Table, so all three ligands are of approximately equal size. **Table B** shows some common examples.

Abbreviated formula	Name	Colour
$[Mn(H_2O)_6]^{2+}$	hexaaquamanganese(II)	very pale pink (usually described as colourless)
$[Fe(H_2O)_4(OH)_2]$	tetraaquadihydroxoiron(II)	pale green
$[Al(OH)_6]^{3-}$	hexahydroxoaluminate(III)	colourless

table B Examples of octahedral complexes.

Note that the last example contains an element that is not a transition metal ion, nor even a d-block element. This example shows that the names and shapes of complexes of non-transition metals such as aluminium can be predicted in the same way as those of transition metal ions.

Fig A shows the shapes of these complexes.

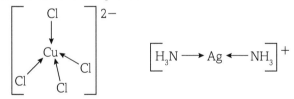

fig A The structures of some octahedral complexes.

Note that:
- Solid and hatched wedge bonds are used to indicate the shapes, although the 3D shape can be shown in other ways.
- The overall charges are all different – this depends on the charge on the original transition metal ion and on how many negatively charged ligands there are.
- All three structures use square brackets, although these are sometimes omitted when the complex is neutral.
- The water ligands are shown as both H_2O and OH_2 – the important point is that the bond must be shown to come from oxygen because it supplies the lone pair of electrons for the dative bond.

Tetrahedral and linear complexes

These are much less common than octahedral complexes. The only tetrahedral complexes that you are likely to come across in your course are those in which chloride ions act as ligands, such as in the $[CuCl_4]^{2-}$ ion. Note that because chlorine is a Period 3 element, its ions are much bigger than water, ammonia and hydroxide ions. So, there is usually insufficient room around the central metal ion for six chloride ions to act as ligands.

The only linear complex you are likely to come across in your course is the reactive ion present in Tollens' reagent (sometimes called ammoniacal silver nitrate). An explanation of why the Ag^+ ion has only two ligands, and not six, is beyond the scope of this book, but note that silver is a transition metal in Period 5 (not Period 4) of the Periodic Table. It therefore behaves differently from the transition metals in Period 4.

Fig B shows the shapes of these two complexes.

fig B The structures of a tetrahedral complex and a linear complex.

Learning tip

You can usually assume that a complex with six ligands is octahedral, one with four ligands is tetrahedral and one with two ligands is linear.

Questions

1. A complex that forms between cobalt ions and nitrite ions (NO_2^-) has the abbreviated formula $[Co(NO_2)_6]^{3-}$. Predict its shape and the oxidation number of the transition metal ion.

2. A complex forms between nickel(II) ions and chloride ions. Predict its shape, name and formula.

Key definition

Six-fold coordination refers to complexes in which there are six ligands forming coordinate bonds with the transition metal ion.

15.1 5 Square planar complexes

By the end of this section, you should be able to...

- understand that square planar complexes are also formed by transition metal ions and that *cis*-platin is an example of such a complex
- understand why *cis*-platin used in cancer treatment is supplied as a single isomer and not in a mixture with the *trans* form

Square planar molecules

You are familiar with the idea that the electron pair repulsion theory can be applied to predict the shapes of simple molecules and ions. However, you may not have come across an example of a **square planar** shape. One example is xenon tetrafluoride, XeF_4. Although xenon is a noble gas, it does form some stable compounds.

An atom of xenon has eight electrons in its outermost energy level, and each fluorine atom uses one of its electrons to form a covalent bond. The outer energy shell now contains 12 electrons. These are arranged in six pairs, forming an octahedral arrangement. Two of these pairs are lone pairs, which repel each other and are therefore located opposite each other. So, the four bonding pairs are in a plane, with the four fluorine atoms at the corners of a square and an F–Xe–F bond angle of 90°, as shown in **fig A**.

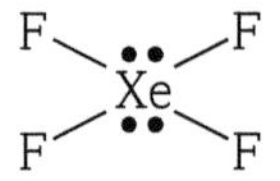

fig A The square planar shape of the xenon tetrafluoride molecule.

Cis-platin

Square planar complexes are much less common than octahedral and tetrahedral complexes. One particular complex of this type, *cis*-platin, has become well known in recent years because of its use as an effective treatment for some types of cancer, especially testicular cancer.

An explanation of why the four ligands in this complex form a square planar shape, and not a tetrahedral shape, is beyond the scope of this book. It is not easily explained by the electron pair repulsion theory. Part of the explanation is that platinum is a transition metal in Period 6, and so behaves differently from the transition metals in Period 4.

Cis-trans isomers

You have learned about *E-Z* and *cis-trans* isomerism in **Book 1 Topic 6**. Your understanding of this type of isomerism can now be applied to inorganic compounds such as *cis*-platin and its isomer, *trans*-platin. These consist of a platinum(II) ion, two ammonia ligands and two chloride ion ligands.

The diagram shows the structures of these isomers and the relationship between them.
- The *cis*- prefix indicates that identical ligands are next to each other.
- The *trans*- prefix indicates that they are opposite each other.

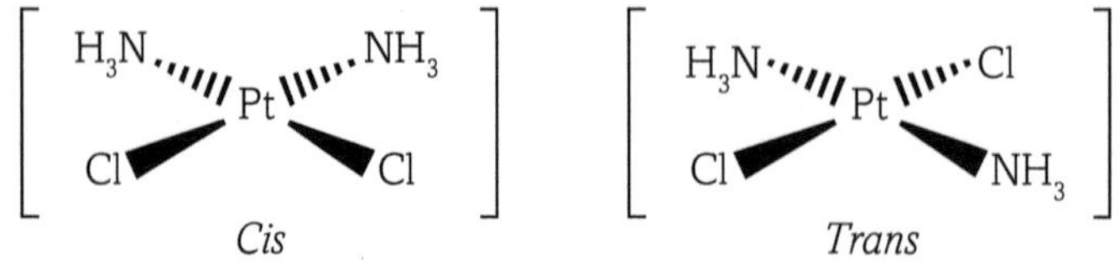

fig B Structures of *cis*-platin and *trans*-platin.

Anti-cancer action

A full understanding of how *cis*-platin kills cancer cells is beyond the scope of this book. Put simply:

- All cells, including cancer cells, contain deoxyribonucleic acid (DNA).
- During cell division, the two strands of DNA must separate from each other to form more DNA.
- The structure of *cis*-platin enables it to form a bond between the two strands of DNA, which prevents them from separating and so prevents the cancer cells from dividing.

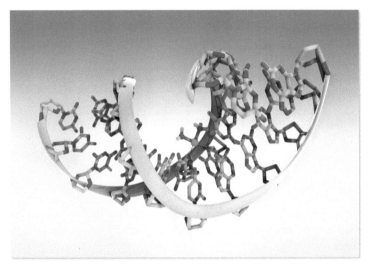

fig C DNA strands are shown in grey and *cis*-platin in pink.

Trans-platin

The isomers *trans*-platin and *cis*-platin have different structures. The difference in structure means that *trans*-platin is a much less effective cancer treatment than *cis*-platin. It is also more toxic. So, it is important to use only the *cis*- isomer in the cancer treatment.

Learning tip
You should know the structures of *cis*-platin and *trans*-platin, but you do not need to know exactly why *cis*-platin is effective at treating cancer but *trans*-platin is not.

Questions

1. Explain why you would expect the ICl_4^- ion to have a square planar shape.

2. (a) Draw the structures of the two *cis-trans* isomers of the $IF_2Cl_2^-$ ion.
   (b) Why are there no *cis-trans* isomers of the $IFCl_3^-$ ion?

Key definition
A **square planar** shape contains a central atom or ion surrounded by four atoms or ligands in the same plane and with bond angles of 90°.

15.1 6 Multidentate ligands

By the end of this section, you should be able to...
- understand why H_2O, Cl^- and NH_3 act as monodentate ligands
- identify bidentate ligands, such as $NH_2CH_2CH_2NH_2$ and multidentate ligands such as $EDTA^{4-}$
- understand that haemoglobin is an iron(II) complex containing a multidentate ligand
- understand that a ligand exchange reaction occurs when an oxygen molecule bound to haemoglobin is replaced by a carbon monoxide molecule
- understand, in terms of the large positive increase in ΔS_{system}, that the replacement of a monodentate ligand by a bidentate or multidentate ligand leads to a more stable complex ion

Denticity

Denticity is a rarely used English word, but it is a property of ligands that you should be aware of. It comes from the Latin word *dentis* (meaning 'tooth'), from which we get the familiar word 'dentist'. There is a connection!

When ligands were introduced earlier in this book, we could have described them as **monodentate ligands**. You can think of monodentate as meaning 'one tooth' or 'one bite', which means that the ligand uses one lone pair of electrons on one atom to form the dative bond with the metal ion.

Now for **bidentate ligands**. *Bi* means two, so a bidentate ligand has two atoms, each of which can use a lone pair of electrons to form a dative bond with the metal ion. You can imagine that a bidentate ligand has two atoms that can 'bite' onto the metal ion.

You can work out for yourself what a **multidentate ligand** is – a ligand with several atoms, each of which uses a lone pair of electrons to form a dative bond with the metal ion. The most common example you are likely to meet has six such atoms, so it could be described as a hexadentate ligand.

Bidentate ligands

The most likely bidentate ligand you will meet in your chemistry course is the organic compound with the structural formula $NH_2CH_2CH_2NH_2$. Its correct name is 1,2-diaminoethane, although for a reason we will soon see, it is sometimes called ethylenediamine – the ethylene comes from the CH_2CH_2 part, and it is correctly described as a diamine because there are two amino groups. Its structure is:

fig A The structure of 1,2-diaminoethane.

Occasionally it is abbreviated to 'en', especially when used in equations representing its reactions as a ligand.

When it acts as a bidentate ligand, it uses the lone pair of electrons on each nitrogen atom to attach to the metal ion. **Fig B** shows an Ni^{2+} ion joined to three molecules of 1,2-diaminoethane.

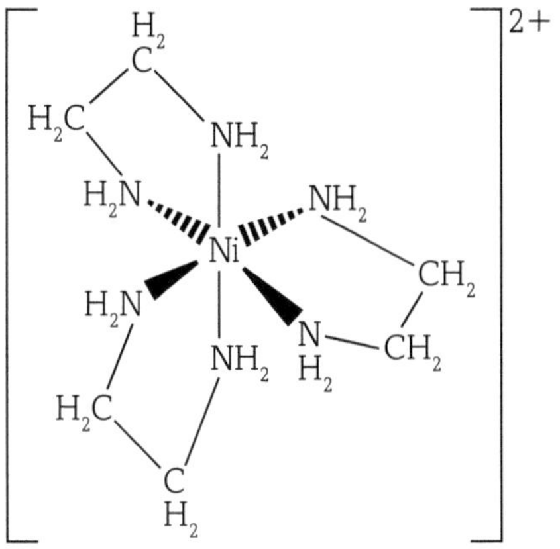

fig B The structure of the complex formed when Ni²⁺ reacts with three molecules of 1,2-diaminoethane.

If this complex were formed by the reaction between a hexaaqua metal ion and 1,2-diaminoethane, this abbreviated equation could be written:

$$[Ni(H_2O)_6]^{2+} + 3en \rightarrow [Ni(en)_3]^{2+} + 6H_2O$$

Multidentate ligands

The most likely multidentate ligand you will meet in your chemistry course is an organic ion with a rather complicated structure.

First, consider the structure of 1,2-diaminoethane. Next, imagine that each of the four hydrogen atoms on the two nitrogens are replaced by $-CH_2COOH$ (this is ethanoic acid bonded to the nitrogens via the CH_3 group).

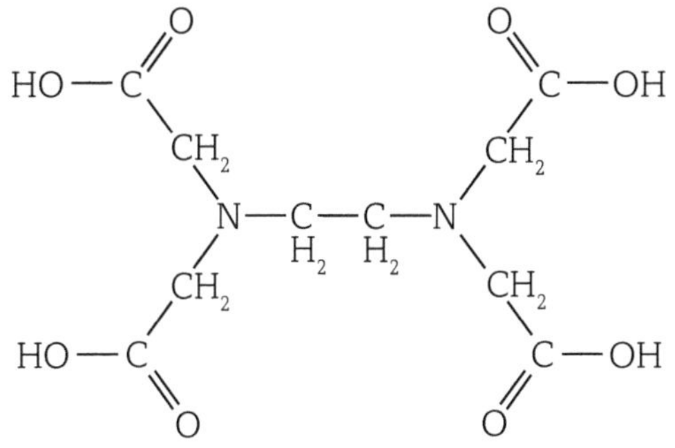

fig C The structure of the EDTA molecule.

You may know that the old name for ethanoic acid is acetic acid. Four of these molecules have been used to form this structure. Now, remembering the alternative name for 1,2-diaminoethane, you should be able to see that one name for this structure could be ethylenediaminetetraacetic acid. This is a bit of a mouthful, so pick out four key letters from the name to get the abbreviation that is normally used for the compound. EthyleneDiamineTetraAcetic acid now becomes 'EDTA'. We will not bother with the IUPAC name!

The final step is to consider the ion formed when each of the ethanoic acid groups loses its H⁺ ion – this gives an ion with four negative charges. It is a hexadentate ligand referred to as EDTA⁴⁻ and has this structure:

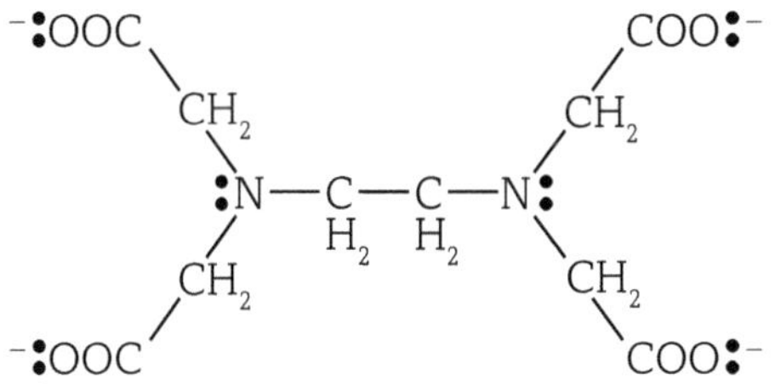

fig D The structure of the EDTA⁴⁻ ion.

The advantage of displaying the EDTA^{4-} ion in this way is that it clearly shows the six lone pairs of electrons – two pairs on the nitrogen atoms and four pairs on the oxygen atoms – that are used to form the dative bonds when the ion acts as a ligand. **Fig E** uses a skeletal formula to show the complex formed between one M^{2+} ion and one EDTA^{4-} ion.

fig E The structure of the complex formed between M^{2+} and EDTA^{4-}.

The six dative bonds formed by EDTA^{4-} are shown as dashed lines.

The stability of complexes

Most complexes are stable, in the sense that they do not decompose readily. In this heading, 'stability' does not refer to complexes containing transition metal ions with unstable oxidation numbers, such as +2 in scandium (Sc^{2+}). Instead, it refers to a comparison of the stabilities of two complexes in which the number of ligands has changed. For a discussion of 'stability' see **Topic 13**. Consider this ligand exchange reaction in which a monodentate ligand is replaced by a bidentate ligand:

$$[Cu(H_2O)_6]^{2+} + 3en \rightarrow [Cu(en)_3]^{2+} + 6H_2O$$

Six water ligands are replaced by three 1,2-diaminoethane ligands, so the total number of species has increased from four to seven. This means that the system is more disordered, and so there is an increase in ΔS_{system}. Ligand exchange reactions of this sort lead to an increase in stability of the products compared to the reactants, and so formation of the products is favoured.

When a monodentate ligand is replaced by a multidentate ligand, the increase in stability is even greater, as in this example with EDTA^{4-}:

$$[Cu(H_2O)_6]^{2+} + EDTA^{4-} \rightarrow [Cu(EDTA)]^{2-} + 6H_2O$$

Here, six water ligands are replaced by one EDTA^{4-} ligand, so the total number of species has increased from two to seven.

Haemoglobin and oxygen transport

You may know that haemoglobin is a protein in red blood cells that plays a vital role in transporting oxygen through the bloodstream in humans and other animals. Here is a simplified explanation of the role of haemoglobin in oxygen transport, without including details of its structure and of the haem group that it contains:

- Haemoglobin consists of three main parts, the largest of which is protein (the 'globin' part).
- Within the protein, there are four haem groups that are made up mostly of carbon and hydrogen atoms.
- Inside each haem group, there are four nitrogen atoms that hold an Fe^{2+} ion by forming dative bonds with it in a square planar structure.
- There is a fifth dative bond from the protein to the Fe^{2+} ion.
- When blood passes through the lungs, haemoglobin collects oxygen molecules and transports them to cells, where it is released.
- When haemoglobin collects oxygen, the oxygen molecule acts as a ligand by using one of its lone pairs of electrons to form a dative bond with one of the Fe^{2+} ions.

Fig F is a very simplified diagram showing the six dative bonds in oxyhaemoglobin.

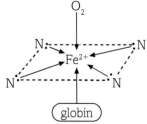

fig F Part of the structure of oxyhaemoglobin, showing the dative covalent bonds to the central metal ion.

Haemoglobin and carbon monoxide

A carbon monoxide molecule has a lone pair of electrons on its carbon atom that enables it to act as a ligand. The strength of the dative bond between oxygen and haemoglobin is not particularly strong, but the advantage of this is that the oxygen molecule is relatively easily released when it is needed.

Unfortunately, a much stronger dative bond forms between carbon monoxide and haemoglobin, and so any carbon monoxide breathed in is very likely to replace the oxygen already bound to haemoglobin – this is a ligand substitution reaction. Once the carbon monoxide has formed carboxyhaemoglobin, the dative bond is so strong that it does not break easily. This means that if enough haemoglobin molecules are converted to carboxyhaemoglobin, there may be too little oxygen transported to support life.

Put very simply, this reaction is reversible:

haemoglobin + oxygen ⇌ oxyhaemoglobin

but this reaction is not:

haemoglobin + carbon monoxide → carboxyhaemoglobin

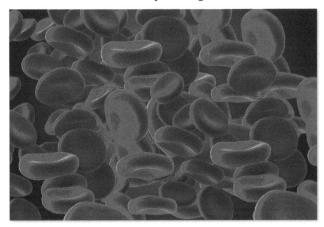

fig G These red blood cells contain haemoglobin that transports oxygen from the lungs to the cells in the body.

Learning tip

In 1,2-diaminoethane, there are only two lone pairs of electrons in the molecule and both are used when it acts as a ligand. In $EDTA^{4-}$ the lone pairs of electrons on the O of the C=O are not used. Only those on the N atoms and the negatively charged O atoms are used.

Questions

1. The water molecule contains three atoms and has two lone pairs of electrons. Explain why it can only act as a monodentate ligand, and not as a bidentate ligand.

2. Write an equation for the ligand substitution reaction that occurs when four of the ligands in the hexaaquairon(III) ion are replaced by 1,2-diaminoethane (en).

Key definitions

A **monodentate ligand** is one that forms one dative bond with a metal ion.
A **bidentate ligand** is one that forms two dative bonds with a metal ion.
A **multidentate ligand** is one that forms several dative bonds with a metal ion.

15.2 1 Different types of reaction

By the end of this section, you should be able to...

- understand that colour changes in d-block metal ions may arise as a result of changes in oxidation number, ligand and coordination number
- record observations and write suitable equations for the reactions of Cu^{2+}(aq) with aqueous sodium hydroxide and aqueous ammonia, including in excess
- understand that the substitution of small, uncharged ligands (such as H_2O) by larger, charged ligands (such as Cl^-) can lead to a change in coordination number
- understand that ligand exchange, and an accompanying colour change, occurs in the formation of:
  (i) $[Cu(NH_3)_4(H_2O)_2]^{2+}$ from $[Cu(H_2O)_6]^{2+}$ via $Cu(OH)_2(H_2O)_4$
  (ii) $[CuCl_4]^{2-}$ from $[Cu(H_2O)_6]^{2+}$
- write ionic equations to show the difference between ligand exchange and amphoteric behaviour

Types of reaction

So far, we have considered the origin of colour in transition metal ions. We can now consider why there are often colour changes when transition metal ions take part in reactions. Four main types of reaction can occur:

- redox – the oxidation number of the transition metal ion changes
- acid–base reaction – one or more of the ligands gains or loses a hydrogen ion
- ligand exchange – one or more of the ligands around the transition metal ion is replaced by a different ligand
- coordination number change – the number of ligands changes.

Any one of these types of reaction can cause a change in the colour of the complex. Some reactions involve more than one of these types of reaction.

Change in oxidation number

An aqueous solution containing Fe^{2+}(aq) ions is pale green, but when it is exposed to air it gradually turns yellow or brown, as the oxidation number of iron increases from +2 to +3. The type and number of ligands remain unchanged in this oxidation reaction, so the formulae of the two complexes are $[Fe(H_2O)_6]^{2+}$ and $[Fe(H_2O)_6]^{3+}$. The colour change in this reaction is best illustrated using solid samples containing these ions.

Equations are not usually written for oxidation reactions in which the only change is the oxidation number of the transition metal ion.

The formation of $[Cu(NH_3)_4(H_2O)_2]^{2+}$ – acid-base and ligand exchange reactions

Consider the reaction that occurs when aqueous sodium hydroxide is added to copper(II) sulfate solution. The observation is that a pale blue solution forms a blue precipitate. The equation for this reaction is:

$$[Cu(H_2O)_6]^{2+} + 2OH^- \rightarrow [Cu(H_2O)_4(OH)_2] + 2H_2O$$

This contains $[Fe(H_2O)_6]^{2+}$

This contains $[Fe(H_2O)_6]^{3+}$

fig A Solid samples showing colour differences between ions.

You might think that this is a ligand substitution reaction – that two hydroxide ions have replaced two water molecules. In fact, it is an acid–base reaction – the two hydroxide ions have removed hydrogen ions from two of the water ligands and converted them into water molecules. The two water molecules that have lost hydrogen ions are now hydroxide ligands. The term **amphoteric behaviour** is sometimes used to describe this acid–base reaction because of its reversible nature. When an acid is added to the blue precipitate, the hydrogen ions from the acid react with the hydroxide ion ligands and convert them back to water molecules.

Exactly the same observations can be made during the careful addition of aqueous ammonia instead of aqueous sodium hydroxide. The equation for this reaction is:

$[Cu(H_2O)_6]^{2+} + 2NH_3 \rightarrow [Cu(H_2O)_4(OH)_2] + 2NH_4^+$

It is much easier to see that this equation involves an acid–base reaction – two of the water ligands transfer a hydrogen ion to the ammonia molecules.

When aqueous ammonia is added to the blue precipitate formed, it dissolves to form a deep blue solution. The equation for this reaction is:

$[Cu(H_2O)_4(OH)_2] + 4NH_3 \rightarrow [Cu(NH_3)_4(H_2O)_2]^{2+} + 2H_2O + 2OH^-$

This is a ligand exchange reaction – four ammonia molecules replace two water molecules and two hydroxide ions. The solutions at the start and end of the reaction, and the intermediate precipitate are shown in **fig B**.

fig B The pale blue solution contains $[Cu(H_2O)_6]^{2+}$, the pale blue precipitate contains $[Cu(H_2O)_4(OH)_2]$ and the deep blue solution contains $[Cu(NH_3)_4(H_2O)_2]^{2+}$.

The formation of $[CuCl_4]^{2-}$ – change in coordination number

These reactions always involve a change of ligand as well. A good example is the reaction between copper(II) sulfate solution and concentrated hydrochloric acid. When the acid is added slowly and continuously, the colour gradually changes from blue to green and finally to yellow. This is the equation for the reaction:

$[Cu(H_2O)_6]^{2+} + 4Cl^- \rightleftharpoons [CuCl_4]^{2-} + 6H_2O$

The state symbols, all (aq), have been omitted for clarity.

You can see that all six water ligands have been substituted by four chloride ions. This reaction is also an example of a change in coordination number, from 6 to 4. Note that although the charge on the complex has changed from 2+ to 2−, there has been no change in oxidation number.

The arrow shows that the reaction is reversible, which helps to explain the colour change observed. The hexaaquacopper(II) ion is blue and the tetrachlorocuprate(II) ion is yellow, so the green colour is due to a mixture of the blue and yellow complex ions. The solutions at the start and end of the reaction, and the intermediate mixture are shown in **fig C**.

fig C The pale blue solution contains $[Cu(H_2O)_6]^{2+}$, the yellow solution on the right contains $[CuCl_4]^{2-}$ and the green solution in the middle contains a mixture of the two.

Learning tip

You should remember the colours of the species in these reactions but you do not need to understand why the species have the actual colours.

Questions

1 The equation for a reaction of a transition metal ion is:

   $[Fe(H_2O)_6]^{3+} + SCN^- \rightarrow [Fe(SCN)(H_2O)_5]^{2+} + H_2O$

   Explain which of the four main types of reaction are illustrated by this equation.

2 The complex $[Co(H_2O)_6]^{2+}$ is converted into $[Co(NH_3)_6]^{3+}$

   Explain which types of reaction occur in this conversion.

Key definition

Amphoteric behaviour refers to the ability of a species to react with both acids and bases.

15.2 2 Reactions of cobalt and iron complexes

By the end of this section, you should be able to...

- record observations and write suitable equations for the reactions of Fe^{2+}(aq), Fe^{3+}(aq) and Co^{2+}(aq) with aqueous sodium hydroxide and aqueous ammonia, including in excess
- understand that ligand exchange, and an accompanying colour change, occurs in the formation of $[CoCl_4]^{2-}$ from $[Co(H_2O)_6]^{2+}$

Reactions involving cobalt complexes

fig A The bright colour of these glass bottles is caused by blue cobalt compounds.

Reaction with alkalis

Consider the reaction that occurs when aqueous sodium hydroxide is added to a solution containing the hexaaquacobalt(II) ion until no further change is seen. The observation is that a pink solution forms a blue precipitate. The equation for this reaction is:

$$[Co(H_2O)_6]^{2+} + 2OH^- \rightarrow [Co(H_2O)_4(OH)_2] + 2H_2O$$

As with copper, this is an acid–base reaction – the two hydroxide ions have removed hydrogen ions from two of the water ligands and converted them into water molecules. The two water molecules that have lost hydrogen ions are now hydroxide ligands. Upon standing, the colour of the precipitate gradually changes to pink.

The same observations can be made when aqueous ammonia is used as the alkali. However, when aqueous ammonia is added to excess, there is an observation not made with aqueous sodium hydroxide – the precipitate dissolves to form a brown solution. The equation for the acid–base reaction is:

$$[Co(H_2O)_6]^{2+} + 2NH_3 \rightarrow [Co(H_2O)_4(OH)_2] + 2NH_4^+$$

The equation for the ligand exchange reaction forming the brown solution is:

$$[Co(H_2O)_4(OH)_2] + 6NH_3 \rightarrow [Co(NH_3)_6]^{2+} + 4H_2O + 2OH^-$$

Some features of transition metal chemistry are impossible to explain without going well beyond an A level understanding. One of these is why six ammonia ligands are involved in the reaction with the cobalt complex, while in the corresponding reaction with copper only four ammonia ligands are involved.

Upon standing, this brown solution changes colour because of oxidation by the oxygen in the atmosphere. The oxidation number of cobalt increases from +2 to +3, and the yellow $[Co(NH_3)_6]^{3+}$ ion forms. Unfortunately, the resulting solution usually looks much darker than yellow, as there are other products whose formation is difficult to explain.

Reaction with concentrated hydrochloric acid

This reaction is very like the one with the $[Cu(H_2O)_6]^{2+}$ ion. When concentrated hydrochloric acid is slowly added to a solution containing the hexaaquacobalt(II) ion, the pink solution gradually changes to blue. This is the equation for the reaction:

$$[Co(H_2O)_6]^{2+} + 4Cl^- \rightarrow [CoCl_4]^{2-} + 6H_2O$$

You can see that all six water ligands have been replaced by four chloride ions. This reaction is also an example of a change in coordination number, from 6 to 4. Note that although the charge on the complex has changed from 2+ to 2−, there has been no change in oxidation number.

Reactions involving iron complexes

Reaction of iron(II) complexes with alkalis

Consider the reaction that occurs when aqueous sodium hydroxide is added to a solution containing the hexaaquairon(II) ion until no further change is seen. The observation is that a pale green solution forms a green precipitate. The equation for this reaction is:

$$[Fe(H_2O)_6]^{2+} + 2OH^- \rightarrow [Fe(H_2O)_4(OH)_2] + 2H_2O$$

fig B Ordinary glass is often very pale green. This is more noticeable if you look through a considerable thickness of glass. The green colour is caused by the presence of impurities containing iron.

As with copper and cobalt, this is an acid–base reaction – the two hydroxide ions have removed hydrogen ions from two of the water ligands and converted them into water molecules. The two water molecules that have lost hydrogen ions are now hydroxide ligands.

The same observations can be made when aqueous ammonia is used as the alkali. The equation for the acid–base reaction is:

$$[Fe(H_2O)_6]^{2+} + 2NH_3 \rightarrow [Fe(H_2O)_4(OH)_2] + 2NH_4^+$$

Upon standing, the colour of the green precipitate gradually changes to brown as oxygen from the atmosphere causes oxidation, forming $[Fe(H_2O)_3(OH)_3]$ – this is the triaquatrihydroxoiron(III) complex.

Reaction of iron(III) complexes with alkalis

Consider the reaction that occurs when aqueous sodium hydroxide is added to a solution containing the hexaaquairon(III) ion until no further change is seen. The observation is that a yellow-brown solution forms a brown precipitate. The equation for this reaction is:

$$[Fe(H_2O)_6]^{3+} + 3OH^- \rightarrow [Fe(H_2O)_3(OH)_3] + 3H_2O$$

As with iron(II), this is an acid–base reaction – the three hydroxide ions have removed hydrogen ions from three of the water ligands and converted them into water molecules. The three water molecules that have lost hydrogen ions are now hydroxide ligands.

The same observations can be made when aqueous ammonia is used as the alkali. The equation for the acid–base reaction is:

$$[Fe(H_2O)_6]^{3+} + 3NH_3 \rightarrow [Fe(H_2O)_3(OH)_3] + 3NH_4^+$$

There are no further reactions when an excess of either aqueous sodium hydroxide or aqueous ammonia is added and no further changes upon standing.

Learning tip

Note that atmospheric oxidation occurs with complexes containing a transition metal ion with a +2 oxidation number, but not for those with a +3 oxidation number.

Questions

1. When aqueous ammonia is added in excess to a solution containing $[Co(H_2O)_6]^{2+}$ ions, an acid–base reaction occurs, followed by a ligand exchange reaction. Write an equation to describe this reaction.

2. A student wrote this equation to explain the formation of the brown precipitate when aqueous ammonia was added to a solution containing Fe^{3+} ions:

    $$[Fe(H_2O)_6]^{3+} + 3OH^- \rightarrow [Fe(H_2O)_3(OH)_3] + 3H_2O$$

    Why is the above reaction less likely to occur than the one shown below?

    $$[Fe(H_2O)_6]^{3+} + 3NH_3 \rightarrow [Fe(H_2O)_3(OH)_3] + 3NH_4^+$$

15.2 3 The chemistry of chromium

By the end of this section, you should be able to...

- record observations and write suitable equations for the reactions of $Cr^{3+}(aq)$ with aqueous sodium hydroxide and aqueous ammonia, including in excess
- understand, in terms of the relevant $E^\ominus$ values, that the dichromate(VI) ion, $Cr_2O_7^{2-}$:
  (i) can be reduced to Cr^{3+} and Cr^{2+} ions using zinc in acidic conditions
  (ii) can be produced by the oxidation of Cr^{3+} ions using hydrogen peroxide in alkaline conditions
- understand that the dichromate(VI) ion, $Cr_2O_7^{2-}$, can be converted into chromate(VI) ions as a result of the equilibrium, $2CrO_4^{2-} + 2H^+ \rightleftharpoons Cr_2O_7^{2-} + H_2O$

Introduction

Some of the reactions of chromium are similar to those we have already met for the ions of copper, cobalt, iron(II) and iron(III). However, chromium has ions in which the metal has an oxidation number of +6, so reactions involving these ions must be considered in a different way.

A word of warning

We have already met the idea that there is an infinite number of colours in the spectrum of visible light, although we sometimes refer to the seven colours of the rainbow. The perception of colours by humans varies considerably. A specific colour might be described as pink or purple by different people, and a colour described as blue-green by one person might be described as blue, green or turquoise by others.

The situation with the colours used to describe chromium compounds is further complicated by these factors:

- Some compounds have different colours as solids and aqueous solutions.
- The colour of a solution depends on concentration.
- The presence of dissolved oxygen in an aqueous solution can affect the colour observed.

This means that the colours used in this book may not always correspond to those in other books, and may be different from those observed in test-tube reactions that you see.

Reactions involving chromium(III) complexes

Reaction of chromium(III) complexes with alkalis

Consider the reaction that occurs when aqueous sodium hydroxide is added to a solution containing the hexaaquachromium(III) ion until no further change is seen. The observation is that a green or violet solution forms a green precipitate. The equation for this reaction is:

$$[Cr(H_2O)_6]^{3+} + 3OH^- \rightarrow [Cr(H_2O)_3(OH)_3] + 3H_2O$$

As with iron(III), this is an acid–base reaction – the three hydroxide ions have removed hydrogen ions from three of the water ligands and converted them into water molecules. The three water molecules that have lost hydrogen ions are now hydroxide ligands.

The same observations can be made when aqueous ammonia is used as the alkali. The equation for the acid–base reaction is:

$$[Cr(H_2O)_6]^{3+} + 3NH_3 \rightarrow [Cr(H_2O)_3(OH)_3] + 3NH_4^+$$

When an excess of aqueous sodium hydroxide is added, the green precipitate dissolves to form a green solution. The equation for this acid–base reaction can be represented as:

$$[Cr(H_2O)_3(OH)_3] + OH^- \rightarrow [Cr(H_2O)_2(OH)_4]^- + H_2O$$

If the aqueous sodium hydroxide is more concentrated, further acid–base reactions occur, such as

$$[Cr(H_2O)_2(OH)_4]^- + 2OH^- \rightarrow [Cr(OH)_6]^{3-} + 2H_2O$$

although there is no further change in colour.

The reactions involving hydroxide ions can be reversed by the addition of acid, illustrating the amphoteric nature of the neutral complex.

When an excess of aqueous ammonia is added to the green precipitate, the precipitate is slow to dissolve, but eventually a violet or purple solution forms:

$$[Cr(H_2O)_3(OH)_3] + 6NH_3 \rightarrow [Cr(NH_3)_6]^{3+} + 3H_2O + 3OH^-$$

In all of the reactions so far, the oxidation number of chromium has remained unchanged at +3. However, providing that the solutions are alkaline, oxidation is easily achieved by the addition of the oxidising agent hydrogen peroxide. In this reaction the solution changes from green to yellow as the chromate(VI) ion, with oxidation number +6, is formed. The equation for this oxidation is:

$$2[Cr(OH)_6]^{3-} + 3H_2O_2 \rightarrow 2CrO_4^{2-} + 2OH^- + 8H_2O$$

Note that although the chromate(VI) ion is a complex, it is not enclosed in square brackets.

Chromate(VI) and dichromate(VI) ions

Chromate(VI) ions are stable in alkaline solution, but in acidic conditions the dichromate(VI) ion is more stable. So, if acid is added there is a colour change from yellow to orange as the following reaction occurs:

$$2CrO_4^{2-} + 2H^+ \rightleftharpoons Cr_2O_7^{2-} + H_2O$$

This reaction is easily reversed by adding alkali. When considering redox reactions, it is often easier to use simplified formulae by omitting square brackets and ligands that do not undergo redox reactions, especially water.

The reduction of dichromate(VI) ions

When zinc metal is added to an acidic solution containing dichromate(VI) ions, reduction reactions occur in which the oxidation number of chromium decreases first to +3 and then to +2.

The first stage of the reduction involves a colour change from orange to green, as this reaction occurs:

$$Cr_2O_7^{2-} + 14H^+ + 3Zn \rightarrow 2Cr^{3+} + 7H_2O + 3Zn^{2+}$$

The second stage of the reduction involves a colour change from green to blue, as this reaction occurs:

$$2Cr^{3+} + Zn \rightarrow 2Cr^{2+} + Zn^{2+}$$

Explanation of redox reactions using $E^\ominus$ values

It is often helpful to explain why redox reactions occur by considering the standard electrode potentials of the different redox systems involved. For chromium, we need to consider these values:

1. $Zn^{2+} + 2e^- \rightleftharpoons Zn$ $E^\ominus = -0.76$ V
2. $Cr^{3+} + e^- \rightleftharpoons Cr^{2+}$ $E^\ominus = -0.41$ V
3. $CrO_4^{2-} + 4H_2O + 3e^- \rightleftharpoons Cr(OH)_3 + 5OH^-$ $E^\ominus = -0.13$ V
4. $H_2O_2 + 2e^- \rightleftharpoons 2OH^-$ $E^\ominus = +1.24$ V
5. $Cr_2O_7^{2-} + 14H^+ + 6e^- \rightleftharpoons 2Cr^{3+} + 7H_2O$ $E^\ominus = +1.33$ V

If you have already learned about standard electrode potentials earlier in this book, you may remember that when half-equations are arranged from high negative $E^\ominus$ values at the top to high positive values at the bottom, then the best reducing agent is at the top and on the right, and the best oxidising agent is at the bottom and on the left.

Explaining oxidation from +3 to +6

The simplified equation for the oxidation of chromium(III) by hydrogen peroxide in alkaline conditions is:

$$2Cr(OH)_3 + 3H_2O_2 + 4OH^- \rightarrow 2CrO_4^{2-} + 8H_2O$$

How can this equation be obtained from the half-equations?

The relevant half-equations are 3 and 4 (see previous page). In these equations, the best reducing agent is $Cr(OH)_3$ and the best oxidising agent is H_2O_2. For these two species to react together, we need to add half-equation 3 (reversed) to half-equation 4. When adding half-equations, they may need to be multiplied so that the number of electrons is the same in both. In this example, we need to multiply equation 3 by 2, and equation 4 by 3, so as to obtain $6e^-$ in both:

$$2Cr(OH)_3 + 10OH^- \rightleftharpoons 2CrO_4^{2-} + 8H_2O + 6e^-$$
$$3H_2O_2 + 6e^- \rightleftharpoons 6OH^-$$

Adding the half-equations and cancelling identical species on both sides gives:

$$2Cr(OH)_3 + 3H_2O_2 + 4OH^- \rightarrow 2CrO_4^{2-} + 8H_2O$$

Since the $E^\ominus$ value for half-equation 3 is more negative than the $E^\ominus$ value for half-equation 4, $Cr(OH)_3$ is electron releasing with respect to H_2O_2 and hence the reaction is thermodynamically feasible.

Explaining reduction from +6 to +3

The simplified equation for the reduction of chromium(VI) by zinc in acidic conditions is:

$$Cr_2O_7^{2-} + 14H^+ + 3Zn \rightarrow 2Cr^{3+} + 7H_2O + 3Zn^{2+}$$

How can this equation be obtained from the half-equations?

The relevant half-equations are 1 and 5 (see previous page). In these equations, the best reducing agent is Zn and the best oxidising agent is $Cr_2O_7^{2-}$. For these two species to react together, we need to add half-equation 1 (reversed) to half-equation 6, bearing in mind multiplying half-equations:

$$3Zn \rightleftharpoons 3Zn^{2+} + 6e^-$$
$$Cr_2O_7^{2-} + 14H^+ + 6e^- \rightleftharpoons 2Cr^{3+} + 7H_2O$$

As in the previous example, adding gives:

$$Cr_2O_7^{2-} + 14H^+ + 3Zn \rightarrow 2Cr^{3+} + 7H_2O + 3Zn^{2+}$$

Since the $E^\ominus$ value for half-equation 1 is more negative than the $E^\ominus$ value for half-equation 5, Zn is electron releasing with respect to $Cr_2O_7^{2-}$ and hence the reaction is thermodynamically feasible.

Explaining reduction from +3 to +2

The simplified equation for the reduction of chromium(III) by zinc in acidic conditions is:

$$2Cr^{3+} + Zn \rightarrow 2Cr^{2+} + Zn^{2+}$$

How can this equation be obtained from the half-equations?

The relevant half-equations are 1 and 2. In these equations, the best reducing agent is Zn and the best oxidising agent is Cr^{3+}. For these two species to react together, we need to add half-equation 1 (reversed) to half-equation 2, bearing in mind multiplying half-equations:

$$Zn \rightleftharpoons Zn^{2+} + 2e^-$$
$$2Cr^{3+} + 2e^- \rightleftharpoons 2Cr^{2+}$$

As in the previous example, adding gives:

$$Cr^{3+} + Zn \rightarrow 2Cr^{2+} + Zn^{2+}$$

Since the $E^\ominus$ value for half-equation 1 is more negative than the $E^\ominus$ value for half-equation 2, Zn is electron releasing with respect to Cr^{3+} and hence the reaction is thermodynamically feasible.

Chromium chemistry summary

Table A summarises the important reactions of chromium complexes in four sequences.

Start	Reagent	Intermediate	Reagent	Finish
$[Cr(H_2O)_6]^{3+}$ green solution	add NaOH(aq)	$[Cr(H_2O)_3(OH)_3]$ green precipitate	add NaOH(aq)	$[Cr(OH)_6]^{3-}$ green solution
$[Cr(H_2O)_6]^{3+}$ green solution	add NH_3(aq)	$[Cr(H_2O)_3(OH)_3]$ green precipitate	add NH_3(aq)	$[Cr(NH_3)_6]^{3+}$ violet solution
$[Cr(OH)_6]^{3-}$ green solution	add H_2O_2/OH^-	CrO_4^{2-} yellow solution	add H^+	$Cr_2O_7^{2-}$ orange solution
CrO_4^{2-} yellow solution	add Zn/H^+	$[Cr(H_2O)_6]^{3+}$ green solution	add Zn/H^+	$[Cr(H_2O)_6]^{2+}$ blue solution

table A

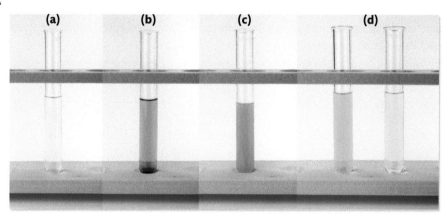

fig A Tube (a) contains the very pale blue $[Cr(H_2O)_6]^{2+}$ ion, produced by reduction using zinc. Tube (b) contains the $[Cr(H_2O)_6]^{3+}$ ion, which is normally green, but can appear violet in the presence of other ions, as here. Tube (c) contains the $[Cr(OH)_6]^{3-}$ ion, which is also green, present in alkaline solutions. The tubes in (d) contain chromium with oxidation number +6. The orange solution contains the $Cr_2O_7^{2-}$ ion and the yellow solution contains the CrO_4^{2-} ion.

Learning tip

Understanding chromium chemistry involves more effort than for some other transition metals because of the number of different ions. As well as all of the complexes involving Cr^{2+} and Cr^{3+}, there are two complexes with oxidation number +6 (CrO_4^{2-} and $Cr_2O_7^{2-}$).

Questions

1. Excess acid is added to a green solution containing $[Cr(OH)_6]^{3-}$ ions. Write an equation for the reaction in which a different green chromium-containing solution is formed.

2. The standard electrode potential for the redox system $Mn^{2+} + 2e^- \rightleftharpoons Mn$ is -1.18 V.
   Explain whether or not zinc can reduce Mn^{2+} ions to the element Mn.

15.2 4 The chemistry of vanadium

By the end of this section, you should be able to...

- recognise the colours of the oxidation states of vanadium (+5, +4, +3 and +2) in its compounds
- understand redox reactions for the interconversion of the oxidation states of vanadium (+5, +4, +3 and +2), in terms of the relevant $E^\ominus$ values

Redox reactions

Like chromium, vanadium is a transition metal that forms ions with several oxidation numbers. **Table A** shows the important ones.

Oxidation number	Formula	Name	Colour of aqueous solution
+2	V^{2+}	vanadium(II)	purple
+3	V^{3+}	vanadium(III)	green
+4	VO^{2+}	oxovanadium(IV)	blue
+5	VO_2^+	dioxovanadium(V)	yellow

table A

Unlike with chromium and the other transition elements, the focus with vanadium is purely on redox reactions. Vanadium is a good choice for this because there is one distinct colour for each of the main oxidation numbers found in its compounds. So, it is relatively easy to demonstrate the change in oxidation number by observing the change in colour during the reaction. As with chromium, we will consider the feasibility of these redox reactions in terms of $E^\ominus$ values. Although most of the species involved are complexes and some contain water ligands, the square brackets and water ligands are omitted for clarity.

Reducing vanadium from +5 to +2

The usual source of vanadium with oxidation number +5 is the compound ammonium vanadate(V), NH_4VO_3. An acidic solution of this compound contains the dioxovanadium(V) ion, VO_2^+. When zinc is added to this solution, reduction begins and there is a gradual colour change from yellow, through blue and green, to purple as the oxidation number decreases from +5 to +2. All of these colours are of solutions – there are no precipitates involved. The test tubes in **fig A** show the colours.

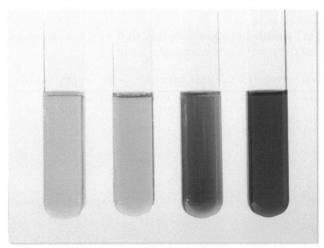

fig A The tube on the left (yellow) contains the VO_2^+ ion. The second tube (blue) contains the VO^{2+} ion. The third tube (green) contains the V^{3+} ion. The tube on the right (purple) contains the V^{2+} ion.

Explanation using $E^\ominus$ values

For vanadium, we need to consider these values:

1	$V^{2+} + 2e^- \rightleftharpoons V$	$E^\ominus = -1.18$ V
2	$Zn^{2+} + 2e^- \rightleftharpoons Zn$	$E^\ominus = -0.76$ V
3	$V^{3+} + e^- \rightleftharpoons V^{2+}$	$E^\ominus = -0.26$ V
4	$VO^{2+} + 2H^+ + e^- \rightleftharpoons V^{3+} + H_2O$	$E^\ominus = +0.34$ V
5	$VO_2^+ + 2H^+ + e^- \rightleftharpoons VO^{2+} + H_2O$	$E^\ominus = +1.00$ V

If you followed the explanations for the redox reactions of chromium in the previous topic, you should be able to understand an abbreviated explanation for each stage that just shows the two relevant half-equations and then the overall equation.

Reduction from +5 to +4

$VO_2^+ + 2H^+ + e^- \rightarrow VO^{2+} + H_2O$

$Zn \rightarrow Zn^{2+} + 2e^-$

Overall reaction:

$2VO_2^+ + 4H^+ + Zn \rightarrow 2VO^{2+} + Zn^{2+} + 2H_2O$

Since the $E^\ominus$ value for half-equation 2 is more negative than the $E^\ominus$ value for half-equation 5, Zn is electron releasing with respect to VO_2^+ and hence the reaction is thermodynamically feasible.

Reduction from +4 to +3

$VO^{2+} + 2H^+ + e^- \rightarrow V^{3+} + H_2O$

$Zn \rightarrow Zn^{2+} + 2e^-$

Overall reaction:

$2VO^{2+} + 4H^+ + Zn \rightarrow 2V^{3+} + Zn^{2+} + 2H_2O$

Since the $E^\ominus$ value for half-equation 2 is more negative than the $E^\ominus$ value for half-equation 4, Zn is electron releasing with respect to VO^{2+} and hence the reaction is thermodynamically feasible.

Reduction from +3 to +2

$V^{3+} + e^- \rightarrow V^{2+}$

$Zn \rightarrow Zn^{2+} + 2e^-$

Overall reaction:

$2V^{3+} + Zn \rightarrow 2V^{2+} + Zn^{2+}$

Since the $E^\ominus$ value for half-equation 2 is more negative than the $E^\ominus$ value for half-equation 3, Zn is electron releasing with respect to V^{3+} and hence the reaction is thermodynamically feasible.

Reduction from +2 to 0

$V^{2+} + 2e^- \rightarrow V$

$Zn \rightarrow Zn^{2+} + 2e^-$

Overall reaction:

$2V^{2+} + Zn \rightarrow 2V + Zn^{2+}$

Since the $E^\ominus$ value for half-equation 2 is less negative than the $E^\ominus$ value for half-equation 1, Zn is not electron releasing with respect to V^{2+} and hence the reaction is not thermodynamically feasible.

Predicting oxidation reactions

A similar method can be used to predict whether a given oxidising agent will oxidise a vanadium species to one with a higher oxidation number. You can see how to do this in Question 2 in this topic.

Learning tip

It is important that you do not get confused between the two oxo ions of vanadium: VO_2^+ and VO^{2+}.

Questions

1. The standard electrode potential for the redox system $Sn^{2+} + 2e^- \rightleftharpoons Sn$ is -0.14 V.

   Explain why the use of tin as a reducing agent will not obtain V^{2+} from a solution containing VO_2^+ ions.

2. The standard electrode potential for the redox system $Cu^{2+} + 2e^- \rightleftharpoons Cu$ is $+0.34$ V.

   Explain whether copper(II) ions can be used to oxidise VO^{2+} ions to VO_2^+ ions.

15.3 1 Heterogeneous catalysis

By the end of this section, you should be able to...

- understand that a heterogeneous catalyst is in a different phase from the reactants and that the reaction occurs at the surface of the catalyst
- understand, in terms of oxidation number, how V_2O_5 acts as a catalyst in the Contact process
- understand how a catalytic converter decreases carbon monoxide and nitrogen monoxide emissions from internal combustion engines by:
  (i) adsorption of CO and NO molecules onto the surface of the catalyst
  (ii) weakening of bonds and chemical reaction
  (iii) desorption of CO_2 and N_2 product molecules from the surface of the catalyst

Different ways to understand catalysis

When you first learned about catalysts, you probably recognised them as substances that increased the rate of a reaction without being chemically changed during the process. This description is fine as an introduction to catalysis, but it is incomplete and not always correct!

In **Book 1**, you learned about catalysis in terms of its effect on the activation energy of a reaction, and how to represent catalytic action using Maxwell–Boltzmann energy distributions and energy profiles.

In this topic and the next one, we will consider catalysis in more detail by looking at the two main types of action: heterogeneous and homogeneous.

Transition metals as heterogeneous catalysts

A **heterogeneous catalyst** is one that is in a different phase from that of the reactants. You can probably remember an example of this type of catalysis from many years ago. Oxygen gas can be made in the laboratory by decomposing a hydrogen peroxide solution. The catalyst used in this reaction is usually manganese(IV) oxide – note that this is a compound of a transition metal. However, what makes this an example of heterogeneous catalysis is that manganese(IV) oxide is a solid and hydrogen peroxide solution is a liquid.

Many transition metals, and their compounds, are used as solid catalysts. Their action can be explained in terms of what happens on the surface of the catalyst. The fact that the action takes place only on the surface explains why many of them are used in a finely divided form, as small particles (including powder) rather than as large lumps. Sometimes, they are used instead as a thin coating on an inert support material.

The Contact process

Sulfuric acid is one of the most widely used chemicals, both in terms of the quantity produced and the number of industries that use it. Its biggest single use is in the manufacture of fertilisers. A complete description of the Contact process used to manufacture sulfuric acid is beyond the scope of this book, but the key reaction in the process is the conversion of sulfur dioxide to sulfur trioxide in this reaction:

$$2SO_2 + O_2 \rightleftharpoons 2SO_3$$

At the temperatures and pressures used in the process, all of the substances are in the gas phase, and the mixture of reactants is passed over a catalyst of vanadium(V) oxide, V_2O_5, usually known in industry as vanadium pentoxide.

fig A Sample of vanadium(V) oxide – used as the catalyst in the Contact process.

Surface adsorption theory

The way that a heterogeneous catalyst works is often explained with reference to this theory, which is usually considered as having three steps:

1. **Adsorption**, in which one or more reactants becomes attached to the surface of the catalyst.

2. Reaction, following the weakening of bonds in the adsorbed reactants.

3. **Desorption**, in which the reaction product becomes detached from the surface of the catalyst.

In the Contact process, the reaction step has two parts:

Part 1: sulfur dioxide adsorbs onto the vanadium(V) oxide and a redox reaction occurs:

$$V_2O_5 + SO_2 \rightarrow V_2O_4 + SO_3$$

Note that the oxidation number of vanadium decreases from +5 to +4. The sulfur trioxide then desorbs.

Part 2: oxygen reacts with the V_2O_4 on the surface of the catalyst and another redox reaction occurs:
$$V_2O_4 + \tfrac{1}{2}O_2 \rightarrow V_2O_5$$
Note that the original catalyst is regenerated as the oxidation number increases from +4 to +5.

Catalytic converters

The problems

One of the major problems associated with the steady growth in road vehicle usage over several decades is the increased emission of pollutant gases from vehicle exhausts. In some cities, the quality of air breathed by humans is below acceptable levels. Without catalytic converters, the situation would now be much worse.

Although many different pollutants come from vehicle exhausts, two of the most significant are carbon monoxide and nitrogen monoxide.

- Carbon monoxide is a toxic gas that interferes with oxygen transport from the lungs through the bloodstream to vital organs in the body (see **Section 15.1.6**).
- Nitrogen monoxide is easily oxidised in the atmosphere to nitrogen dioxide, where it can act as a respiratory irritant and contribute to the formation of acid rain.

Carbon monoxide forms through the incomplete combustion of hydrocarbon fuels, and nitrogen monoxide forms through the reaction between nitrogen and oxygen at the high temperatures that exist in an internal combustion engine.

The solutions

For over 20 years, all cars sold in EU countries have been fitted with catalytic converters in an attempt to alleviate the effect of vehicle emissions on air quality. The main transition metals used in catalytic converters are platinum and rhodium, and sometimes palladium. The method of action can be explained by the surface adsorption theory described above. In one type of catalyst, molecules of carbon monoxide and nitrogen monoxide are adsorbed onto the surface. Then, because their bonds are weakened, they react together to form carbon dioxide and nitrogen, which are then desorbed from the surface of the catalyst. The overall reaction can be represented by this equation:

$$2CO + 2NO \rightarrow 2CO_2 + N_2$$

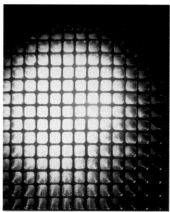

fig B View through the element of a catalytic converter from a car exhaust. The inner surface is coated with an alloy containing platinum, rhodium and palladium.

Learning tip
Try not to confuse absorb/absorption with adsorb/adsorption. Very simply, absorption involves one substance becoming distributed throughout another (for example water in a sponge), while adsorption only happens at the surface of a substance.

Questions

1. How does the description of the catalytic role of vanadium(V) oxide show that this statement is not true? 'A catalyst increases the rate of a chemical reaction but does not take part in the reaction'.

2. Write an equation to show how nitrogen monoxide can form acid rain.

Key definitions
A **heterogeneous catalyst** is one that is in a different phase from the reactants.
Adsorption is the process that occurs when reactants form weak bonds with a solid catalyst.
Desorption is the process that occurs when products leave the surface of a solid catalyst.

15.3 / 2 Homogeneous catalysis

By the end of this section, you should be able to...

- understand that a homogeneous catalyst is in the same phase as the reactants and appreciate that the catalysed reaction will proceed via an intermediate species
- understand the role of Fe^{2+} ions in catalysing the reaction between I^- and $S_2O_8^{2-}$ ions
- understand the role of Mn^{2+} ions in autocatalysing the reaction between MnO_4^- and $C_2O_4^{2-}$ ions

Transition metals as homogeneous catalysts

A **homogeneous catalyst** is one that is in the same phase as the reactants. This means that they are either all gases, or more often, all in aqueous solution. You will not have come across many examples of this type of catalyst, compared with those that take part in heterogeneous catalysis. Homogeneous catalysis is also much less common in industry.

The key feature of homogeneous catalysis is the formation of an intermediate species for which a specific formula can be written.

A reaction of the $S_2O_8^{2-}$ ion

The ion with this formula has several names, the simplest of which is the 'persulfate ion', although it is often known as the 'peroxodisulfate ion'. We will not use the IUPAC name, which is very complicated!

It acts as an oxidising agent in its reaction with iodide ions, the equation for which is:

$$S_2O_8^{2-} + 2I^- \rightarrow 2SO_4^{2-} + I_2$$

One reason why this reaction is slow at room temperature is that the two reactant ions are both negatively charged and so repel each other. The reaction is much faster in the presence of Fe^{2+} ions, which act as a catalyst. All of the reactants and products, and the catalyst, are in the aqueous phase, so this reaction is an example of homogeneous catalysis.

The steps in the catalysed reaction

Step 1: The Fe^{2+} ions are not repelled by the $S_2O_8^{2-}$ ions because they have the opposite charge. They react together as follows:

$$S_2O_8^{2-} + 2Fe^{2+} \rightarrow 2SO_4^{2-} + 2Fe^{3+}$$

Step 2: The Fe^{3+} ions formed in Step 1 now react with the I^- ions (these also have opposite charges) as follows:

$$2Fe^{3+} + 2I^- \rightarrow 2Fe^{2+} + I_2$$

The iron(II) ions are used in Step 1 and regenerated in Step 2, so the two steps can repeat continuously.

An alternative mechanism

The reaction is also catalysed by Fe^{3+} ions, for which the following mechanism can be written.

Step 1: $2Fe^{3+} + 2I^- \rightarrow 2Fe^{2+} + I_2$

Step 2: $S_2O_8^{2-} + 2Fe^{2+} \rightarrow 2SO_4^{2-} + 2Fe^{3+}$

The same two reactions are involved, but they occur in a different order.

Transition metals as catalysts 15.3

The oxidation of ethanedioate ions

You may be familiar with titrations in which potassium manganate(VII) in acidic conditions acts as an oxidising agent. For these titrations to work satisfactorily, it is important that the reactions are fast, so that the end point of the reaction can be accurately observed.

One example of a redox reaction that can be used in titrations is the oxidation of ethanedioate ions by potassium manganate(VII). The equation for this reaction is:

$$2MnO_4^- + 5C_2O_4^{2-} + 16H^+ \rightarrow 2Mn^{2+} + 5CO_2 + 8H_2O$$

As with the $S_2O_8^{2-}/I^-$ reaction, the reacting species are both negatively charged. Therefore, the reaction is slow. However, as more potassium manganate(VII) solution is added, the reaction rate increases.

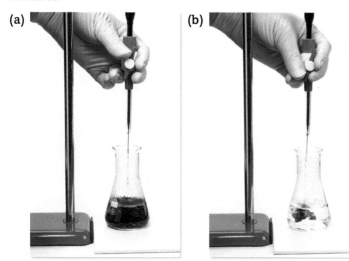

fig A (a) At the start of the titration, the colour of the potassium manganate(VII) solution takes some time to disappear; (b) The colour of the potassium manganate(VII) solution now disappears more quickly because there are Mn^{2+} ions to catalyse the reaction; (c) At the end of the titration, there are no more ethanedioate ions left to react, so the remaining potassium manganate(VII) solution gives the mixture a pink colour.

Note that one of the reaction products is the Mn^{2+} ion. This can act as a catalyst in the reaction, which explains why the reaction rate increases as the titration proceeds. The formation of a reaction product that increases the rate of the reaction is sometimes referred to as **autocatalysis**. This effect is sometimes represented using a graph.

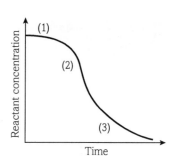

fig B (1) The slow decrease in the reactant concentration is because the reaction is uncatalysed at first. (2) The reactant concentration decreases more rapidly as the reaction rate increases because of catalysis. (3) The reactant concentration then decreases more slowly because there is little reactant left.

Learning tip

Note that in **Section 15.3.1**, all of the transition metal catalysts were solids, but that in this section they are all ions in aqueous solution.

Questions

1. State why Mg^{2+} ions do not act as a catalyst in the reaction between $S_2O_8^{2-}$ and I^- ions.

2. State why autocatalysis of the reaction between MnO_4^- and $C_2O_4^{2-}$ ions does not occur under alkaline conditions.

Key definitions

A **homogeneous catalyst** is one that is in the same phase as the reactants.

Autocatalysis occurs when a reaction product acts as a catalyst for the reaction.

THINKING BIGGER

POWERFUL PIGMENTS

Octopuses are complex and intelligent creatures. One of their many fascinating attributes is the blue colour of their blood. What is the chemical that causes this colour and what is its function? Read the article below to find out.

BLUE BLOOD HELPS OCTOPUS SURVIVE BRUTALLY COLD TEMPERATURES

Posted by Stefan Sirucek in *Weird & Wild* on 10 July 2013

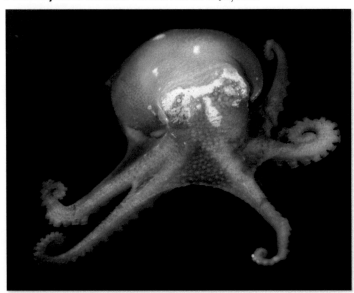

fig A A Pareledone charcoti octopus caught near the South Orkney Islands.

Researchers at the Alfred Wegener Institute for Polar and Marine Research in Germany have found that a specialized pigment in the blood of Antarctic octopods allows them to survive temperatures that often drop below freezing.

It's all down to a respiratory pigment in their blood called haemocyanin that allows the octopus 'to maintain an aerobic lifestyle at sub-zero temperatures,' said Michael Oellermann, who took part in the research. Haemocyanin also has the effect of making the octopods' blood blue.

'Haemocyanin contains copper, which is the responsible ion for binding oxygen,' explained Oellermann in an email. 'Haemoglobins contain iron instead, which gives our blood a red color.'

Blood chilling

It's not a question of blood freezing.

'The reason why octopus blood does not freeze at −1.9 degrees C [28.58 degrees Fahrenheit] is that they are isosmotic, which means that their blood shares the same salinity as the surrounding seawater,' explained Oellermann.

The problem in very cold temperatures is that, without special adaptations, the aerobic process would, in most cases, shut down.

'Cold temperatures increase oxygen affinity to the extent that oxygen cannot be released in the tissue anymore,' explained Oellermann, making survival in waters that can drop to as low as 29 degrees Fahrenheit a challenge to say the least.

Comparing octopods

The researchers looked at a particular species of Antarctic octopod called *Pareledone charcoti* and compared it with temperate and warm-adapted octopods to observe the difference in how cold-water octopods transport oxygen in their blood.

'*Pareledone charcoti* decreased the oxygen affinity of its haemocyanin to counter the adverse effect of temperature on oxygen binding accompanied by changes in their protein sequence, to assure sufficient oxygen supply to its tissues and organs,' explained Oellermann.

Octopods likely developed the adaptation because they require a lot of oxygen compared to other invertebrates, notes Oellermann, and because they are largely non-migratory and must instead adapt to their environment.

In contrast, said Oellermann, Antarctic icefish survive in the same frigid environment by dint of their lower oxygen demand, rather than through the adapted system of oxygen transport employed by their blue-blooded neighbors.

Interestingly, the same adaptation may be what allows octopods to survive temperatures on the other end of the thermometer, such as the 86-degree F heat often found near thermal vents.

Where else will I encounter these themes?

Book 1 | 11 | 12 | 13

Thinking Bigger

Let us start by considering the nature of the writing in the article.

1. How might an article like the one opposite be used to highlight the importance of species diversity and environmental conservation?

> Why might it be more effective to focus on one specific example of a fascinating species than to talk about species diversity in general?

Now we will look at the chemistry in detail. Some of these questions will link to topics earlier in this book or in **Book 1**, so you may need to combine concepts from different areas of chemistry to work out the answers.

2. Oxygenated haemocyanin is blue in colour whereas deoxygenated haemocyanin is colourless. Explain this observation using your knowledge of copper chemistry.

3. Describe the type of the bond between copper and nitrogen in oxyhaemocyanin, as shown below in **fig B**.

fig B An oxyhaemocyanin molecule.

4. Suggest why zinc is unlikely to be able to form a protein complex that is capable of reversibly binding oxygen.

5. a. Use the two half equations below to show that the reaction between oxygen and copper(I) ions under standard conditions is feasible:

$$Cu^{2+}(aq) + e^- \rightleftharpoons Cu^+(aq)$$
$$O_2(g) + 4H^+(aq) + 4e^- \rightleftharpoons 2H_2O(l)$$

b. Calculate the total entropy change for this reaction under standard conditions (298 K) (F = 96 500 C).

Activity

Design an A3 poster for the general public to illustrate the role of transition metals and their compounds as catalysts. You could choose to focus on their role in biological systems (enzymes) or their role in industrial chemistry. In each case, choose a specific example to illustrate the general principles.

- From the website of the magazine *National Geographic*

15 Exam-style questions

1. Scandium and titanium are both d-block elements, but only titanium is a transition metal.
   (a) Complete the electronic configuration of each of these ions.

   Sc^{3+} $1s^2 2s^2 2p^6$ ....

   Ti^{3+} $1s^2 2s^2 2p^6$ .... [2]

   (b) Explain why titanium is a transition metal, but scandium is not. [2]

   (c) What is the oxidation number of the d-block element in each of these compounds?

   $Sc(OH)_3$ ....

   $CaTiO_3$ .... [2]

   (d) Which ion is isoelectronic with Ti^{3+}?
   - A Cu^+
   - B Cr^{2+}
   - C Fe^{3+}
   - D V^{4+} [1]

   [Total: 7]

2. (a) What is the meaning of the term 'ligand'? [2]

   (b) Which molecule is *not* able to act as a ligand?
   - A CO
   - B H_2
   - C H_2O
   - D O_2 [1]

   (c) (i) Give the formula of the complex with the name tetraaquadihydroxoiron(II). [1]
   (ii) Give the name of the complex with the formula $[CoCl_4]^{2-}$. [1]

   (d) The arrangement of d-electrons in the complex with the formula $[Cr(H_2O)_6]^{3+}$ can be represented using this diagram.

   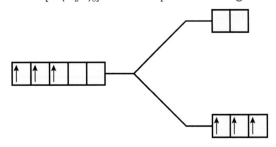

   Explain, with reference to this diagram, why a solution containing this complex is coloured. [5]

   [Total: 10]

3. Most transition metal complexes have octahedral or tetrahedral shapes.

   (a) Explain why the complex formed between one copper(II) ion and six water molecules has an octahedral shape. [2]

   (b) Concentrated hydrochloric acid is added to a solution containing the complex in part (a). During the reaction that occurs, there is a colour change from blue to yellow.
   (i) Explain why the number of ligands around the metal ion decreases during this reaction. [2]
   (ii) State one other feature of the complex that changes during this reaction, and one feature that does *not* change. [2]

   (c) A complex with a linear shape has the formula $[Ag(NH_3)_2]^+$, and is used in a test to distinguish between aldehydes and ketones.
   (i) State the name of this complex and the reagent that contains it. [2]
   (ii) State the type of change that this complex undergoes when it reacts with an aldehyde, and the name of the metal species formed. [2]

   (d) A complex with a different shape, used in cancer treatment, is *cis*-platin, and has the formula $[Pt(NH_3)_2Cl_2]$. Draw the structure of *cis*-platin and state the name of its shape. [2]

   (e) A multidentate ligand has a formula abbreviated to $EDTA^{4-}$. Its structure is:

   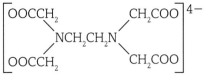

   (i) Explain why $EDTA^{4-}$ is described as a multidentate ligand. [1]
   (ii) Draw the structure of $EDTA^{4-}$ showing the lone pairs of electrons used in the formation of bonds with a transition metal ion. [1]
   (iii) Using the abbreviation EDTA, give the formula of the complex formed when $[Fe(H_2O)_6]^{3+}$ reacts with $EDTA^{4-}$. [1]

   [Total: 15]

4 The following sequence summarises some reactions, 1, 2, 3 and 4, of chromium compounds.

$[Cr(H_2O)_6]^{3+} \xrightarrow{1} [Cr(H_2O)_3(OH)_3] \xrightarrow{2} [Cr(OH)_6]^{3-} \xrightarrow{3} CrO_4^{2-} \xrightarrow{4} Cr_2O_7^{2-}$

(a) State the appearance of the reactant and product in reaction 1. [2]

(b) State the reagent used in both reactions 1 and 2. [1]

(c) With reference to oxidation numbers, identify which reaction is a redox reaction. [2]

(d) Describe the colour change in reaction 4. [1]

(e) Write an equation for reaction 4. [1]

[Total: 7]

5 The heterogeneous catalyst used in the Contact process to manufacture sulfuric acid has the formula V_2O_5.

Explain, with the help of suitable equations and with reference to oxidation states of vanadium, the steps in this catalysed reaction. [6]

[Total: 6]

TOPIC 16

Further kinetics

Introduction

Your health depends on a complex interplay of a large number of chemical reactions taking place in the cells of your body. In a healthy body, these reactions will take place at the correct rate, in the right place and at the right time. These reactions are controlled by enzymes. The branch of chemistry concerned with the rates of chemical reactions is called chemical kinetics. The term 'kinetics' implies motion (derived from the ancient Greek word for movement – kinesis). We can use the information obtained from the study of chemical kinetics to understand the body's metabolism and to model the effects of pollutants in the Earth's atmosphere. The development of new catalysts is another branch of chemical kinetics, and is one that is of immense importance to the chemical industry.

In **Book 1 Topic 9**, we used a qualitative approach to the understanding of reaction rates. In this topic we shall develop this further by adopting a quantitative approach. This will enable us to begin to understand what is happening during chemical reactions. We know how atoms can bond together to form molecules, but why do the atoms change partners during chemical reactions? What exactly happens when a hydrogen molecule meets an iodine molecule and they subsequently react to produce hydrogen iodide?

It has been reported by scientists that the herb St John's Wort (*Hypericum perforatum*) may be just as effective as Prozac at treating depression. Herbs and plant extracts have been used for centuries to cure diseases and to relieve pain. In many cases they are effective because they control the rates of reactions within the body. In this topic we shall study the rates of chemical reactions and the mechanisms by which they take place.

All the maths you need

- Recognise and make use of appropriate units in calculations
- Recognise and use expressions in decimal and ordinary form
- Use calculators to find and use power functions
- Use an appropriate number of significant figures
- Change the subject of an equation
- Substitute numerical values into algebraic expressions using appropriate units for physical quantities
- Solve algebraic expressions
- Translate information between graphical, numerical and algebraic forms
- Plot two variables from experimental or other data
- Determine the slope and intercept of a linear graph
- Calculate rate of change for a graph showing a linear relationship
- Draw and use the slope of a tangent to a curve as a measure of rate of change

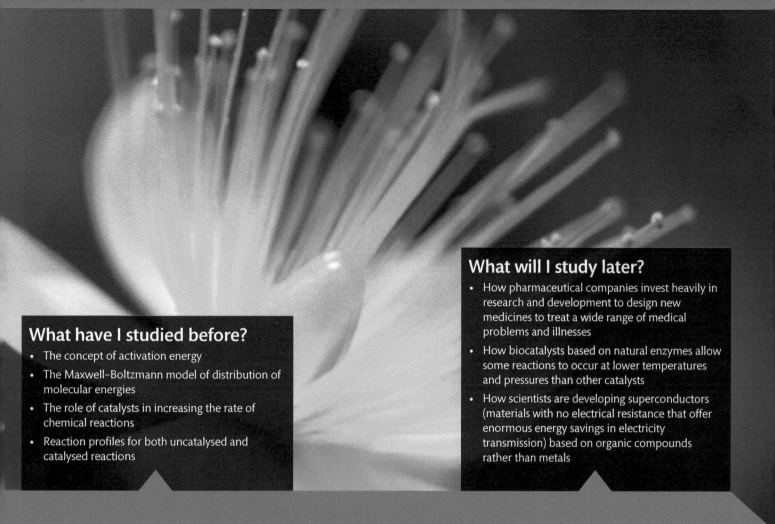

What have I studied before?
- The concept of activation energy
- The Maxwell–Boltzmann model of distribution of molecular energies
- The role of catalysts in increasing the rate of chemical reactions
- Reaction profiles for both uncatalysed and catalysed reactions

What will I study later?
- How pharmaceutical companies invest heavily in research and development to design new medicines to treat a wide range of medical problems and illnesses
- How biocatalysts based on natural enzymes allow some reactions to occur at lower temperatures and pressures than other catalysts
- How scientists are developing superconductors (materials with no electrical resistance that offer enormous energy savings in electricity transmission) based on organic compounds rather than metals

What will I study in this topic?
- Order of reaction and rate equations
- Selection of an appropriate technique to follow the rate of a reaction
- Initial rate and continuous rate methods for following reactions
- Reaction mechanisms
- Homogeneous and heterogeneous catalysis

16.1 1 Methods of measuring the rate of reaction

By the end of this section, you should be able to...

- understand the term 'rate of reaction'
- select and justify a suitable experimental technique to obtain rate data for a given reaction, including:
  - (i) volume of gas evolved
  - (ii) mass change
  - (iii) colorimetry
  - (iv) titration
  - (v) other suitable technique(s) for a given reaction.

Rate of reaction

The rate of a reaction can be expressed either in terms of how the concentration of a product *increases* with time, or how the concentration of a reactant *decreases* with time.

So,

$$\text{rate} = \frac{\text{change in concentration of product}}{\text{time}}$$

or

$$\text{rate} = -\frac{\text{change in concentration of reactant}}{\text{time}}$$

The negative sign in the second expression reflects the fact that the concentration of the reactant is decreasing and therefore gives a positive value for the rate.

Rate is measured in units of concentration per unit time, and the most common unit is mol dm^{-3} s^{-1}.

For the mathematicians amongst you, the expressions in calculus notation are:

$$\text{rate} = \frac{d[\text{product}]}{dt} = -\frac{d[\text{reactant}]}{dt}$$

This **rate of reaction** is sometimes called the 'overall rate of reaction'.

Methods of measuring the rate of reaction

Before investigating the rate of a particular reaction, it is necessary to know the overall equation, including state symbols, for the reaction so that we can decide what method to use to follow the reaction.

There are various methods available, such as:

- measuring the volume of a gas evolved
- measuring the change in mass of a reaction mixture
- monitoring the change in intensity of colour of a reaction mixture (colorimetry)
- measuring the change in concentration of a reactant or product using titration
- measuring the change in pH of a solution
- measuring the change in electrical conductivity of a reaction mixture.

The method chosen to follow the reaction will depend on the nature of the reactants and products, as well as the conditions under which the reaction is carried out.

For example, the reaction between calcium carbonate and dilute hydrochloric acid,

$$CaCO_3(s) + 2HCl(aq) \rightarrow CaCl_2(aq) + H_2O(l) + CO_2(g)$$

could conveniently be followed by measuring the volume of gas given off at regular time intervals, or by measuring the change in mass of the reaction mixture with time.

However, the reaction between propanone and iodine in aqueous solution,

$$CH_3COCH_3(aq) + I_2(aq) \rightarrow CH_3COCH_2I(aq) + H^+(aq) + I^-(aq)$$

could not be followed by measuring change in mass because all products of the reaction remain in solution. It would be possible, however, to follow the reaction by monitoring the decrease in intensity of colour of the reaction mixture, since $I_2(aq)$ is the only coloured species present.

Measuring the volume of a gas evolved

The two most common techniques for collecting and measuring the volume of a gas evolved during a reaction are:

1. collection over water into a measuring cylinder, and
2. collection using a gas syringe.

The technique chosen will depend partly on the level of precision required. The gas syringe has greater degree of precision, but if a large volume of gas is being measured, the difference in the degree of measurement uncertainty becomes so small that either instrument is sufficiently precise.

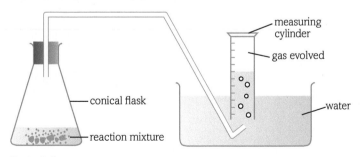

fig A Collecting a gas over water.

Further kinetics 16.1

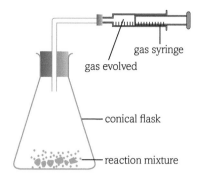

fig B Collecting a gas in a gas syringe.

Measuring the change in mass of a reaction mixture

This is another technique applicable to reactions in which a gas is evolved.

The reaction flask and contents are placed on a digital balance and the decrease in mass is measured as the reaction proceeds.

This technique is most precise when the gas given off has a relatively high density, such as with carbon dioxide. With a low-density (i.e. low molecular weight) gas such as hydrogen, the mass changes are so small that the measurement uncertainties become significant.

fig C Cotton wool is placed in the neck of the flask to prevent the loss of liquid spray.

Monitoring a colour change (colorimetry)

Monitoring a colour change can sometimes be done with sufficient precision using the naked eye. However, under certain circumstances greater precision can be obtained by using a colorimeter. A colorimeter can detect far more subtle changes than the human eye can observe, and provides a quantitative (rather than a subjective) measurement.

fig D A colorimeter.

Analysis by titration

This technique involves using a pipette to remove small samples (aliquots) from a reaction mixture at regular intervals. The reaction in the aliquot is either stopped by adding another substance or slowed down to almost zero by immersing it in an ice bath. The aliquot is then titrated to determine the concentration of a reactant or product species.

The process of stopping or slowing down the reaction in an aliquot is known as 'quenching'.

For example, if the reaction involves an acid, the aliquot, after quenching, could be titrated against standard sodium hydroxide to determine the concentration of the acid. This technique is used to investigate the reaction between iodine and propanone, which is catalysed by acid. Sodium hydrogencarbonate is added to the aliquot to remove the acid catalyst and hence stop the reaction. The remaining iodine is then titrated against a standard solution of sodium thiosulfate.

$$CH_3COCH_3(aq) + I_2(aq) \rightarrow CH_3COCH_2I(aq) + H^+(aq) + I^-(aq)$$
$$I_2(aq) + 2S_2O_3^{2-}(aq) \rightarrow 2I^-(aq) + S_4O_6^{2-}(aq)$$

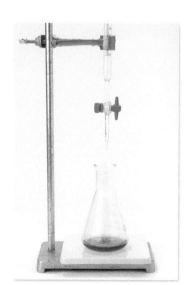

fig E Titrating iodine against sodium thiosulfate.

Measuring the electrical conductivity

If the total number or type of ions in solution changes during a reaction it might be possible to follow the reaction by measuring changes in electrical conductivity of the solution using a conductivity meter.

Measuring any other physical property that shows a significant change

Possible physical properties that have not already been mentioned include changes in the volume of liquid ('dilatometry'), chirality and refractive index.

Questions

1. State suitable techniques to obtain rate data for each of the following reactions.
   (a) Magnesium with dilute sulfuric acid:
   $$Mg(s) + 2H^+(aq) \rightarrow Mg^{2+}(aq) + H_2(g)$$
   (b) Ethyl ethanoate with sodium hydroxide:
   $$CH_3COOCH_2CH_3(l) + OH^-(aq) \rightarrow CH_3COO^-(aq) + CH_3CH_2OH(aq)$$
   (c) Hydrogen gas with iodine gas:
   $$H_2(g) + I_2(g) \rightarrow 2HI(g)$$

2. Why would the method of measuring the change in mass of a reaction vessel and contents not work well in the reaction between magnesium and dilute sulfuric acid?

3. The reaction between calcium carbonate and dilute hydrochloric acid can be followed by collecting and measuring the volume of gas produced. The gas could be collected over water in a measuring cylinder or in a gas syringe. Which method would be the more suitable for this reaction? Explain your answer.

Key definition

The (overall) **rate of reaction** is the change in concentration of a species divided by the time it takes for the change to occur. All reaction rates are positive.

16.1 2 Rate equations, rate constants and orders of reaction

By the end of this section, you should be able to...

- understand the terms:
  (i) rate equation
  (ii) order of reaction with respect to a substance in a rate equation
  (iii) overall order of reaction
  (iv) rate constant
  (v) rate-determining step

Rate equation

What is a rate equation?

The usual relationship between the rate of reaction and the concentration of a reactant is that the rate of reaction is directly proportional to the concentration. That is, as the concentration is doubled, the rate of reaction doubles.

Unfortunately, this is not always the case. Sometimes the rate will double, but sometimes it will increase by a factor of four. In some cases, the rate of reaction does not increase at all, or it increases in an unexpected way.

Let us consider the simple relationship where the rate is directly proportional to the concentration of a reactant, say A. We can represent this by the expression:

$$\text{rate} \propto [A]$$

or:

$$\text{rate} = k[A]$$

where k is the proportionality constant.

This is called the *first order* **rate equation**. The constant k is called the *rate constant*.

Every reaction has its own particular rate equation and its own rate constant.

Rate constants will only change their value with a change in temperature.

Other common rate equations with respect to an individual reactant are:

second order rate equation: rate = $k[A]^2$
zero order rate equation: rate = $k[A]^0$ or rate = k

Zero order reactions do not occur very often and it might be difficult at this stage for you to imagine why they should occur at all. However, all will soon be revealed.

If two or more reactants are involved, then it is possible to have a third order rate equation:

third order rate equation: rate = $k[A]^2[B]$

The units of rate constants

Table A shows the units for rate constants, using mol dm^{-3} as the unit of concentration and seconds as the unit of time. You find the units by rearranging the rate equation.

Order	Unit
Zero	mol dm^{-3} s^{-1}
First	s^{-1}
Second	dm^3 mol^{-1} s^{-1}
Third	dm^6 mol^{-2} s^{-1}

table A

The units are obtained by rearranging the rate equation. For example, for a second order reaction:

$$k = \frac{\text{rate}}{[A]^2}$$

Inserting the units we obtain:

$$\frac{\text{mol dm}^{-3}\,\text{s}^{-1}}{\text{mol dm}^{-3} \times \text{mol dm}^{-3}}$$

This cancels down to:

$$\frac{\text{s}^{-1}}{\text{mol dm}^{-3}}$$

which equates to dm^3 mol^{-1} s^{-1}.

The majority of reactions involve two or more reactants. If we call the reactants A, B and C, then the reaction may be first **order** with respect to A, first order with respect to B and second order with respect to C. The *overall* rate equation will be:

$$\text{rate} = k[A][B][C]^2$$

and the **overall order** of the reaction is four (1 + 1 + 2). Note that you are adding the powers.

For a general reaction in which the orders are *m*, *n* and *p*, we have:

$$\text{rate} = k[A]^m[B]^n[C]^p$$

The overall order of the reaction is $m + n + p$.

Reaction mechanisms

Many reactions can be represented by a stoichiometric equation containing many reactant particles. For example, the reaction between manganate(VII) ions and iron(II) ions in acidic solution can be represented by:

$$MnO_4^-(aq) + 8H^+(aq) + 5Fe^{2+}(aq) \rightarrow Mn^{2+}(aq) + 4H_2O(l) + 5Fe^{3+}(aq)$$

If this reaction actually proceeded in a single step, then the reaction would be very slow indeed. The probability of 14 particles simultaneously colliding, all with the correct orientation and energy, is so small that we can say it is effectively zero. The reaction is, however, very fast indeed even at room temperature. It must, therefore, proceed via a series of steps, all of which follow on quickly from one another. It is important to recognise that a reaction involving simultaneous collision of more than two particles is very rare.

The orders of reaction of the individual reactants can help us to suggest a possible mechanism for a reaction. The mechanism cannot be inferred from the stoichiometric equation, since the mathematical relationship between the rate of reaction and the concentration of reactants (i.e. the orders of reaction) can only be determined *through experiments*.

Consider the reaction:

$$A + B + C \rightarrow D + E$$

for which the experimentally determined rate equation is:

rate = $k[A][B]$

This could mean that C was present in such a large excess that changes in its concentration were negligible and therefore had no measurable effect on the rate of reaction.

If, however, this is not the case, and changes to [C] really do not have any effect on the overall rate of reaction, then a different explanation must be sought for why [C] does not appear in the rate equation.

In this case, there must be a step involving a reaction between A and B that has an effect on the rate of reaction. There must also be another step in which C reacts, but this reaction has *no effect* on the overall rate of reaction. This could be explained by assuming that the reaction between A and B takes place *before* C has a chance to react, and that the reaction between A and B is significantly *slower* than the reaction involving C. If this were the case, then the mechanism for the reaction could be:

Step 1: $A + B \rightarrow Z$ SLOW

Step 2: $Z + C \rightarrow D + E$ FAST

Since only Step 1 is **rate-determining**, then only changes in [A] and [B] will affect the overall rate of reaction. Changes in the rate at which Step 2 occurs, owing to changes in [C], will be negligible.

Important points to remember are that:

The slowest step in a reaction determines the overall rate of the reaction.

The slowest step is known as the *rate-determining step* of the reaction.

We shall return to the concept of reaction mechanisms, and consider them in much more detail, in **Section 16.1.4**.

WORKED EXAMPLE

A useful way of visualising the idea of a rate-determining step is to imagine that three students are arranging some sheets of notes into sets (**fig A**). The notes are arranged into ten piles and the first student collects a sheet from each of the piles (Step 1). The second student takes the set of ten papers and shuffles them so that they are tidy and ready for stapling (Step 2). The third student staples the set of notes together (Step 3).

The overall rate of this process (i.e. the rate at which the final sets of notes are prepared) depends on the rate of Step 1, the collecting of the sheets of notes, since this is by far the slowest step. It does not matter, within reason, how quickly the tidying up for stapling is done. For most of the time the second and third students will be doing nothing while they wait for the first student to collect the sheets. The mechanism of the process is therefore:

Step 1: Student 1 collects sheets   SLOW
Step 2: Student 2 tidies set of sheets FAST
Step 3: Student 3 staples set of sheets FAST

fig A Analogy for rate-determining step.

Learning tip

All reactant species involved either in or before the rate-determining step have an effect on the rate of the reaction and will appear in the rate equation.

Questions

1. The rate equation for the reaction between peroxydisulfate ions and iodide ions is:
    $$\text{rate} = k[S_2O_8^{2-}][I^-]$$
    (a) What is the order of reaction with respect to (i) peroxydisulfate ions and (ii) iodide ions?
    (b) What is the overall order of reaction?

2. The rate equation for the reaction between P and Q is:
    $$\text{rate} = k[P][Q]^2$$
    What will be the increase in rate if:
    (a) [P] is doubled, while [Q] is kept constant?
    (b) [Q] is doubled, while [P] is kept constant?
    (c) [P] and [Q] are both doubled?

3. In the reaction between R, S and T:
    - when the concentration of R is doubled, the rate increases by four times
    - when the concentration of S is doubled, the rate does not change
    - when the concentration of T is doubled, the rate doubles.
    (a) Deduce the orders of reaction with respect to R, S and T.
    (b) Write the rate equation for the reaction.
    (c) What is the overall order of reaction?

Key definitions

The **rate equation** is an equation expressing the mathematical relationship between the rate of reaction and the concentrations of the reactants.

The **order** of a reactant species is the power to which the concentration of the species is raised in the rate equation.

The **overall order** of a reaction is the sum of all the individual orders.

The **rate-determining step** of a reaction is the slowest step in the mechanism for the reaction.

16.1 3 Determining orders of reaction

By the end of this section, you should be able to...
- understand experiments that can be used to investigate reaction rates by:
  (i) a continuous monitoring method
  (ii) an initial-rate method
- understand the term 'half-life'
- calculate the rate of reaction and the half-life of a first order reaction using data from a concentration–time graph or a volume–time graph
- deduce the order (0, 1 or 2) with respect to a substance in a rate equation using data from a concentration-time graph
- deduce the order (0, 1 or 2) with respect to a substance in a rate equation using data from an initial-rate method
- determine and use rate equations of the form: rate = $k[A]^m[B]^n$, where m and n are 0, 1 or 2
- deduce the order (0, 1 or 2) with respect to a substance in a rate equation using data from a rate-concentration graph.

How can we determine the rate equation?

This question can equally be phrased 'How can we determine the order of reaction with respect to each reactant?'

There are two methods for determining orders of reaction. They are both experimental. Indeed, orders of reaction can only be determined by experiment.

The first is sometimes called the 'continuous method'.

The second is sometimes called the 'initial-rate method'.

The continuous method

In this method, *one* reaction mixture is made up and samples of the reaction mixture are withdrawn at regular time intervals. The reaction in the sample is stopped, if necessary, by quenching. The concentration of the reactant is then determined by an appropriate experimental technique, such as titration.

- The first step is to draw a 'concentration–time' graph.
- The second step is to find out the **half-life** for the reaction at different concentrations.

If the half-life has a constant value, then the reaction is first order with respect to the reactant.

The 1st half-life, for the change in concentration from 120 to 60 units, is 100 s (**fig A**).

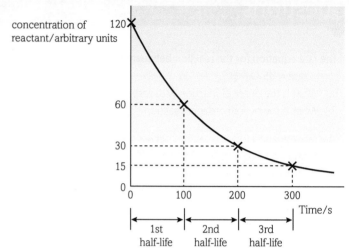

fig A A graph of concentration of reactant against time for a first order reaction.

The 2nd half-life, for the change in concentration from 60 to 30 units, is 100 s.

The 3rd half-life, for the change in concentration from 30 to 15 units, is 100 s.

Since all three half-lives are the same, the reaction is first order with respect to the concentration of the reactant plotted.

If the half-life doubles as the reaction proceeds, then the reaction is second order.

If the graph is a straight line with a negative gradient, then the rate of reaction is constant no matter what the concentration of the reactant. In other words, the reaction is zero order with respect to the reactant.

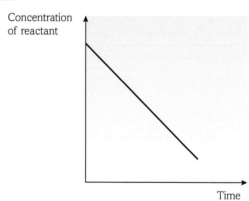

fig B A graph of concentration against time for a zero order reaction.

A typical set of results is shown in **table A** for the reaction:

$$2N_2O_5(g) \rightarrow 4NO_2(g) + O_2(g)$$

148

Time/min	0	10	20	30	40	50	60	70	80	90
$[N_2O_5]/10^{-3}$ mol dm^{-3}	22.90	16.27	12.29	9.35	6.89	4.88	3.68	2.74	2.16	1.85

table A

If you plot a graph of $[N_2O_5]$ against time using the above data, you will find that the line is curved and the half-life is constant; hence, the reaction is first order with respect to N_2O_5.

Calculating rate from a concentration–time graph

The rate of reaction at any given time can be determined from a concentration–time or volume–time graph by drawing a tangent to the curve at the given time and calculating the gradient of the tangent.

Fig C shows the change in concentration of a reactant with time.

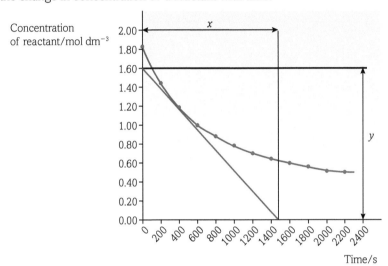

fig C Determining an instantaneous rate of reaction from a graph of concentration of reactant against time.

A tangent to the curve has been drawn at time = 400 s.

To find the rate at this time point, draw as large a triangle as possible and then measure x and y.

$x = 1470$ s and $y = -1.60$ mol dm^{-3}.

$$\text{rate} = -\frac{\text{change in concentration of reactant}}{\text{time}}$$

$$= -\frac{-1.60 \text{ mol dm}^{-3}}{1470 \text{ s}} = 1.09 \times 10^{-3} \text{ mol dm}^{-3} \text{ s}^{-1}$$

Calculating rate from a volume–time graph

The procedure is exactly the same for a volume–time graph for a gas evolved. This time the curve will slope upwards not downwards.

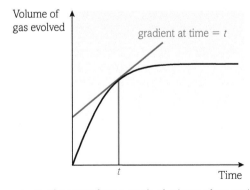

fig D Determining an instantaneous rate of reaction from a graph of volume of gas evolved against time.

The rate of reaction obtained in this way is sometimes called the **instantaneous reaction rate**.

The initial-rate method for determining the rate equation

In this method, *several* reaction mixtures are made up and the initial rate (i.e. the time taken for a fixed amount of reactant to be used up or for a fixed amount of product to be formed) is measured. From these times, it is then possible to calculate the mathematical relationship between the rate of the reaction and the concentration of the reactant.

A typical set of results is shown in **table B** for the reaction:

$$A + B \rightarrow C$$

Experiment	[A]/mol dm^{-3}	[B]/mol dm^{-3}	Initial rate of formation of C/mol dm^{-3} s^{-1}
1	0.1	0.1	0.02
2	0.1	0.2	0.04
3	0.2	0.1	0.04
4	0.2	0.2	0.08

table B

If we now look at experiments 1 and 3, we can see that [A] has doubled whilst [B] has remained constant. The rate of reaction has also doubled. This indicates that the reaction is first order with respect to A and hence we can write:

$$\text{rate} \propto [A]$$

If we now look at experiments 1 and 2, or experiments 3 and 4, we can see that [B] has doubled while [A] has remained constant. The initial rate of reaction has also doubled. This indicates that the rate of reaction is directly proportional to [B]. That is, the reaction is first order with respect to B. Hence we can write:

$$\text{rate} \propto [B]$$

The overall order of the reaction is two and the complete rate law (rate equation) is:

$$\text{rate} = k[A][B]$$

You may suspect that the numbers in **table B** were deliberately kept simple to illustrate a point. Real experimental results rarely fit such an exact pattern and you may have to look for the nearest whole numbers to obtain orders.

Determining order from a rate–concentration graph

The method described above to determine initial rates with changing concentrations results in only approximate values for the initial rates. However, since most orders of reaction with respect to individual reactants have integer values, this approximation is usually acceptable.

In an initial rates experiment, the time measured is that for a fixed amount of product to be formed, or that for a fixed amount of reactant to be used up. Since the amount of product formed, or reactant used up, is kept constant, the initial rate of reaction is proportional to the reciprocal of the time, t, measured. That is:

$$\text{rate} \propto \frac{1}{t}$$

This assumes that the rate is constant for the whole time period, t. However, this is not true because as soon as the reaction starts the rate begins to decrease. The rate calculated from the expression is, therefore, the *mean (average)* rate over time, t, and not the *true* initial rate of reaction. The approximation becomes poorer with larger values of t. However, the approximation is good enough to determine integer order.

It might be possible to determine orders of reaction from just a few measurements. However, it is usual to record a range of results and plot a graph of $\frac{1}{t}$ against concentration of reactant.

The shape of the graph indicates the order of reaction.

Further kinetics 16.1

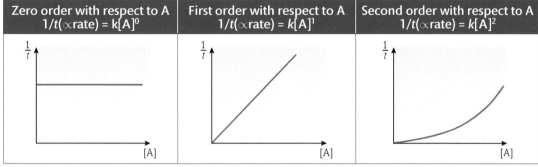

table C

It is impossible to determine directly by sight from a rate–concentration graph that the reaction is second order. If the graph is a curve as shown, then it is necessary to then plot $\frac{1}{t}$ against $[A]^2$. If this produces a straight line passing through the origin, then the reaction is second order with respect to A.

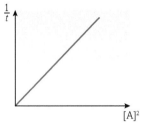

fig E A graph of $\frac{1}{t}$ against the square of the concentration of a reactant for a second order reaction.

Questions

1 A compound P decomposes when heated. The graph shows the change in concentration when a sample of P is heated.

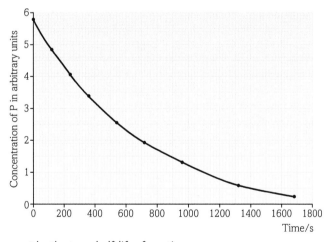

(a) State what is meant by the term *half-life of reaction*.
(b) Use the graph to show that the decomposition of P is a first order reaction.
(c) Explain the effect on the half-life of doubling the initial concentration of P.
(d) Calculate the rate constant, k, for this reaction using the following expression:

$$k = \frac{0.693}{\text{half-life}}$$

Include units in your answer.
(e) Write the rate equation for this reaction.
(f) (i) Use the graph to determine the concentration of P at 800 s.
(ii) Use the rate equation from (e) and your answers to (d) and (f)(i) to calculate the rate of reaction at 800 s. Include units in your answer.
(g) Describe how could you determine the reaction rate at 800 s directly from the graph.

151

2 The equation for the reaction between bromide ions and bromate(V) ions in acidified aqueous solution is:
$$5Br^-(aq) + BrO_3^-(aq) + 6H^+(aq) \rightarrow 3Br_2(aq) + 3H_2O(l)$$
The table shows the results of four experiments carried out using different concentrations of the three reactants.

Experiment	$[Br^-(aq)]$/ mol dm^{-3}	$[BrO_3^-(aq)]$/ mol dm^{-3}	$[H^+(aq)]$/ mol dm^{-3}	Initial rate/ mol dm^{-3} s^{-1}
1	0.10	0.10	0.10	1.2×10^{-3}
2	0.10	0.20	0.10	2.4×10^{-3}
3	0.30	0.10	0.10	3.6×10^{-3}
4	0.10	0.20	0.20	9.6×10^{-3}

(a) Deduce the order of reaction with respect to:
  (i) Br^-
  (ii) BrO_3^-
  (iii) H^+
(b) Write the rate equation for the reaction.
(c) Using the results from experiment 1, calculate the rate constsnt, k, for the reaction. Include units in your answer.

3 Bromine can be formed by the oxidation of hydrogen bromide with oxygen.
A proposed mechanism for this reaction is:
  Step 1: $HBr + O_2 \rightarrow HBrO_2$
  Step 2: $HBrO_2 + HBr \rightarrow 2HBrO$
  Step 3: $HBrO + HBr \rightarrow Br_2 + H_2O$
  Step 4: $HBrO + HBr \rightarrow Br_2 + H_2O$ (a repeat of Step 3)
The rate equation for this reaction is:
  rate = $k[HBr][O_2]$

(a) Explain which of the above four steps is the rate-determining step for this reaction.
(b) Write the overall equation for the reaction.

Key definitions

The **half-life** of a reaction is the time taken for the concentration of the reactant to fall to one-half of its initial value.

The **instantaneous reaction rate** is the gradient of a tangent drawn to the line of the graph of concentration against time. The instantaneous rate varies as the reaction proceeds (except for a zero order reaction).

16.1 4 Rate equations and mechanisms

By the end of this section, you should be able to...

- deduce a rate-determining step from a rate equation and vice versa
- deduce a reaction mechanism, using knowledge from a rate equation and the stoichiometric equation for the reaction
- understand that knowledge of the rate equations for the hydrolysis of halogenoalkanes can be used to provide evidence of S_N1 or S_N2 mechanisms for tertiary and primary halogenoalkane hydrolysis
- understand how to:
  (i) obtain data to calculate the order with respect to the reactants (and the hydrogen ion) in the acid catalysed iodination of propanone
  (ii) use these data to make predictions about species involved in the rate-determining step
  (iii) deduce a possible mechanism for the reaction.

Reaction mechanisms

You will remember from **Book 1** that the basic view as to how a reaction takes place at a particulate level is that particles (atoms, molecules, ions or radicals) first have to collide in the correct orientation and with sufficient energy for products to be formed.

For the following reaction:

$$P + Q \rightarrow \text{products}$$

we expect the rate law to be:

$$\text{rate} = k[P][Q]$$

That is, we might expect the reaction to be first order with respect to each reactant and second order overall.

A reaction taking place in this manner, that is a single collision between the two reactant particles, is described as being *elementary*. If we know that a reaction is elementary, then we can deduce the rate law directly from the stoichiometric equation. For example, the following reaction is known to be elementary:

$$NO(g) + O_3(g) \rightarrow NO_2(g) + O_2(g)$$

so the rate equation is:

$$\text{rate} = k[NO][O_3]$$

The reaction takes place when a molecule of NO collides with a molecule of O_3.

If the reaction is not elementary, it is *not* possible to deduce the rate equation by simply looking at the stoichiometric equation for the reaction.

For example, the decomposition of dinitrogen pentoxide into nitrogen dioxide and oxygen is first order with respect to dinitrogen pentoxide, not second.

$$2N_2O_5(g) \rightarrow 4NO_2(g) + O_2(g)$$

The experimentally determined rate equation is:

$$\text{rate} = k[N_2O_5]$$

A reaction that is not elementary takes place via a series of interconnected elementary reactions that are collectively called the *mechanism* for the reaction. You will have already come across a number of such mechanisms in your study of organic chemistry. For example, the radical substitution reaction between methane and chlorine to form chloromethane (CH_3Cl) is thought to have the following mechanism:

Step 1: $Cl_2 \rightarrow 2Cl\bullet$

Step 2: $Cl\bullet + CH_4 \rightarrow HCl + \bullet CH_3$

Step 3: $Cl_2 + \bullet CH_3 \rightarrow Cl\bullet + CH_3Cl$

The overall stoichiometric equation for the reaction is:

$CH_4 + Cl_2 \rightarrow CH_3Cl + HCl$

The species $Cl\bullet$ and $\bullet CH_3$ are called *intermediates*. They do not appear in the overall equation for the reaction, but are involved in reactions that ultimately result in the reactants being converted into the products.

If the experimentally determined rate equation does not match the overall stoichiometry, then it is almost certain that the reaction is not elementary. For example, the rate equation for the following reaction:

$NO_2(g) + CO(g) \rightarrow NO(g) + CO_2(g)$

is:

rate = $k[NO_2]^2$

This suggests that only molecules of NO_2 are involved in the rate-determining step, and that *two* molecules of NO_2 are involved in this step.

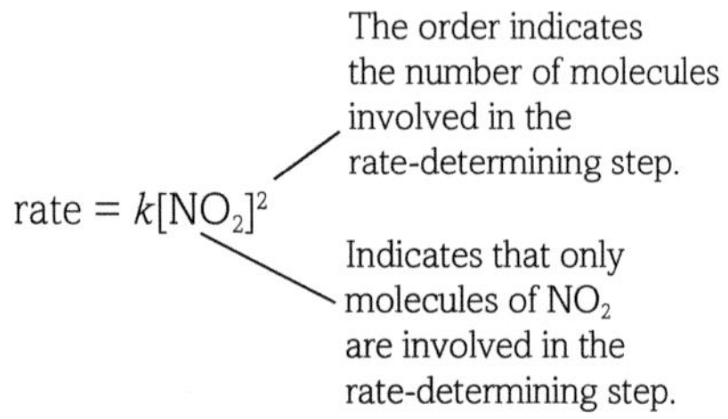

rate = $k[NO_2]^2$

- The order indicates the number of molecules involved in the rate-determining step.
- Indicates that only molecules of NO_2 are involved in the rate-determining step.

fig A Molecules involved in the rate-determining step.

Using our knowledge of molecules that do exist, two possible rate-determining steps are:

1. $NO_2 + NO_2 \rightarrow N_2O_4$, or
2. $NO_2 + NO_2 \rightarrow 2NO + O_2$

Either of these two reactions are equally valid and we have no way of knowing, without carrying out further investigations, which is the more likely to be taking place.

We also have no way of knowing what is involved in the remaining steps, but we do know that the sum of all the steps must add up to the overall stoichiometric equation. Using the second of the two possible rate-determining steps, the following mechanism is consistent with the data:

$NO_2 + NO_2 \rightarrow 2NO + O_2$ SLOW

$CO + O_2 \rightarrow CO_2 + O$ FAST

$NO + O \rightarrow NO_2$ FAST

This shows how the particles in the proposed mechanism cancel to produce the overall equation:

$$NO_2 + \cancel{NO_2} \rightarrow \cancel{NO} + NO + \cancel{O_2}$$
$$CO + \cancel{O_2} \rightarrow CO_2 + \cancel{O}$$
$$\cancel{NO} + \cancel{O} \rightarrow \cancel{NO_2}$$
$$\overline{NO_2 + CO \rightarrow NO + CO_2}$$

As it happens, further investigations into this reaction have identified the mechanism as:

$NO_2 + NO_2 \rightarrow NO_3 + NO$ SLOW

$NO_3 + CO \rightarrow NO_2 + CO_2$ FAST

This is not what we would immediately suspect as the mechanism because the existence of NO_3 is not something with which we would be familiar.

Even if the experimentally determined rate equation is simple, it does not necessarily follow that the reaction proceeds in a single elementary step. For example, the rate expression for the following gas-phase reaction:

$2N_2O_5(g) \rightarrow 4NO_2(g) + O_2(g)$

is:

rate = $k[N_2O_5]$

Yet, the reaction is thought to involve several steps and a number of intermediates.

Alkaline hydrolysis of halogenoalkanes

The hydrolysis of halogenoalkanes by hydroxide ions is a reaction we introduced in **Book 1**. The hydroxide ion acts as a nucleophile and replaces the halogen in the halogenoalkanes. The reaction, therefore, can also be described as *nucleophilic substitution*.

Hydrolysis of a tertiary halogenoalkane

The equation for the alkaline hydrolysis of 2-chloromethylpropane is:

$(CH_3)_3CCl + OH^- \rightarrow (CH_3)_3COH + Cl^-$

The experimentally determined rate equation for this reaction is:

rate = $k[(CH_3)_3CCl]$

The reaction is first order with respect to 2-chloromethylpropane, but zero order with respect to the hydroxide ion.

The sensible conclusion to reach is that the 2-chloromethylpropane undergoes slow ionisation as the rate-determining step. This is then followed by a very fast step involving attack by the hydroxide ion on the carbocation formed in Step 1.

Step 1: $(CH_3)_3CCl \rightarrow (CH_3)_3C^+ + Cl^-$ SLOW

Step 2: $(CH_3)_3C^+ + OH^- \rightarrow (CH_3)_3COH$ FAST

This type of mechanism is named S_N1, i.e. Substitution *N*ucleophilic *uni*molecular.

The rate-determining step is said to be *uni*molecular because there is only *one* reactant particle present, $(CH_3)_3CCl$.

The carbocation, $(CH_3)_3C^+$, formed in Step 1 is an *intermediate* (see **Section 16.1.6**).

Further kinetics 16.1

Hydrolysis of a primary halogenoalkane

The equation for the alkaline hydrolysis of 1-chlorobutane is:

$$CH_3CH_2CH_2CH_2Cl + OH^- \rightarrow CH_3CH_2CH_2CH_2OH + Cl^-$$

The experimentally determined rate equation for this reaction is:

rate = $k[CH_3CH_2CH_2CH_2Cl][OH^-]$

This time the reaction is first order with respect to each reactant, so it is reasonable to suggest that one particle of each reactant is present in the rate-determining step of the mechanism.

The accepted mechanism for the reaction is:

fig B S_N2 mechanism of the alkaline hydrolysis of 1-chlorobutane.

This type of mechanism is named S_N2, i.e. *S*ubstitution *N*ucleophilic *bi*molecular.

The rate-determining step is *bi*molecular because there are *two* reactant particles present.

It is a continuous single one-step reaction. The complex shown in square brackets is not an intermediate (like the carbocation formed in the S_N1 mechanism), but is a *transition state* (see **Section 16.1.6**).

Investigation

A study of the acid catalysed iodination of propanone

The reaction between propanone and iodine in aqueous solution may be acid catalysed:

$$I_2(aq) + CH_3COCH_3(aq) + H^+(aq) \rightarrow CH_3COCH_2I(aq) + 2H^+(aq) + I^-(aq)$$

The influence of the iodine on the reaction rate may be studied if the concentrations of propanone and hydrogen ions effectively remain constant during the reaction. This is achieved by using a *large excess* of both propanone and acid in the original reaction mixture.

Procedure

1. Mix 25 cm³ of 1 mol dm⁻³ aqueous propanone with 25 cm³ of 1 mol dm⁻³ sulfuric acid.
2. Start the stop clock *the moment you add* 50 cm³ of 0.02 mol dm⁻³ iodine solution. Shake well.
3. Using a pipette, withdraw a 10 cm³ sample and place it in a conical flask. Stop the reaction by adding a 'spatula-measure' of sodium hydrogencarbonate. Note the exact time at which the sodium hydrogencarbonate is added.
4. Titrate the remaining iodine present in the sample with 0.01 mol dm⁻³ sodium thiosulfate(VI) solution, using starch indicator.
5. Withdraw further 10 cm³ samples at suitable time intervals (approx. 5 to 7 minutes) and treat them similarly, always noting the exact time at which the sodium hydrogencarbonate is added. Waste should be contained in a fume cupboard.

Treatment of results

Plot a graph of titre against time. (The titre is proportional to the concentration of iodine.)
Deduce from the graph the order of reaction with respect to iodine.

Analysis of results

The graph produced shows that the reaction is zero order with respect to iodine.
Similar experiments show that the reaction is first order with respect to both propanone and hydrogen ions.
This gives us the following rate equation:

rate = $k[CH_3COCH_3][H^+]$

This would suggest the following reaction as the first step of the reaction:

$$CH_3-\underset{O}{\underset{\|}{C}}-CH_3 + H^+ \rightarrow CH_3-\underset{^+O-H}{\underset{\|}{C}}-CH_3$$

Similar reactions in other mechanisms are very fast, so this reaction is unlikely to be the rate-determining step of this reaction.

The second step probably controls the rate of the reaction and produces H⁺ that, being a catalyst, is not used up in the reaction:

$$CH_3-\underset{^+O-H}{\underset{\|}{C}}-CH_3 \rightarrow CH_3-\underset{O-H}{\underset{|}{C}}=CH_2 + H^+ \quad \text{SLOW}$$

This rearrangement is likely to be very slow and hence is probably the rate-determining step of the reaction.

Iodine can now react in a fast step as follows:

$$CH_3-\underset{O-H}{\underset{|}{C}}=CH_2 + I_2 \rightarrow CH_3-\underset{O}{\underset{\|}{C}}-CH_2I + I^- + H^+ \quad \text{FAST}$$

Testing the mechanism

We have a proposed mechanism for the reaction that is consistent with the kinetic data obtained from experiment. This is not the same as saying that the mechanism is the correct one, or indeed the most likely one. We now need to carry out further experiments to confirm, or deny, our proposed mechanism.

Here are three techniques that can be employed in this case.

- *Use a wider range of concentrations*
  The experimentally determined rate equation may hold over only a limited range of concentrations. For example, the mechanism we have proposed for the iodination of propanone predicts that, at very low concentrations of iodine, the order of reaction with respect to iodine will change from zero to first. This is because the rate of the final step in the mechanism is given by the expression rate = $k[I_2][CH_3C(OH)=CH_2]$ and so the rate of the step will decrease as $[I_2]$ decreases. This has been shown to be the case, so supports the proposed mechanism.

- *Use instrumental analysis*
  This may be able to detect the presence of intermediates that have been proposed. Nuclear magnetic resonance can be used to show that acidified propanone contains about one molecule in a million in the form $CH_3C(OH)=CH_2$. This is an intermediate stated in the proposed mechanism.

- *Carry out the reaction with deuterated propanone, CD_3COCD_3*
  Deuterium ^2_1H or (simply D) behaves slightly differently to hydrogen ^1_1H or (simply H).
  A C–D bond is slightly stronger, and therefore harder to break, than a C–H bond. If this is the bond that breaks in the rate-determining step, as suggested by our proposed mechanism, then the iodination of CD_3COCD_3 will be slower than that of CH_3COCH_3. This is found to be the case.

Questions

1. The equation for the reaction between ethanal, CH_3CHO, and hydrogen cyanide, HCN, is:
$$CH_3CHO + HCN \rightarrow CH_3CH(OH)CN$$
Two mechanisms that have been proposed for this reaction are:
Mechanism 1
Step 1: $CH_3CHO + H^+ \rightarrow [CH_3CHOH]^+$
Step 2: $[CH_3CHOH]^+ + CN^- \rightarrow CH_3CH(OH)CN$
Mechanism 2
Step 1: $CH_3CHO + CN^- \rightarrow [CH_3CHOCN]^-$
Step 2: $[CH_3CHOCN]^- + H^+ \rightarrow CH_3CH(OH)CN$
The rate equation for the reaction is:
$$\text{rate} = k[CH_3CHO][CN^-][H^+]^0$$
(a) Explain which of the two mechanisms is consistent with the rate equation.
(b) Which step in this mechanism is the rate-determining step?

2. The equation for the reaction between hydrogen and nitrogen monoxide is:
$$2H_2(g) + 2NO(g) \rightarrow 2H_2O(g) + N_2(g)$$
The rate equation for the reaction is:
$$\text{rate} = k[H_2][NO]^2$$
A proposed mechanism for this reaction is:
Step 1: $2NO(g) \rightarrow N_2O_2(g)$ FAST
Step 2: $N_2O_2(g) + H_2(g) \rightarrow H_2O(g) + N_2O(g)$ SLOW
Step 3: $N_2O(g) + H_2(g) \rightarrow N_2(g) + H_2O(g)$ FAST
Is this mechanism consistent with the rate equation? Explain your answer.

3. Assume the following proposed reaction mechanism is correct:
$Cl_2 \rightarrow 2Cl\bullet$ SLOW
$H_2 + Cl\bullet \rightarrow HCl + H\bullet$ FAST
$H\bullet + Cl\bullet \rightarrow HCl$ FAST
(a) Write the overall equation for the reaction.
(b) Write a rate equation for the reaction that is consistent with the mechanism.
(c) What would be the effect on the rate of reaction of doubling the concentration of Cl_2?
(d) What would be the effect on the rate of reaction of doubling the concentration of H_2?

4. The nucleophilic substitution reaction between equimolar quantities of CH_3Cl and OH^- is second order overall. However, if the reaction is carried out using a large excess of OH^-, the reaction becomes first order overall. Suggest an explanation for these observations.

16.1 5 Activation energy and catalysis

By the end of this section, you should be able to...
- understand the terms:
    (i) activation energy
    (ii) homogeneous catalyst and heterogeneous catalyst
    (iii) autocatalysis

Activation energy, E_a

In **Book 1 Section 9.1.1**, we defined activation energy, E_a, as the minimum energy that colliding particles must possess for a reaction to occur.

The activation energy represents the energy that the colliding particles must obtain in order to reach the energy level of the *transition state* (see **Section 16.1.6** for more details). Once the energy level of the transition state has been reached, the particles can react to form the products and release energy as they do so. The energy profile diagram for an exothermic reaction is shown below.

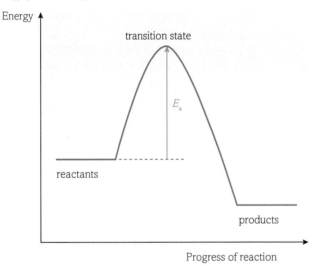

fig A A reaction profile for an elementary exothermic reaction.

Catalysts

In **Book 1 Section 9.1.4**, we defined a catalyst as a substance that increases the rate of a chemical reaction but is chemically unchanged at the end of the reaction.

We then went on to explain that a catalyst works by providing an alternative route for the reaction, and that this alternative route has a lower activation energy than the original route. Although the original route is still available for the reactants, most collisions resulting in reaction will occur via the alternative route since the fraction of particles possessing this lower activation energy will be greater.

Catalysts can be divided into two groups – homogeneous catalysts and heterogeneous catalysts.

Homogeneous catalysts

A *homogeneous catalyst* is in the same phase (solid, liquid, solution or gas) as the reactants.

Many reactions in aqueous solution are catalysed by the hydrogen ion, $H^+(aq)$. An example is the iodination of propanone, a reaction we discussed in detail in **Section 16.1.4**:

$$CH_3COCH_3(aq) + I_2(aq) \rightarrow CH_3COCH_2I(aq) + H^+(aq) + I^-(aq)$$

The production of chlorine radicals from chlorofluorohydrocarbons (CFCs) is responsible for the destruction of ozone in the upper atmosphere. Ultraviolet radiation from the Sun produces chlorine radicals, Cl•, from CFCs such as dichlorodifluoromethane, CCl_2F_2:

$$CCl_2F_2(g) \xrightarrow{\text{ultraviolet radiation}} \bullet CClF_2(g) + Cl\bullet(g)$$

The chlorine radicals then take part in a chain reaction with ozone:

$$Cl\bullet(g) + O_3(g) \rightarrow ClO\bullet(g) + O_2(g)$$
$$ClO\bullet(g) + O_3(g) \rightarrow 2O_2(g) + Cl\bullet(g)$$

The chlorine radical is regenerated and so is acting as a catalyst. Since the catalyst is in the same phase, i.e. the gas phase, as the reacting species, it is classified as a homogeneous catalyst.

The overall reaction is:

$$2O_3(g) \rightarrow 3O_2(g)$$

Another reaction involving a homogeneous catalyst is the one between peroxydisulfate ions and iodide ions in aqueous solution:

$$S_2O_8^{2-}(aq) + 2I^-(aq) \rightarrow 2SO_4^{2-}(aq) + I_2(aq)$$

This reaction is catalysed by either $Fe^{2+}(aq)$ or $Fe^{3+}(aq)$.

With $Fe^{2+}(aq)$, the reaction mechanism is:

Step 1: $S_2O_8^{2-}(aq) + 2Fe^{2+}(aq) \rightarrow 2SO_4^{2-}(aq) + 2Fe^{3+}(aq)$
Step 2: $2Fe^{3+}(aq) + 2I^-(aq) \rightarrow 2Fe^{2+}(aq) + I_2(aq)$

With $Fe^{3+}(aq)$ as the catalyst, Steps 1 and 2 occur in the reverse order

Learning tip

The action of $Fe^{2+}(aq)$ or $Fe^{3+}(aq)$ as a catalyst in the reaction between $S_2O_8^{2-}(aq)$ and $I^-(aq)$ is sometimes explained in terms of standard electrode potentials ($E^\ominus$ values). However it is important to note that since standard electrode potentials can predict only the thermodynamic feasibility of a reaction, and not the kinetics. The reason why the reactions are faster in the presence of either Fe^{2+} or Fe^{3+} is that the activation energies for both Steps 1 and 2 are lower than the activation energy for the overall reaction.

Heterogeneous catalysts

A *heterogeneous* catalyst is in a different phase to that of the reactants.

Two important uses of heterogeneous catalysts in industry are in the Haber process and the Contact process. The use of solid vanadium(V) oxide (V_2O_5), in which the vanadium changes its oxidation number, in the Contact process is described in the section on transition elements (**Section 15.3.1**). We will now describe the action of solid iron as a catalyst in the reaction between nitrogen gas and hydrogen gas to form ammonia gas in the Haber process.

The equation for the formation of ammonia in the Haber process is:

$$N_2(g) + 3H_2(g) \rightleftharpoons 2NH_3(g)$$

Iron is able to act as a catalyst because it can form an *interstitial hydride* with hydrogen molecules. In this hydride, hydrogen atoms are held in spaces between the metal ions in the lattice. The atoms are then able to react with nitrogen molecules that are adsorbed onto the metal surface nearby. There are three stages in catalysis involving surface adsorption. These are:

- *Adsorption* – the reactants are first adsorbed onto the surface of the catalyst.
- *Reaction* – the reactant molecules are held in positions that enable them to react together.
- *Desorption* – the product molecules leave the surface.

The rate of reaction is controlled by how fast the reactants are adsorbed and how fast the products are desorbed. As mentioned in **Book 1 Sections 9.1.2** and **10.2.2**, once the surface of the iron is covered with molecules, there is no further increase in reaction rate even if the pressure of the reactants is increased.

Did you know?

The efficiency of a heterogeneous catalyst depends on the surface of the catalyst. In particular the efficiency of the catalyst can be affected significantly by poisoning and by the use of promoters.

Poisoning
Many catalysts are made ineffective by trace impurities. For example, catalysts used in hydrogenation reactions are poisoned by sulfur impurities. This is one reason why nickel is preferred to platinum as a catalyst in the hydrogenation of alkenes. Nickel is relatively inexpensive, so a large quantity can be used. If some of it is deactivated by poisoning, enough will remain for it still to be effective. On the other hand, platinum is expensive and so small quantities would have to be used. In this case, it is possible for all of the catalyst to become poisoned.

Promoters
The spacing on the surface of the catalyst is important. For example, only some surfaces of the iron crystals act as effective catalysts in the Haber process. The addition of traces of potassium oxide and aluminium oxide act as promoters by producing *active sites* where the reaction takes place most readily.

Did you know?

An interstitial hydride, sometimes called a metallic hydride, is not strictly a compound as such. It is more like an alloy than a compound. The hydrogen absorbs into the metal and can exist in the form of either atoms or diatomic molecules. For example, palladium absorbs up to 900 times its own volume of hydrogen at room temperature, and forms palladium hydride. This material has been considered as a method to carry hydrogen in fuel cells for use in vehicles (see **Section 14.2.1**).

A heterogeneous catalyst is used in a three-way catalytic converter, which is employed in the exhaust of cars. This converts unburned hydrocarbons into water and carbon dioxide, and carbon and carbon monoxide into carbon dioxide. It also converts oxides of nitrogen into oxygen and nitrogen. The reactions involved are discussed in more detail in **Topic 15**.

Did you know?

The electrophilic addition reaction between ethene and bromine vapour (**Book 1 Sections 6.2.6** and **6.2.7**) was once thought to be entirely a homogeneous process in which the molecules collided in the gas phase and reacted. It is now known that the reaction takes place almost entirely on the surface of the walls of the reaction vessel. The reaction takes place very quickly on a glass surface and more quickly when the surface of the vessel is covered with a polar substance such as cetyl alcohol (hexadecan-1-ol, $CH_3(CH_2)_{15}OH$). It takes place even more quickly on a surface of stearic acid (octadecanoic acid, $CH_3(CH_2)_{16}COOH$), presumably because stearic acid molecules are more polar than molecules of cetyl alcohol. However, if the walls of the reaction vessel are coated with a non-polar substance such as paraffin wax, very little reaction, if any, occurs. So, a process that was once thought to be a homogeneous gas phase reaction is now known to be heterogeneously catalysed.

It is sometimes stated that a catalyst does not change the products of a reaction, but only increases the rate at which the reaction takes place. This is not entirely true, as demonstrated by the different products obtained when hot ethanol vapour is passed over different solid catalysts.

With copper or nickel as the catalyst, at 400 °C, a dehydrogenation reaction takes place in which the products are ethanal and hydrogen:

$$CH_3CH_2OH(g) \rightarrow CH_3CHO(g) + H_2(g)$$

If, at the same temperature, aluminium oxide is used as the catalyst, the more familiar dehydration reaction into ethene and water takes place:

$$CH_3CH_2OH(g) \rightarrow CH_2=CH_2(g) + H_2O(g)$$

It is interesting to note that the dehydrogenation catalysts, copper and nickel, adsorb hydrogen very strongly, whereas the dehydration catalyst, aluminium oxide, adsorbs water in preference.

These two examples illustrate that there is a lot more to heterogeneous catalysis than is mentioned at A level. This might encourage you to do your own research into the topic and learn even more about it.

The oxidation of ethanedioic acid by manganate(VII) ions

In autocatalysis, the reaction is catalysed by one of its products. One of the simplest examples of this is the oxidation of ethanedioic acid by acidified potassium manganate(VII). The equation for the reaction is:

$$5(COOH)_2(aq) + 2MnO_4^-(aq) + 6H^+(aq) \rightarrow 10CO_2(g) + 2Mn^{2+}(aq) + 8H_2O(l)$$

The reaction is very slow at room temperature, but is catalysed by manganese(II) ions, Mn^{2+}. The Mn^{2+} ions are not present initially, so the reaction starts off extremely slowly at room temperature. However, Mn^{2+} is a product of the reaction. As soon as it is produced in a catalytic amount, the reaction rate increases.

You can show this effect by plotting the concentration of one of the reactants against time. The graph obtained is unlike the normal rate curve for a reaction.

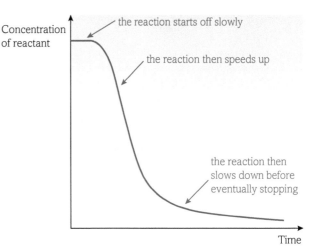

fig B A graph of concentration against time for an autocatalysed reaction.

You can see the slow (uncatalysed) reaction at the beginning. As catalyst begins to be formed in the mixture, the reaction speeds up – it gets faster and faster as more and more catalyst is formed. Eventually, the rate decreases in the normal way as the concentrations of the reactants decrease.

Learning tip

Do not assume that a rate curve that looks like this *necessarily* shows an example of autocatalysis. There are other effects that might produce a similar graph. For example, if the reaction involves a solid reacting with a liquid, there might be a substance coating the surface of the solid that the liquid has to penetrate before the expected reaction can happen. Another possibility is that the reaction is strongly exothermic and the temperature is not being controlled. The heat evolved during the reaction speeds up the reaction.

Questions

1. Explain why only a small quantity of a homogeneous catalyst is required in order for it to be effective.

2. Cars in some parts of the world still run on leaded petrol. Leaded petrol contains a compound called tetraethyl lead, $(CH_3CH_2)_4Pb$. Tetraethyl lead reacts in the engine with oxygen and forms lead and lead(II) oxide. These substances remove radical intermediates in the combustion reactions, thus increasing the octane rating of the petrol. Excess lead and lead(II) oxide is removed by adding 1,2-dibromoethane. This forms volatile lead(II) bromide that escapes through the exhaust.

   Suggest why a catalytic converter could not be used in a car running on leaded petrol.

16.1 6 The effect of temperature on the rate constant

By the end of this section, you should be able to...
- use the Arrhenius equation to explain the effect of temperature on the rate constant of a reaction
- use graphical methods to find the activation energy from experimental data

The relationship between temperature and rate of reaction

In **Book 1** we discussed qualitatively why an increase in temperature increased the rate of a reaction. There are two reasons for this:

- an increase in the fraction of molecules with energy equal to or greater than the activation energy for the reaction.
- an overall increase in the frequency of collisions between the reacting molecules

The second effect is considerably more significant than the first – so much so that we can effectively ignore the overall increase in frequency of collisions.

In 1889, Svante Arrhenius, a Swedish chemist, proposed a quantitative relationship between temperature and the rate constant, k, for a reaction. This is described by the *Arrhenius equation* and is usually expressed in the form:

$$k = Ae^{\left(-\frac{E_a}{RT}\right)}$$

- A is a constant known as the pre-exponential factor, which is a measure of the rate at which collisions occur irrespective of their energy. It also includes other factors, the most important of which is that reactions can only occur when the molecules are correctly orientated at the time of collision.
- E_a is the activation energy of the reaction.
- R is the gas constant.
- T is the absolute temperature (i.e. the temperature in Kelvin).

If we take natural logarithms (i.e. logarithms to the base 'e') of the Arrhenius equation. we obtain:

$$\ln k = -\frac{E_a}{R}\frac{1}{T} + \ln A$$

If a graph of $\ln k$ is plotted against $\frac{1}{T}$, a straight line is obtained with a gradient of $-\frac{E_a}{R}$

This provides an experimental method for determining the activation energy of a reaction.

The intercept with the vertical axis gives $\ln A$.

Example

Table A shows some data related to the reaction between peroxydisulfate and iodide ions in aqueous solution:

$$S_2O_8^{2-}(aq) + 2I^-(aq) \rightarrow 2SO_4^{2-}(aq) + I_2(aq)$$

Temperature/K	Magnitude of rate constant, k	$\ln k$	$1/T/K^{-1}$
300	0.00513	−5.27	0.00333
310	0.00833	−4.79	0.00323
320	0.0128	−4.36	0.00313
330	0.0201	−3.91	0.00303
340	0.0301	−3.50	0.00294

table A

Further kinetics | 16.1

Fig A shows $\ln k$ plotted against $\frac{1}{T}$.

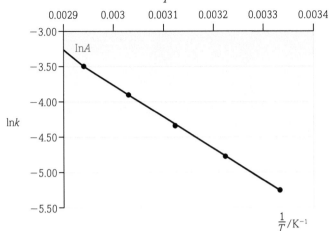

fig A A graph of $\ln k$ against $1/T$.

The gradient of the line, $-\frac{E_a}{R}$

$$= -\frac{(-3.50 - -5.27)}{(0.00333 - 0.00294)}$$

$$= -4538$$

$[R = 8.31 \text{ J mol}^{-1} \text{ K}^{-1}]$

$E_a = -(-4538 \times 8.31) = 37\,700 \text{ J mol}^{-1} = +37.7 \text{ kJ mol}^{-1}$ (to 3 s.f.)

Did you know?

In stating that the relationship between $\ln k$ and $-\frac{1}{T}$ is linear, we have assumed that both E_a and A remain constant over a range of temperatures. This is not strictly true. However, the change in the values of E_a and A with temperature are insignificant compared with the effect of temperature on the rate constant, and so can be ignored.

Did you know?

The significance of the factor $e^{-\frac{E_a}{RT}}$

The factor $e^{-\frac{E_a}{RT}}$ represents the fraction of collisions that have energy equal to or greater than the activation energy, E_a.
For example, a reaction with an E_a of 60 kJ mol^{-1} gives a fraction, at 298 K, of:

$$e^{-(60\,000/8.31 \times 298)} = 3 \times 10^{-11}$$

So, only one collision in 3×10^{11} has sufficient energy to react.

It is often assumed that the fraction of collisions with energy equal to or greater than E_a is the same as the fraction of molecules with this energy. This is not strictly true, but the difference is insignificant at high energies, where the fraction is very small. It is, therefore, not unreasonable to draw the molecular energy distribution curve, instead of the collision frequency curve, when demonstrating the effect of temperature on collision frequency. (See **Book 1 Section 9.1.3**).

Learning tip

The effect of E_a on reaction rate:
- reactions with a large E_a are slow, but the rate increases rapidly with an increase in temperature
- reactions with a small E_a are fast, but the rate does not increase as rapidly with an increase in temperature
- catalysed reactions have small values of E_a.

Did you know?

Reaction profiles

When drawing reaction profile diagrams for multi-step reactions, it is important to distinguish between an *intermediate* and a *transition state*.

An intermediate has an energy minimum, whereas a transition state occurs at the top of the energy curve and therefore has an energy maximum.

An intermediate is a definite chemical species that exists for a finite length of time. By comparison, a transition state has no significant, permanent lifetime of its own – it exists for a period of the order 10^{-15} seconds, when the molecules are in contact with one another. Even a very reactive intermediate, with a lifetime of only 10^{-6} seconds, has a long lifetime in comparison with the period that colliding molecules are in contact with one another.

A simple one-step reaction, such as the S_N2 hydrolysis of a primary halogenoalkane, has a single maximum energy. The reaction profile for the hydrolysis of a primary halogenoalkane, represented as RHal, is therefore:

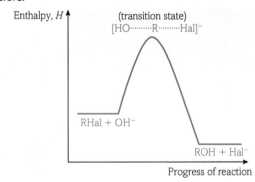

fig B The reaction profile diagram for the hydrolysis of a primary halogenoalkane.

The reaction profile for the S_N1 hydrolysis of a tertiary halogenoalkane, R_3CHal, which is a two-step reaction involving the intermediate R_3C^+, is:

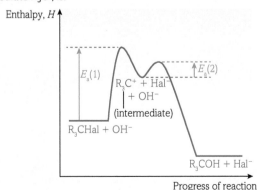

fig C The reaction profile diagram for the hydrolysis of a tertiary halogenoalkane.

161

Learning tip

In the S_N1 hydrolysis of a tertiary halogenoalkane, the first step of the mechanism is the rate-determining step of the reaction and it therefore has the higher activation energy. That is, $E_a(1) > E_a(2)$.

If the second step in a reaction mechanism is rate-determining, $E_a(1) < E_a(2)$.

Did you know?

The relative rates of S_N1 and S_N2 reactions depend on structure

You learned in **Book 1** that alkyl groups donate electrons by the inductive effect. This means that the stability of carbocations increases in the order 1° < 2° < 3°, as the number of alkyl groups donating electrons towards the positive carbon atom increases.

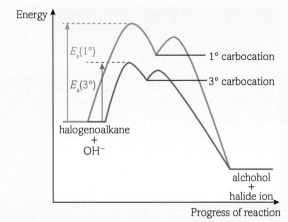

As the stability of the carbocation increases, the activation energy for the reaction leading to its formation also decreases.

fig D Reaction profiles for the S_N1 hydrolysis of primary and tertiary halogenoalkanes.

We therefore expect the rate of the S_N1 reaction to increase in the order 1° < 2° < 3°.

No carbocations are formed during the S_N2 reaction. Other factors come into play. In an S_N2 reaction, the transition state has five groups arranged around the central carbon atom. It is therefore more crowded than either the starting halogenoalkane or the alcohol product, each of which have only four groups around the central carbon atom.

Alkyl groups are much larger than hydrogen atoms. Therefore, the more alkyl groups around the central carbon atom, the more crowded will be the transition state, and the higher the activation energy for its formation.

We therefore expect the rate of the S_N2 reaction to increase in the order 3° < 2° < 1°.

The two effects reinforce one another. The S_N1 reaction is fastest with tertiary halides and slowest with primary halides, whilst the S_N2 reaction is fastest with primary halides and slowest with tertiary halides. Overall, primary halides react predominantly via the S_N2 mechanism and tertiary halides react predominantly via the S_N1 mechanism. Secondary halides tend to react via a mixture of the two mechanisms.

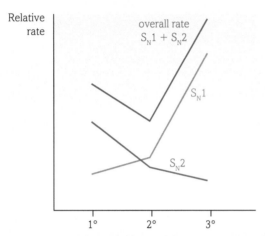

fig E S_N1 and S_N2 hydrolysis reactions for primary, secondary and tertiary halogenoalkanes.

Questions

1. Use the Arrhenius equation to explain why an increase in temperature results in an increase in the rate of reaction.

2. The rate constant, k, for a reaction increases from $10.0\,s^{-1}$ to $100.0\,s^{-1}$ when the temperature is increased from 300 K to 400 K.
   Calculate the activation energy, E_a, for this reaction. [$R = 8.31\,J\,mol^{-1}\,K^{-1}$].

THINKING BIGGER

CATALYST CRAFT

In this article you will look at an example of how an engineered metalloenzyme can improve both the kinetics and specificity of a reaction. The extract is from *Chemistry World*, the print and online magazine of the Royal Society of Chemistry.

ENGINEERED METALLOENZYME CATALYSES FRIEDEL–CRAFTS REACTION

3 November 2014, Debbie Houghton

Reprogramming the genetic code of bacteria to incorporate an unnatural amino acid has allowed scientists in the Netherlands to create a new metalloenzyme capable of catalysing an enantioselective reaction.

'Nature is extremely good at catalysing reactions with very high rate accelerations and very high selectivity. But it does so, from our perspective, with a relatively limited set of reactions,' explains Gerard Roelfes from the University of Groningen, the Netherlands, who led the study. His group is targeting existing reactions that use traditional catalysts, but fail to achieve the same rate acceleration and selectivity as enzyme catalysed reactions.

Metalloenzymes combine the flexibility of metal catalysts with the high activity and selectivity of enzymes. Artificial metalloenzymes are produced by inserting a catalytically active transition metal complex into a biomolecular scaffold, like a protein.

Roelfes' team engineered *Escherichia coli* cells to include a copper-binding amino acid into one of its proteins. This method requires no further chemical modification or purification steps, giving it an advantage over existing methods. The resulting metalloenzyme was tested on a catalytic asymmetric Friedel–Crafts alkylation reaction, achieving an enantiomeric excess of up to 83%.

Takafumi Ueno, who researches the mechanisms of chemical reactions in living cells at the Tokyo Institute of Technology in Japan, is impressed by the work. 'It could be applied not only for rapid screening of artificial metalloenzymes but also for in vivo use of them to govern cell fate in future'.

The group are now looking to develop new artificial metalloenzymes with the capability to perform chemistry that traditional transition metal catalysts cannot. Ultimately, they hope to integrate these enzymes into biosynthetic pathways.

fig A The artificial metalloenzymes were applied in a catalytic asymmetric Friedel–Crafts alkylation reaction.

Where else will I encounter these themes?

Book 1 | 11 | 12 | 13

Thinking Bigger 16

Let us start by considering the nature of the writing in the article.

1. **a.** After reading the article for the first time, write a one-paragraph summary of what you think the article is about.

   b. Now write down a list of any words from the article of which you do not know the meaning. Do some research into what they mean. Is there anything about your summary from (a) that you would change?

Now we will look at the chemistry in detail. Some of these questions will link to topics earlier in this book or **Book 1**, so you may need to combine concepts from different areas of chemistry to work out the answers.

2. Assuming the reaction is exothermic, sketch an energy profile for the catalysed reaction showing:

   a. the overall reaction enthalpy change

   b. the activation energy for the formation of the enzyme-substrate complex.

3. Under certain conditions the rate equation for an enzyme catalysed reaction takes the form: Rate = $k[E]^n$ were $[E]$ is the enzyme concentration and k and n are experimentally determined constants. What does this rate equation suggest about the reaction mechanism under these conditions?

4. Name the type of bond that is formed between the copper ion and the unusual amino acid in the metalloenzyme shown in **fig A**.

5. Suggest why the use of a metalloenzyme of this type might be a better alternative than the traditional Friedel–Crafts methods for alkylation.

6. Suggest why this method might be expected to increase the enantioselectivity of this reaction.

7. Suggest why the Zn^{2+} ion may not be a suitable replacement for the Cu^{2+} ion in this enzyme.

> You may find that it is easier than you think to get an overall understanding of what a detailed scientific article is about. You do not need to understand every single word to learn something new and interesting.

> When you are asked to draw something, make sure you use a sharp HB pencil so that your drawing is clear and use a ruler if you are drawing straight lines.

Activity

Although the metalloenzyme detailed above is genetically engineered, there are many naturally occurring metalloenzymes that carry out a range of reactions in organisms. Most of these can be found in a freely accessible online database called the PDB (alternatively, the PDBe).

Choose one of the following metals: Fe, Zn, Cu, Mo, Co.

- Find a protein that contains the metal.
- Find out what the protein does.
- Find out how much of the metal is present in a typical healthy adult.

Give a five-minute presentation on your chosen metal, featuring no more than five slides. You will have the opportunity to display your chosen molecules using some of the freely available software packages such as Jmol.

Did you know?

Although a typical adult has only about 1.5 mg of copper per kg of body mass, an inability to regulate copper has disastrous effects. For example, the inability to excrete excess copper from the body (Wilson's disease) can be fatal if left untreated. Menkes' syndrome manifests itself if the body is unable to retain copper. Although not fatal, it can lead to developmental delay and neurological problems.

- From *Engineered Metalloenzyme Catalyses Friedal-Crafts Reaction* by Debbie Houghton, *Chemistry World*, 3 November 2014

16 Exam-style questions

1 Nitrogen monoxide, NO, is a pollutant gas emitted from the exhaust of cars. It is oxidised in the atmosphere to form nitrogen dioxide, NO_2.

The overall equation for this reaction is:

$NO(g) + \frac{1}{2}O_2(g) \rightarrow NO_2(g)$

(a) An investigation into the rate of this reaction produced the results shown in the table.

Experiment	Initial [NO(g)] /mol dm^{-3}	Initial [O$_2$(g)] /mol dm^{-3}	Initial rate of formation of NO$_2$ /mol dm^{-3} s^{-1}
1	1.0×10^{-3}	1.0×10^{-3}	6.0×10^{-4}
2	1.0×10^{-3}	2.0×10^{-3}	1.2×10^{-3}
3	2.0×10^{-3}	1.0×10^{-3}	2.4×10^{-3}
4	2.0×10^{-3}	2.0×10^{-3}	4.8×10^{-3}

(i) Deduce the order of reaction with respect to (i) NO(g) and (ii) O$_2$(g). Justify your reasoning. [4]

(ii) Write a rate equation for the reaction between nitrogen monoxide and oxygen. [1]

(b) A proposed mechanism for the reaction between nitrogen monoxide and oxygen is:

$2NO(g) \rightarrow N_2O_2(g)$ FAST
$N_2O_2(g) + O_2(g) \rightarrow 2NO_2(g)$ SLOW

Comment on this proposed mechanism in light of your answers to part (a). [3]

[Total: 8]

2 Propanone and iodine react in aqueous acidic solution according to the following overall equation:

$CH_3COCH_3 + I_2 + H^+ \rightarrow CH_3COCH_2I + 2H^+ + I^-$

The experimentally determined rate equation for this reaction is:

rate = $k[CH_3COCH_3][H^+]$

(a) With initial concentrations as shown, the initial rate of reaction was 1.43×10^{-6} mol dm^{-3} s^{-1}.

	Initial concentration/mol dm^{-3}
CH$_3$COCH$_3$	0.400
H$^+$	0.200
I$_2$	4.00×10^{-4}

Calculate a value for the rate constant, k, for the reaction. [3]

(b) Explain the effect on the rate of reaction of doubling the concentration of iodine, but keeping the concentrations of propanone and hydrogen ions constant. [2]

(c) The proposed mechanism for the overall reaction is:

Step 1 $CH_3COCH_3 + H^+ \rightarrow CH_3-\overset{+}{\underset{OH}{C}}-\underset{H}{\overset{H}{C}}-H$

Step 2 $CH_3-\overset{+}{\underset{OH}{C}}-\underset{H}{\overset{H}{C}}-H \rightarrow CH_3-\underset{OH}{C}=\underset{H}{\overset{H}{C}} + H^+$

Step 3 $CH_3-\underset{OH}{C}=\underset{H}{\overset{H}{C}} + I_2 \rightarrow CH_3-\underset{\overset{+}{O}-H}{\overset{\|}{C}}-CH_2I + I^-$

Step 4 $CH_3-\underset{\overset{|}{O^+}-H}{\overset{\|}{C}}-CH_2I \rightarrow CH_3-\underset{O}{\overset{\|}{C}}-CH_2I + H^+$

Explain which of the four steps could be the rate-determining step. [3]

(d) Add one or more curly arrows to the carbocation to complete the mechanism for Step 2. [1]

Step 2 $CH_3-\overset{+}{\underset{OH}{C}}-\underset{H}{\overset{H}{C}}-H \rightarrow CH_3-\underset{OH}{C}=\underset{H}{\overset{H}{C}} + H^+$

[Total: 9]

3 Most chemical reactions involve two or more steps. The experimentally determined rate equation indicates which species are involved either before or in the rate-determining step.

(a) State what is meant by the term **rate-determining step**. [1]

(b) Hydrogen reacts with iodine monochloride in a two-step mechanism according to the following overall equation:

$H_2(g) + 2ICl(g) \rightarrow 2HCl(g) + I_2(g)$

The experimentally determined rate equation for this reaction is:

rate = $k\,[H_2(g)][ICl(g)]$

The rate-determining step is the first step in the mechanism for the reaction.

(i) Write an equation for the rate-determining step. [1]

(ii) Write an equation for the second step. [1]

(c) A series of experiments were carried out on the reaction A + B + C → products. The results are shown in the table.

Experiment	Initial concentrations/ mol dm^{-3}			Initial rate/ mol dm^{-3} s^{-1}
	[A]	[B]	[C]	
1	0.100	0.100	0.100	6.20×10^{-4}
2	0.100	0.200	0.100	6.20×10^{-4}
3	0.100	0.100	0.200	2.48×10^{-3}
4	0.200	0.100	0.100	1.24×10^{-3}

(i) Determine the order of reaction with respect to A, B and C. Show how you obtain your answers. [3]

(ii) Write the rate equation for the reaction. [1]

(iii) Which reactant is unlikely to be in the rate determining step? [1]

[Total: 8]

4 Nitrogen pentoxide decomposes on heating to form nitrogen tetroxide and oxygen. The equation for the reaction is:

$N_2O_5(g) \rightarrow N_2O_4(g) + \frac{1}{2}O_2(g)$

The progress of the reaction can be followed by measuring the concentration of N_2O_5 present. The graph shows the results obtained in an experiment conducted at constant temperature.

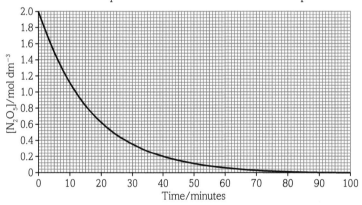

(a) (i) Use the graph to predict the rate of reaction after 20 minutes and after 90 minutes. [3]

(ii) Deduce the rate of production of oxygen after 20 minutes? [1]

(b) (i) Plot on the graph two successive half-lives for this reaction. [2]

(ii) Deduce the order of reaction with respect to N_2O_5. Justify your answer. [2]

(iii) Write a rate equation for the reaction. [1]

(c) (i) Calculate a value for the rate constant, k, for this reaction. [3]

(ii) Deduce the initial rate of reaction when the initial concentration of N_2O_5 is 1.50 mol dm^{-3}. Assume the reaction is carried out under the same conditions of temperature and pressure. [2]

[Total: 14]

5 Nitrogen dioxide, NO_2, is decomposed on heating into nitrogen monoxide, NO, and oxygen. The proposed mechanism for this reaction is:

$NO_2 + NO_2 \rightarrow NO_3 + NO$ SLOW
$NO_3 \rightarrow NO + O_2$ FAST

(a) (i) Write an overall equation for the decomposition of nitrogen dioxide. [1]

(ii) Write a rate equation for the reaction. [1]

(b) The rate constant, k, for the decomposition was determined at several different temperatures. The results obtained were used to plot the graph shown below.

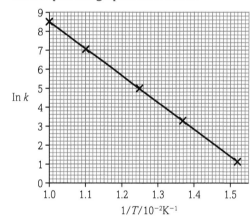

(i) Use the graph and the expression

$\ln k = -\frac{E_a}{R}\frac{1}{T} + \ln A$

to calculate the activation energy, E_a, for the thermal decomposition of nitrogen dioxide. Give your answer to an appropriate number of significant figures. [4]

(ii) A vessel of volume 2.00 dm^3 is filled with 4.00 mol of ntitrogen dioxide at a temperature of 650 K. The initial rate of reaction was found to be 12.64 mol dm^{-3} s^{-1}. Calculate a value for the rate constant, k, at this temperature. [3]

[Total: 9]

TOPIC 17
Further organic chemistry

Introduction

In **Book 1**, you learned about the simplest hydrocarbons – alkanes and alkenes – and how to name them using IUPAC rules. You then considered halogenoalkanes and alcohols, including important practical techniques used to convert one organic compound into another. You learned how the analytical techniques of mass spectrometry and infrared spectroscopy are used to determine the structures of organic compounds.

In this topic, you will extend your understanding of organic chemistry by considering compounds with functional groups such as aldehydes, ketones, carboxylic acids, acyl chlorides and esters. You will also learn about aromatic compounds and amines, and condensation polymers such as polyesters and polyamides. Amino acids and proteins will be introduced, together with reaction schemes that enable one compound to be converted into another in a series of steps. Two more modern analytical techniques – nuclear magnetic resonance spectroscopy and chromatography – will be introduced.

Here are some important areas of organic chemistry you will encounter:

- How measurements of optical activity can be used as evidence for organic reaction mechanisms
- How simple chemical tests can be used to identify different functional groups
- The uses of esters in solvents, food flavourings and perfumes
- How common medicines such as aspirin and paracetamol can be made.

All the maths you need

- Use percentages and ratios to calculate empirical formulae
- Translate information between graphical, numerical and algebraic forms when analysing spectra
- Visualise and represent 2D and 3D forms when distinguishing between the different types of isomerism, identifying chiral centres and drawing structures of isomers

What have I studied before?

- How to use different kinds of formula to represent organic compounds
- Using IUPAC rules to name organic compounds
- Recognising different types of isomerism, including geometrical isomerism
- How to convert one organic compound into another
- How to write reaction mechanisms
- How to use mass spectrometry and infrared spectroscopy to determine the structures of organic compounds

What will I study later?

- How pharmaceutical companies invest heavily in research and development to design new medicines to treat a wide range of medical problems and illnesses
- How biocatalysts based on natural enzymes allow some reactions to occur at lower temperatures and pressures than other catalysts
- How scientists are developing superconductors (materials with no electrical resistance that offer enormous energy savings in electricity transmission) based on organic compounds rather than metals
- Developments involving plastics in the ever more important field of solar cell technology, which may lead to the cost of manufacture falling and efficiency rising, with particular importance for developing countries

What will I study in this topic?

- Chirality and optical isomerism
- Examples of converting one organic compound into another
- Different types of reaction mechanism
- How aromatic compounds are different from aliphatic compounds
- The similarities between manufacturing polyamides and the formation of proteins from amino acids
- The analytical technique of nuclear magnetic resonance spectroscopy

17.1 1 Chirality and enantiomers

By the end of this section, you should be able to...

- recognise that optical isomerism is a result of chirality in molecules with a single chiral centre
- understand that optical isomerism results from chiral centre(s) in a molecule with asymmetric carbon atom(s) and that optical isomers are object and non-superimposable mirror images

Different types of isomerism

In **Book 1**, you learned about structural isomerism and its division into chain isomerism and position isomerism. You then considered a different type of isomerism called 'geometric isomerism', including the terms *E-Z* and *cis-trans* isomerism – this is one type of stereoisomerism.

This chapter looks at a second type of stereoisomerism – optical isomerism. Before we introduce this, it is useful to consider all the types of isomerism that you need to be familiar with, in the form of a family tree.

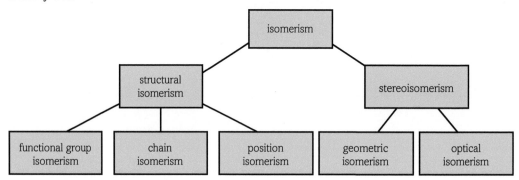

fig A The isomerism family tree.

Structural isomers are compounds with the same molecular formula but with different structural formulae.

Functional group isomers differ because they have different functional groups (e.g. aldehydes and ketones, alcohols and ethers).

Chain isomers differ because they have different patterns of branching in their carbon chains.

Positional isomers differ because the same functional group is attached to different carbon atoms in the chain.

Stereoisomers have the same structural formula but differ because their atoms or groups are arranged differently in three dimensions.

Geometric isomers differ because their atoms or groups are attached at different positions on opposite sides of a C=C double bond.

Optical isomerism

How does optical isomerism fit into this family tree? Optical isomers are non-superimposable mirror images of each other but are otherwise the same. This needs careful explanation!

Chirality

This is a term derived from the Greek word for 'hand', and could be translated as 'handedness'. Many objects, including human hands, can be described as **chiral**. Your hands have the same features and you could describe both hands using the same words. Both consist of a palm with a thumb on one side, and four fingers of different sizes. However, if you place one hand on top of the other, you can see that they are different – you cannot superimpose one hand on the other.

Now put your hands together – you can see that the thumbs are together, the little fingers are together, and so on – everything corresponds. Next, put your left hand in front of a mirror, then look at the image in the mirror and your right hand at the same time – they should look identical, so your hands can be described as mirror images, or sometimes as an object and its mirror image.

Chirality in simple molecules

The key to understanding chirality in molecules is to visualise them in 3D, and ideally use models to help you. Consider a molecule consisting of a single carbon atom joined to four different groups or atoms (represented by W, X, Y and Z). If you represent the molecule in two dimensions, with bond angles of 90°, you cannot easily see why there should be two different arrangements. However, you may remember from **Topic 2** on the shapes of molecules that you can attempt to show the 3D nature of molecules using different styles to represent the bonds. Ordinary lines are used to represent bonds in the plane of the paper, tapered or wedge-shaped lines indicate those above the plane of the paper, and dashed lines indicate those below the plane of the paper. Drawing the two structures side-by-side clearly shows how they relate to each other as object and mirror image.

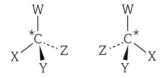

fig B This is one way to show optical isomers in 3D.

The asterisk indicates that the carbon atom next to it is a chiral centre, also known as an **asymmetric** carbon atom.

Notice that if the attached atoms or groups are W, W, X and Y (so two are the same), then although it is possible to show that they relate to each other in the same way (they are mirror images), there is no chirality because one can also be superimposed on the other.

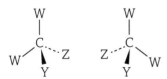

fig C These structures are not optical isomers because there are only three different groups joined to the central carbon atom.

How to identify chiral centres in structural formulae

You need to be able to identify chiral centres in molecules such as the examples shown in the following table – all you need to consider is whether there is a carbon atom joined to four different atoms or groups.

Structure	Comments
CBr_2ClF	The single carbon atom has only three different atoms attached, so there is no chiral centre.
$CH_3CHBrCH_2CH_3$	Carbons 1, 3 and 4 in the chain have two or more hydrogens attached so are not chiral. Carbon 2 is joined to four different groups (methyl, hydrogen, bromine and ethyl) so it is a chiral centre.
$CH_3CH_2CHBrCH_2CH_3$	Carbons 1, 2, 4 and 5 in the chain have two or more hydrogens attached so are not chiral. Carbon 3 is joined to only three different groups (two ethyl, one hydrogen and one bromine) so it is not a chiral centre.
$CH_3CH_2CHBrCHBrCH_3$	Carbons 1, 2, and 5 in the chain have two or more hydrogens attached so are not chiral. Carbon 3 is joined to four different groups (ethyl, hydrogen, bromine and $CHBrCH_3$), so it is a chiral centre. Carbon 4 is joined to four different groups (methyl, hydrogen, bromine and CH_3CH_2CHBr), so it is also a chiral centre.

table A

If a molecule has a chiral centre, then it can exist as optical isomers – each one is known as an **enantiomer**.

Learning tip

It is really worthwhile using molecular models to make some models of simple structures to check the idea of mirror images and superimposability.

Questions

1. What is the main similarity and the main difference between optical isomers and geometric isomers?

2. How many chiral centres are there in each of these molecules?
   (a) $CH_3CCl_2CH_3$
   (b) $CH_3CH(OH)COOH$
   (c) $CH_2Cl–CHCl–CHFCH_3$

Key definitions

Chiral refers to an atom in a molecule that allows it to exist as non-superimposable forms. It can also refer to the molecule itself.

Asymmetric refers to a carbon atom in a molecule that is joined to four different atoms or groups.

Enantiomers are isomers that are related as object and mirror image.

17.1 2 Optical activity

By the end of this section, you should be able to...

- understand that optical activity is the ability of a single optical isomer to rotate the plane of polarisation of plane-polarised monochromatic light in molecules containing a single chiral centre
- understand the nature of a racemic mixture

Plane-polarised light

In the previous section, we looked at the causes of optical isomerism. Before we consider **optical activity**, we must have some understanding of **plane-polarised light**. One way to consider light is as electromagnetic radiation that travels as a transverse wave. This means that the oscillations exist in planes at right angles to the direction of travel. **Fig A** shows 'normal' **unpolarised light** and plane-polarised light with the oscillations in only one plane.

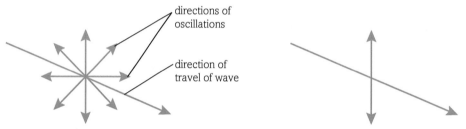

fig A In an unpolarised wave, oscillations may occur in any plane, while in a plane-polarised wave they occur in only one plane.

Some natural materials, and synthetic materials such as Polaroid (the material used in some sunglasses), can absorb all of the oscillations except those in a single plane, and so convert unpolarised light into plane-polarised light. For convenience, the single plane that remains is often assumed to be the vertical plane, as shown in **fig B**.

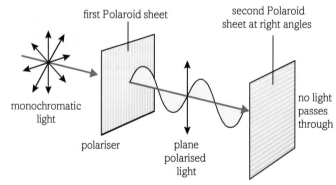

fig B Polarisation of light.

Fig B shows that the second sheet of Polaroid has horizontal lines, which means that only horizontally plane-polarised light can pass through. As the first sheet has produced vertically plane-polarised light, then no light can pass through the second sheet.

Polarimetry

Polarimetry is the use of a **polarimeter** to measure the amount of optical activity, if any, of a substance. **Fig C** shows how a polarimeter works. A monochromatic light source (light of only one colour or frequency) passes through a polarising filter (such as a piece of Polaroid). This is called the **polariser** because it converts unpolarised light into (vertically) plane-polarised light. The plane-polarised light then passes through a sample tube containing some of the substance in solution. If the substance is optically active (because it contains an enantiomer), then the plane of polarisation will be rotated so that it is no longer vertical. In this example, the rotation is clockwise, and the substance can therefore be described as 'dextrorotatory'. If the rotation is anticlockwise, then the substance is 'laevorotatory'.

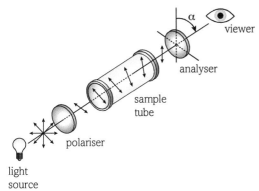

fig C You can see how the plane of polarisation of the polarised light is rotated clockwise as it passes through the sample tube.

The second polarising filter, known as the **analyser**, is rotated to a position where the maximum light intensity can be seen. The angle of rotation (α) is measured, and is quoted as a positive value if the rotation is clockwise and as a negative value if the rotation is anticlockwise. In **fig C**, the rotation, α, is approximately +60°.

Properties of enantiomers

Two enantiomers have identical physical properties, with one exception. The enantiomers rotate the plane of polarisation of plane-polarised light by equal angles, but in opposite directions. So, if the rotation for one enantiomer is +60°, then the rotation for the other enantiomer is −60°.

Two enantiomers have identical chemical properties, with one exception – the way in which they react with enantiomers of other substances. This chemical property may be different for each enantiomer.

Racemic mixtures

What happens if a compound with a chiral centre is present as a mixture of both enantiomers? You can imagine that the dextrorotatory and laevoratatory enantiomers have equal but opposite effects on plane-polarised light, and so the analyser does not need to be rotated to allow the maximum light intensity to be seen. Without knowing anything about the substance, it is not possible to distinguish between a substance that has no optical activity and one that has optically active enantiomers whose effects cancel out. A mixture containing equal amounts of two enantiomers is called a **racemic mixture**.

We cannot explain the origin of every scientific term in this book, but let us try to provide a brief explanation of this one! The Latin word *racemus* means a bunch of grapes. What has this to do with optical activity? The French chemist Louis Pasteur achieved much in his lifetime (1822–1895), and his name lives on in the English word 'pasteurisation', often associated with milk production. Another of his achievements was to realise that tartaric acid produced during winemaking showed optical activity, but that tartaric acid produced in other ways did not. He eventually realised that a single enantiomer of tartaric acid was produced from winemaking and that a mixture of both enantiomers of tartaric acid was produced in the other ways. Since tartaric acid is present in grapes, the term 'racemic acid' was used, and the 'racemic' part of this term was then used to describe an equimolar mixture of the two enantiomers. Voilà!

Here are the structures of the two enantiomers of tartaric acid:

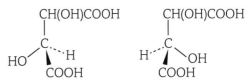

fig D The two enantiomers of tartaric acid.

Learning tip

It is worth spending some time making sure you understand the differences between these terms:

unpolarised, polariser, plane-polarised, plane of polarisation.

Questions

1 Outline how a polariser works.

2 A dextrorotatory enantiomer has a rotation of +43°. A mixture of this enantiomer and its laevorotatory enantiomer has a rotation of −10°. What does this information indicate about the composition of this mixture?

Key definitions

A substance shows **optical activity** if it rotates the plane of polarisation of plane-polarised light.

Plane-polarised light is monochromatic light that has oscillations in only one plane.

Unpolarised light has oscillations in all planes at right angles to the direction of travel.

A **polarimeter** is the apparatus used to measure the angle of rotation caused by a substance.

A **polariser** is a material that converts unpolarised light into plane-polarised light.

An **analyser** is a material that allows plane-polarised light to pass through it.

A **racemic mixture** is an equimolar mixture of two enantiomers that has no optical activity.

17.1 | 3 | Optical activity and reaction mechanisms

By the end of this section, you should be able to...

- use data on optical activity of reactants and products as evidence for S_N1 and S_N2 mechanisms

Nucleophilic substitution in halogenoalkanes

In **Book 1**, you learned to write mechanisms for these reactions, but the story was taken only so far. There was a suggestion that the explanation given in **Book 1** was incomplete and that there was more to come.

Now is the appropriate time to develop the explanation in greater depth. First, let us revise what you may remember from **Book 1**. **Fig A** shows how we represented the mechanism of the reaction between the primary halogenoalkane bromoethane and hydroxide ions.

fig A This is a simple way to represent a nucleophilic substitution mechanism.

We must now accept that this mechanism is an oversimplification and that we need to consider two alternative mechanisms. This mechanism shows the simultaneous movement of two pairs of electrons – those from the O of OH^- ions and those from the C—Br bond. In fact, there are two distinct mechanisms.

The S_N2 mechanism

The individual symbols in 'S_N2' have the following meanings.

S = substitution   N = nucleophilic   2 = **bimolecular** and second order

The term 'bimolecular' means that there are two species involved in the rate-determining step – in this example, the OH^- ion and the halogenoalkane.

Second order reactions were considered in **Topic 16** in this book. In this topic, we are going to revisit the mechanism provided in **Book 1**, by looking at it in more detail. Here is the mechanism for a primary halogenoalkane that is also an enantiomer.

fig B This is the S_N2 mechanism showing the transition state.

Notice that the curly arrows are used in the same way. The mechanism now includes a transition state showing the HO—C bond forming and the C—Br bond breaking. The dashed lines represent partial bonds – those being broken and formed. The most important thing to notice is that the arrangement of the three groups that remain attached to the central carbon atom has been inverted (turned inside out).

Learning tip

Remember to include square brackets and a negative charge outside the brackets when writing the transition state in the S_N2 mechanism. Using $\delta+$ and $\delta-$ to indicate the polar bonds is good practice, but in this section, they have been omitted for clarity.

Evidence from optical activity

If the reaction actually occurs according to the S_N2 mechanism, then, because of the inversion that occurs, the optical activity of the product is different from that of the reactant. If the reactant is dextrorotatory, then the product will be laevorotatory, and *vice versa*. So, by measuring the optical activity of the original halogenoalkane and the alcohol formed, we can show whether the reaction has occurred by the S_N2 mechanism.

The S_N1 mechanism

The individual symbols in 'S_N1' have the following meanings.

S = substitution   N = nucleophilic   1 = **unimolecular** and first order

The term 'unimolecular' means that only one species is involved in the rate-determining step – in this example, just the halogenoalkane. We have used the same halogenoalkane as for the S_N2 mechanism so that you can focus on the differences between the two mechanisms.

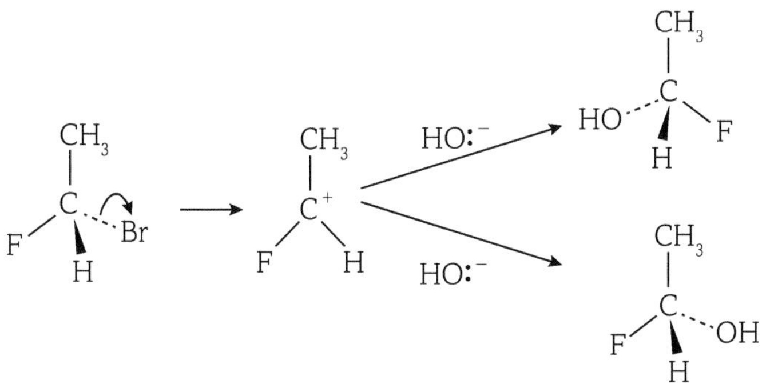

fig C In the S_N1 mechanism, the planar carbocation can be attacked from the left or from the right, giving a pair of enantiomers.

The original halogenoalkane has a tetrahedral shape, but the product of the first step of the reaction is a carbocation with a planar shape. This means that in the second step of the reaction, there is an equal chance that the attacking HO⁻ ion can approach from the left or from the right. There are therefore two products, both of which are enantiomers. These are present in equal numbers, so a racemic mixture is formed.

Evidence from optical activity

If the reaction actually occurs according to the S_N1 mechanism, then the optical activity of the product is different from that of the reactant. Whether the reactant is dextrorotatory or laevorotatory, the product has no optical activity. So, by measuring the optical activity of the original halogenoalkane and the alcohol formed, we can show whether the reaction has occurred by this mechanism.

Which mechanism?

In general, the main mechanism for primary halogenoalkanes is S_N2, and for tertiary halogenoalkanes is S_N1. For secondary halogenoalkanes, both mechanisms can occur, although the main one depends on the actual halogenoalkane and the conditions.

Questions

1. 2-bromobutane and cyanide ions react together by an S_N2 mechanism. Draw the structure of the transition state formed in this reaction.

2. 2-bromopropane and cyanide ions react together by an S_N1 mechanism. Explain why there is no change in optical activity in this reaction.

Key definitions

A mechanism described as **bimolecular** has two species reacting in the rate-determining step.

A mechanism described as **unimolecular** has one species reacting in the rate-determining step.

17.2 1 Carbonyl compounds and their physical properties

By the end of this section, you should be able to...

- identify the aldehyde and ketone functional groups
- understand that aldehydes and ketones:
  (i) do not form intermolecular hydrogen bonds, and this affects their physical properties
  (ii) can form hydrogen bonds with water, and this affects their solubility

Nomenclature

We met carbonyl compounds briefly in **Book 1**, but only as the products of the oxidation of alcohols.

The carbonyl group (C=O) is a carbon atom joined by a double bond to an oxygen atom. Many types of compounds contain this group (such as carboxylic acids which contain COOH), but only two types are classified as carbonyl compounds – these are aldehydes and ketones.

Aldehydes

A carbonyl compound is described as an aldehyde if there is a hydrogen atom joined to the carbonyl group. The naming system uses the suffix -*al*, with the carbon chain being named as for alkanes. If locants are used, then the carbon of the carbonyl group is carbon 1 in the chain. **Table A** shows some examples.

Structural formula	Name	Common name
HCHO	methanal	formaldehyde
CH_3CHO	ethanal	acetaldehyde
CH_3CH_2CHO	propanal	(propionaldehyde)
$(CH_3)_2CHCHO$	methylpropanal	(isobutyraldehyde)
$CH_3CH_2CH_2CHO$	butanal	(butyraldehyde)
C_6H_5CHO	benzenecarbaldehyde	benzaldehyde

table A

The general formula is RCHO.

Ketones

A carbonyl compound is described as a ketone if there are only hydrocarbon groups joined to the carbonyl group. The naming system uses the suffix -*one*, with the carbon chain being named as for alkanes. If locants are used, then one of the end carbon atoms in the chain of the carbonyl group is carbon 1 in the chain. The rules about using the lowest possible numbers, as with alkanes, apply to ketones. **Table B** shows some examples.

Structural formula	Name	Common name
CH_3COCH_3	propanone	acetone
$CH_3COCH_2CH_3$	butanone	(methyl ethyl ketone)
$CH_3COCH_2CH_2CH_3$	pentan-2-one	(methyl propyl ketone)
$CH_3CH_2COCH_2CH_3$	pentan-3-one	(diethyl ketone)
$(CH_3)_2CHCOCH_3$	3-methylbutan-2-one	(methyl isopropyl ketone)

table B

The general formula is RCOR'.

For both aldehydes and ketones, the common names in brackets are rarely used.

Bonding

You are familiar with the nature of the C=C double bond in alkenes. It is composed of a combination of sigma and pi bonding, but the bond is non-polar because there are two identical (carbon) atoms involved. The situation in the carbonyl group is very similar, except that the bond is polar because of the differing electronegativities of carbon and oxygen. **Fig A** compares the two.

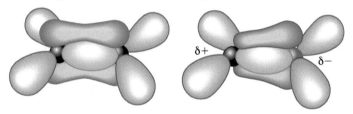

fig A The bonding in an alkene is on the left. The bonding in a carbonyl compound is on the right. Green is used to represent sigma bonding and blue is used to represent pi bonding.

Note that the electron density in the pi bond of the alkene is evenly distributed across both carbon atoms, but that in the carbonyl compound the electron density is greater near the oxygen atom than near the carbon atom.

Physical properties

Aldehydes have distinctive smells. Those with short carbon chains have unpleasant smells (butanal smells of rancid butter). Some of those with long carbon chains have very pleasant smells and are used in expensive perfumes (nonanal, with nine carbon atoms per molecule, smells of roses). The smells of the lower ketones, especially propanone, remind most people of solvents. In fact, propanone is widely used as a solvent, including as nail varnish remover.

fig B Cinnamon (which contains the aldehyde cinnamaldehyde) comes from the dried bark of the tree *Cinnamomum zeylanicum*. It is widely used as a spice in cookery.

Boiling temperatures

Alkanes have only weak intermolecular forces (London forces) because they are non-polar. Alcohols have strong intermolecular forces (hydrogen bonding) because they have a very polar O—H group.

Aldehydes and ketones have intermolecular forces of intermediate strength. They contain the polar C=O group and so have permanent dipole–dipole attractions. They do not have hydrogen bonding because all of their hydrogen atoms are joined to carbon atoms. As with alkanes and alcohols, boiling temperatures increase with increasing molar mass as the extent of London forces increases. At room temperature, methanal is a gas. The other carbonyl compounds are liquids. **Fig C** compares the boiling temperatures of some carbonyl compounds.

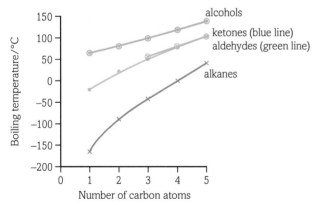

fig C The boiling temperatures of some alkanes, aldehydes, ketones and alcohols. The effect of the intermolecular forces on the boiling temperatures can be seen clearly, with dipole–dipole attractions, which result from the carbonyl groups of the aldehydes and ketones, causing higher boiling temperatures than the equivalent alkanes but lower than the alcohols with their hydrogen bonding.

Solubility in water

The lower aldehydes and ketones are soluble in water because they can form hydrogen bonds (shown as a dashed line) with water molecules. **Fig D** shows hydrogen bonding between ethanal and water.

fig D Hydrogen bonding between water molecules and ethanal molecules is stronger than the intermolecular bonds in the pure aldehyde.

The solubility of aldehydes and ketones decreases with increasing chain length, as the hydrocarbon part of the molecules becomes more significant.

Learning tip

When naming ketones with a chain of five or more carbon atoms, a locant is needed to show the position of the carbonyl group. When naming aldehydes, a locant is never used to show the position of the carbonyl group because the CHO group must be at the end of the chain.

Questions

1. Write the displayed formula for:
   (a) propanal
   (b) propanone.

2. Draw a diagram to explain why propanone is soluble in water.

17.2 2 Redox reactions of carbonyl compounds

By the end of this section, you should be able to...

- understand the reactions of carbonyl compounds with:
  (i) Fehling's or Benedict's solution, Tollens' reagent and acidified dichromate(VI) ions
  (ii) lithium tetrahydridoaluminate (lithium aluminium hydride) in dry ether
  (iii) iodine in the presence of alkali

Which reactions occur?

You may remember from **Book 1** that carbonyl compounds (formed by the oxidation of alcohols) differ in their ease of oxidation – aldehydes (but not ketones) are easily oxidised to carboxylic acids. In this section, we will look at these reactions in more detail, and also consider the reduction reactions of carbonyl compounds. **Fig A** summarises these reactions.

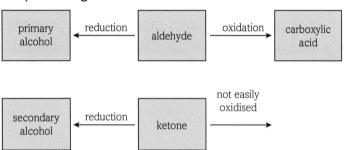

fig A Aldehydes can be reduced and oxidised, but ketones can only be reduced.

Reduction reactions

Both aldehydes and ketones can be reduced to alcohols by the same reagent, lithium tetrahydridoaluminate. This reagent is also known as lithium aluminium hydride, and its formula is $LiAlH_4$. The reduction reactions are carried out with both the carbonyl compound and reducing agent dissolved in dry ether. To simplify the equations, the reducing agent is represented by [H], which is the symbol for an atom of hydrogen provided by the reducing agent in an unspecified way. Note that this is a different use of square brackets – it does not indicate a complex (as with transition metals), and it does not indicate a molar concentration (as in kinetics). Here are some examples of equations.

Reduction of propanal to propan-1-ol:
$$CH_3CH_2CHO + 2[H] \rightarrow CH_3CH_2CH_2OH$$

Reduction of butanone to butan-2-ol:
$$CH_3COCH_2CH_3 + 2[H] \rightarrow CH_3CH(OH)CH_2CH_3$$

In all of these reduction reactions, a hydrogen atom becomes attached to each of the atoms in the carbonyl group.

Oxidation reactions

Ketones are very difficult to oxidise, and as far as this book is concerned, ketones cannot be oxidised. **Table A** shows the reagents that can oxidise aldehydes to carboxylic acids.

Reagent	Colour change	Notes
Acidified potassium dichromate(VI)	orange solution → green solution	The colour change is due to the reduction of chromium from +6 to +3.
Fehling's solution	deep blue solution → red precipitate	In both of these solutions, the colour change is due to the conversion of a copper(II) complex to copper(I) oxide. It is not necessary to understand the difference between the chemical compositions of these two reagents.
Benedict's solution	deep blue solution → red precipitate	
Tollens' reagent	colourless solution → silver mirror	The colour change is due to the conversion of a silver(I) complex to metallic silver, which often sticks to the inside of the tube. The reagent is also described as ammoniacal silver nitrate.

table A

All of these oxidation reactions can be carried out in the laboratory in a test tube, by mixing the carbonyl compound with the oxidising agent and leaving the mixture in a beaker of hot water.

As for the reduction reaction equations, all of the oxidising agents in **table A** can be represented by [O]. This is the symbol for an atom of oxygen provided by the oxidising agent in an unspecified way. Here are some examples of equations.

Oxidation of propanal to propanoic acid:
$$CH_3CH_2CHO + [O] \rightarrow CH_3CH_2COOH$$

Oxidation of butanal to butanoic acid:
$$CH_3CH_2CH_2CHO + [O] \rightarrow CH_3CH_2CH_2COOH$$

Distinguishing between aldehydes and ketones

The oxidation reactions using Fehling's solution, Benedict's solution and Tollens' reagent are often used to distinguish between aldehydes and ketones. A positive result, as described in **table A**, indicates the presence of an aldehyde. A negative result indicates the absence of an aldehyde. If it is also known that the substance being tested is a carbonyl compound, then the negative result can be taken as confirmation of the presence of a ketone.

fig B Colour change of Fehling's solution from a blue solution to a red precipitate in the test for an aldehyde.

fig C Colour change of Tollens' reagent from a colourless solution to a silver mirror in the test for an aldehyde.

Reactions with iodine

Another redox reaction of some carbonyl compounds is rather different – it used to be known as the iodoform reaction, but is now described as the triiodomethane reaction. This is because the product is CHI_3, a yellow insoluble solid.

If a carbonyl compound is added to an alkaline solution of iodine and the mixture warmed and then cooled, a pale yellow precipitate sometimes forms. This result is a positive test for carbonyl compounds containing the CH_3CO group. This group is found in only one aldehyde (ethanal) and in all methyl ketones (those with the carbonyl group as the second carbon in the chain). This means that the reaction occurs with propanone, butanone and pentan-2-one, but not pentan-3-one.

Because oxidising conditions are used in the reaction, this means that some alcohols will also give positive results. **Table B** shows the results for some common alcohols.

Alcohol	Oxidation product	Result
ethanol	CH_3CHO	positive
propan-1-ol	CH_3CH_2CHO	negative
propan-2-ol	CH_3COCH_3	positive
butan-1-ol	$CH_3CH_2CH_2CHO$	negative
butan-2-ol	$CH_3COCH_2CH_3$	positive
2-methylpropan-1-ol	$(CH_3)_2CHCHO$	negative

table B

Learning tip

The reduction reactions are included to show the conversion of aldehydes and ketones to alcohols.

The oxidation reactions are included to show how to distinguish between aldehydes and ketones.

Questions

1. What are the names of the organic products of these reactions?
   (a) The reduction of $CH_3CH_2CH_2CHO$.
   (b) The oxidation of $CH_3CH_2CH_2CH_2CHO$.

2. What colour changes would be observed in these tests?
   (a) Heating a mixture of propanone and acidified potassium dichromate(VI).
   (b) Heating a mixture of ethanal and Tollens' reagent.

17.2 3 Nucleophilic addition reactions

By the end of this section, you should be able to...

- understand the reactions of carbonyl compounds with:
  (i) HCN, in the presence of KCN, as a nucleophilic addition reaction, using curly arrows, relevant lone pairs, dipoles and evidence of optical activity to show the mechanism
  (ii) 2,4-dinitrophenylhydrazine (2,4-DNPH), as a qualitative test for the presence of a carbonyl group and to identify a carbonyl compound given melting temperatures of derivatives data

The reaction with hydrogen cyanide

The reagent is hydrogen cyanide, and the reaction is carried out in an aqueous alkaline solution containing potassium cyanide (KCN). The equations and mechanisms are the same for both aldehydes and ketones. Here are some examples of equations.

Addition to propanal:

$CH_3CH_2CHO + HCN \rightarrow CH_3CH_2CH(OH)CN$

Addition to butanone:

$CH_3COCH_2CH_3 + HCN \rightarrow CH_3C(CN)(OH)CH_2CH_3$

These addition reactions involve a hydrogen atom attaching to the oxygen atom of the carbonyl group and a cyanide group attaching to the carbon atom of the carbonyl group.

These products contain a carbon atom joined to the functional groups OH and CN. They are named somewhat differently from other organic compounds we have met so far. According to IUPAC rules, the OH group should be shown as a prefix and the CN group should be shown as a suffix. When this happens, the OH group is shown as *hydroxy-* and not as *-ol*, and the CN group is shown as the ending *-nitrile*. In general, these products are described as hydroxynitriles.

Applying these naming rules:

$CH_3CH_2CH(OH)CN$ is 2-hydroxybutanenitrile

$CH_3C(CN)(OH)CH_2CH_3$ is 2-hydroxy-2-methylbutanenitrile

The nucleophilic addition mechanism

Nucleophilic addition is a two-step mechanism, illustrated using ethanal as an example.

Step 1 involves the nucleophilic attack by a cyanide ion on the carbon atom of the carbonyl compound.

Step 2 involves the reaction between the intermediate and a hydrogen cyanide molecule.

Using optical activity as evidence

A consideration of optical activity can be used as evidence for this mechanism. The arrangement of the two atoms or groups joined to C=O is planar, which means that in Step 1 there is an equal chance of the cyanide ion attacking from each side of the plane. So, although the product of this reaction contains an asymmetric carbon atom (a chiral centre), there are equal amounts of both dextrorotatory and laevorotatory enantiomers formed, giving a product mixture with no optical activity.

Using ethanal as the example, the product is 2-hydroxypropanenitrile. **Fig A** shows the formation of a racemic mixture containing equal numbers of both enantiomers.

fig A Attack by the cyanide ion from the top leads to the product shown higher up, while attack from the bottom leads to the product shown lower down. Each ethanal molecule is attacked by only one of these cyanide ions.

The cyanide ion can attack from above or below, producing equal amounts of both enantiomers.

The reaction with 2,4-dinitrophenylhydrazine

This compound is sometimes known as Brady's reagent, and its name is often abbreviated as 2,4-DNPH. You do not need to write equations or mechanisms for any reactions involving this reagent, but it is useful to show its structure, and the structure of a product of its reaction with a carbonyl compound. **Fig B** shows its reaction with propanal.

Carbonyl compounds | 17.2

fig B The reaction between 2,4-dinitrophenylhydrazine and propanal.

The reagent reacts with most carbonyl compounds to form a brightly coloured orange solid, so the appearance of an orange precipitate is a positive result that indicates the presence of a carbonyl compound.

Melting temperatures of derivatives

As well as providing a visual indication of a positive result for the presence of a carbonyl compound, these reactions can also be used to identify individual carbonyl compounds. This is because the compounds formed, known as **derivatives** (because they are derived from particular carbonyl compounds), can be filtered, purified and dried, and their melting temperatures measured. These derivatives have the ending *-one*, so the derivative of butanal is butanal 2,4-dinitrophenylhydrazone.

By comparing the experimental values of melting temperatures of these derivatives with data book values, the identity of the original carbonyl compound can often be confirmed. For example, the three carbonyl compounds pentanal, pentan-2-one and pentan-3-one all have the same boiling temperatures and all form coloured precipitates with 2,4-dinitrophenylhydrazine. **Table A** shows the boiling temperatures of the carbonyl compounds and the melting temperatures of their derivatives.

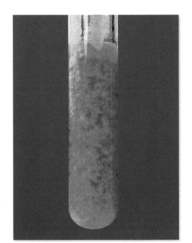

fig C This is the bright orange precipitate formed when a carbonyl compound reacts with 2,4-dinitrophenylhydrazine.

Carbonyl compound	Boiling temperature/°C	Melting temperature of 2,4-dinitrophenylhydrazone derivative/°C
pentanal	102	104
pentan-2-one	102	144
pentan-3-one	102	156

table A

You can see that the original carbonyl compound can now be identified by matching the experimental melting temperature of its derivative to a value in the table.

Learning tip

Remember that when a cyanide group becomes part of an organic compound, the number of carbon atoms increases by one, which affects the name. So, for example, propanal becomes 2-hydroxybutanenitrile.

Questions

1. What are the names of the organic products of these reactions?
   (a) methanal + hydrogen cyanide →
   (b) propanone + hydrogen cyanide →

2. Compound X reacts with 2,4-dinitrophenylhydrazine and with iodine in aqueous alkali in separate tests. In both cases a coloured precipitate forms. Explain which of these could be compound X:
   propanal, propan-1-ol, propan-2-ol, propanone

Key definitions

Nucleophilic addition is a type of mechanism in which a molecule containing two atoms or groups is added across a polar double bond (usually C=O), and the attacking species in the first step is a nucleophile.

Derivatives are compounds formed from other compounds, especially when the properties of the derivatives can be used to identify the original compound.

17.3 1 Carboxylic acids and their physical properties

By the end of this section, you should be able to...

- identify the carboxylic acid functional group
- understand that hydrogen bonding affects the physical properties of carboxylic acids, in relation to their boiling temperatures and solubility

Nomenclature

We met carboxylic compounds briefly in **Book 1**, as the products of the oxidation of primary alcohols.

The carboxylic acid group consists of a carbonyl group (C=O) with a hydroxyl group on the same carbon atom. Both groups influence each other, so carboxylic acids have some properties that are not shared with either carbonyl compounds or alcohols. The group is usually abbreviated to COOH (and sometimes to CO_2H, although this is not recommended).

The suffix *-oic acid* is used, with the carbon chain being named as for alkanes. If locants are used, then the carbon of the carboxylic acid group is carbon 1 in the chain. **Table A** shows some examples.

Structural formula	Name	Common name
HCOOH	methanoic acid	formic acid
CH_3COOH	ethanoic acid	acetic acid
CH_3CH_2COOH	propanoic acid	(propionic acid)
$CH_3CH_2CH_2COOH$	butanoic acid	(butyric acid)
$(CH_3)_2CHCOOH$	methylpropanoic acid	(isobutyric acid)
HOOCCOOH	ethanedioic acid	oxalic acid
C_6H_5COOH	benzenecarboxylic acid	benzoic acid

table A

The general formula is RCOOH; the common names in brackets are rarely used.

Bonding

You are familiar with the nature of the C=O double bond in carbonyl compounds, and of the single bonds in alcohols. The bonding in the carboxylic acid group is a combination of the two. There are now three polar bonds present.

Many reactions of carboxylic acids involve the loss of the hydrogen atom as H^+. This produces the carboxylate ion (COO^-) in which the two carbon–oxygen bonds are identical. **Fig A** compares the two.

fig A The structures of a carboxylic acid and its carboxylate ion.

Note that the charge and double bond character are evenly distributed across both oxygen atoms in the carboxylate ion.

Physical properties

Carboxylic acids have distinctive smells and sour tastes. The smells of some of them are considered unpleasant. For example butanoic acid is responsible for the smell of stale sweat, and hexanoic acid for the characteristic smell of goats. Methanoic acid is found in ants and nettle stings (*formica*, the origin of the name formic acid, is the Latin word for ant). Citric acid gives lemons their very sour taste.

The Latin word for sour is *acidus*, and *acetum* means vinegar, the sour taste of which comes from the acetic acid it contains.

fig B Methanoic acid, HCOOH, makes up more than half of an ant's body mass!

Boiling temperatures

The presence of three polar bonds, including a polar O—H bond, in the carboxylic acid group means that they have strong intermolecular forces (hydrogen bonding). This means that they have high boiling temperatures compared to other organic compounds with a similar molar mass. As the carbon chain lengthens, the London forces between the non-polar hydrocarbon chains increase. Therefore, boiling temperature increases with increasing molar mass, as shown in **table B**.

Name	Formula	Molar mass/ g mol^{-1}	Boiling temperature /°C
methanoic acid	HCOOH	46	101
ethanoic acid	CH$_3$COOH	60	118
propanoic acid	CH$_3$CH$_2$COOH	74	141
butanoic acid	CH$_3$CH$_2$CH$_2$COOH	88	164

table B

The extent of hydrogen bonding in the shorter-chain carboxylic acids means that they form dimers (double molecules) in the absence of a solvent such as water. **Fig C** shows the hydrogen bonding (dashed lines) between two molecules of ethanoic acid.

fig C Hydrogen bonding in ethanoic acid.

Solubility in water

The shorter-chain carboxylic acids are soluble in water because they can form hydrogen bonds with water molecules. **Fig D** shows hydrogen bonding between ethanoic acid and water.

fig D Hydrogen bonding between ethanoic acid and water.

Solubility decreases with increasing chain length, as the hydrocarbon part of the molecules becomes larger.

Learning tip

Note that the hydrogen bonding between carboxylic acid and water molecules involves

$$-\underset{\underset{OH}{|}}{C}=O \text{ ---- } H-O-H \text{ and } -O-H \text{ ---- } O-H$$

where hydrogen bonds are represented by dashed lines.

Questions

1. Write the displayed formula for:
   (a) methanoic acid
   (b) ethanedioic acid.

2. Explain why hexanoic acid is much less soluble than ethanoic acid in water.

17.3 2 Preparations and reactions of carboxylic acids

By the end of this section, you should be able to...

- understand that carboxylic acids can be prepared by the oxidation of alcohols or aldehydes, and the hydrolysis of nitriles
- understand the reactions of carboxylic acids with:
  (i) lithium tetrahydridoaluminate (lithium aluminium hydride) in dry ether
  (ii) bases to produce salts
  (iii) phosphorus(V) chloride (phosphorus pentachloride)
  (iv) alcohols in the presence of an acid catalyst

Preparation by oxidation

Although carboxylic acids are formed in many chemical reactions, two main methods are used to prepare them in the laboratory.

Oxidation uses either a primary alcohol or an aldehyde as the starting material. The usual oxidising agent is acidified potassium dichromate(VI), and the method is to heat the mixture under reflux. If a primary alcohol is used, it first oxidises to an aldehyde, then to a carboxylic acid, but both oxidations occur inside the apparatus used.

When the oxidation is complete, the reaction mixture is fractionally distilled to obtain a pure sample of the carboxylic acid.

Examples

Propanoic acid can be prepared from either propan-1-ol or propanal. The equations use [O] to represent the oxygen supplied by acidified potassium dichromate(VI).

From propan-1-ol:
$$CH_3CH_2CH_2OH + 2[O] \rightarrow CH_3CH_2COOH + H_2O$$

From propanal:
$$CH_3CH_2CHO + [O] \rightarrow CH_3CH_2COOH$$

Preparation by hydrolysis

Nitriles are organic compounds containing the CN group. They can be hydrolysed by heating under reflux with either a dilute acid or aqueous alkali. The same apparatus is used for preparation and purification as for oxidation.

In both cases, the C≡N triple bond breaks. The carbon atom remains part of the organic product and the nitrogen atom becomes either ammonia or the ammonium ion.

Acidic hydrolysis

Propanoic acid can be prepared from propanenitrile. The equation for the reaction is:
$$CH_3CH_2CN + H^+ + 2H_2O \rightarrow CH_3CH_2COOH + NH_4^+$$

Alkaline hydrolysis

Butanoic acid can be prepared from butanenitrile. The equation for the reaction is:
$$CH_3CH_2CH_2CN + OH^- + H_2O \rightarrow CH_3CH_2CH_2COO^- + NH_3$$

The product is actually the butanoate ion, but this is easily converted to butanoic acid by adding dilute acid:
$$CH_3CH_2CH_2COO^- + H^+ \rightarrow CH_3CH_2CH_2COOH$$

Introduction to reactions

There are four main reactions to consider. These are summarised in **fig A**.

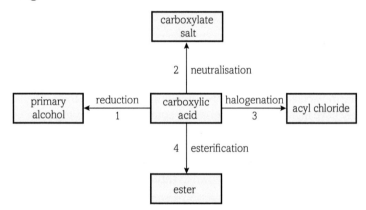

fig A These are the four main reactions of carboxylic acids to form other organic compounds.

Some of the products of these reactions, especially acyl chlorides and esters, will be discussed in more detail later on.

1 Reduction

You will remember that carboxylic acids are products of oxidation. They can be reduced in the same way as aldehydes and ketones. Carboxylic acids can be reduced to primary alcohols, but not to aldehydes. This is because aldehydes are more easily reduced than carboxylic acids, and so any aldehyde produced during the reduction will be immediately reduced to a primary alcohol. The reducing agent is lithium tetrahydridoaluminate, and is used in a solvent of dry ether. As an example, an equation using butanoic acid is:
$$CH_3CH_2CH_2COOH + 4[H] \rightarrow CH_3CH_2CH_2CH_2OH + H_2O$$

The product is butan-1-ol.

2 Neutralisation

Although carboxylic acids are weak acids, they can be completely neutralised by mixing with aqueous alkali. The products are carboxylate salts, which have a wide range of uses. The commonest, sodium ethanoate, is used in hand warmers and to make the additive that gives the flavour to salt and vinegar potato crisps. An equation for the reaction to make sodium ethanoate occurs between ethanoic acid and sodium hydroxide:

$$CH_3COOH + NaOH \rightarrow CH_3COONa + H_2O$$

Carboxylate salts are ionic, and are sometimes shown with formulae such as $CH_3COO^-Na^+$. Their general formula is RCOONa.

3 Halogenation

In this reaction, the OH group is replaced by a halogen atom, usually chlorine, so the functional group becomes COCl. These products are known as acyl chlorides. They are highly reactive compounds with many uses in organic synthesis, as we will see later on. The reagent is phosphorus(V) chloride (old name 'phosphorus pentachloride'). It must be used in anhydrous conditions because both the reagent and the acyl chloride product react with water. The reaction is vigorous, so no heating is required. An equation for the reaction, using propanoic acid as the example is:

$$CH_3CH_2COOH + PCl_5 \rightarrow CH_3CH_2COCl + POCl_3 + HCl$$

Acyl chlorides are named using the suffix -*oyl chloride*, so this product is propanoyl chloride. Their general formula is RCOCl.

The phosphorus-containing product is phosphorus trichloride oxide (also known as phosphorus oxychloride). It is a liquid that mixes with the acyl chloride, which has therefore to be separated by fractional distillation. The hydrogen chloride gas produced escapes, appearing as misty fumes.

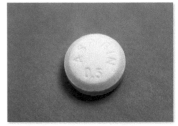

fig B Aspirin is one of the most familiar and inexpensive medications, but is useful in treating many medical conditions. It is an example of a carboxylic acid.

4 Esterification

In this reaction, a carboxylic acid is mixed with an alcohol and a small amount of an acid catalyst, often concentrated sulfuric acid. Even with a catalyst, the reactions are slow and reversible.

Esters are used in industry for many purposes, including as solvents and in making polymers (polyesters). They also occur very commonly in nature – they are responsible for the characteristic smells of fruits, and most animal fats and vegetable oils are esters.

An equation for the reaction between methanoic acid and ethanol is:

$$HCOOH + CH_3CH_2OH \rightleftharpoons HCOOCH_2CH_3 + H_2O$$

The names of esters always contain two words – the first is the alkyl group from the alcohol and the second is the carboxylate from the carboxylic acid. So, this product is ethyl methanoate.

Learning tip

With an ester, be careful to make sure that the name and formula match. CH_3COOCH_3 is methyl ethanoate but $HCOOCH_2CH_3$ is ethyl methanoate.

Questions

1 Write equations for the preparation of methylpropanoic acid:
   (a) by the oxidation of an alcohol
   (b) by the hydrolysis of a nitrile.

2 Write an equation for the conversion of methylpropanoic acid:
   (a) into an acyl chloride
   (b) into a methyl ester.

17.3 3 Acyl chlorides

By the end of this section, you should be able to...

- identify the acyl chloride functional group
- understand the reactions of acyl chlorides with:
    (i) water
    (ii) alcohols
    (iii) concentrated ammonia
    (iv) amines

fig A Adding ethanoyl chloride to water produces fumes of hydrogen chloride.

What are acyl chlorides?

These used to be known as acid chlorides, and have the general formula RCOCl. They are derivatives of carboxylic acids, and can be imagined as the result of replacing the OH group in COOH with Cl. Although, like carboxylic acids, aldehydes and ketones, they contain a carbonyl group, they are not classed as carbonyl compounds. As with carboxylic acids, having two functional groups sharing the same carbon atom means that the properties are different from both carbonyl compounds and halogenoalkanes.

The commonest example is ethanoyl chloride, CH_3COCl, which is still called by its old name acetyl chloride. You may come across others, such as:

	propanoyl chloride	CH_3CH_2COCl
and	butanoyl chloride	$CH_3CH_2CH_2COCl$

Reactions of acyl chlorides

The carbon atom in RCOCl is joined to two electronegative atoms and so is electron-deficient ($\delta+$). You can therefore predict that it will be readily attacked by nucleophiles, such as molecules containing O or N atoms. You need to know about four reactions of acyl chlorides, although no mechanisms are needed for any of them.

Reaction with water

Acyl chlorides react vigorously with cold water, forming a carboxylic acid and releasing hydrogen chloride gas, which appears as misty fumes. The equation for the reaction of ethanoyl chloride with water is as follows:

$$CH_3-COCl + H-OH \longrightarrow CH_3-COOH + H-Cl$$

Reaction with alcohols

Acyl chlorides react readily with ethanol to form an ester and hydrogen chloride gas. The equation for the reaction of ethanoyl chloride with ethanol is as follows:

$$CH_3-COCl + CH_3CH_2-OH \longrightarrow CH_3-COOCH_2CH_3 + H-Cl$$

Notice the similarity between these two reactions. In both cases, the H of the second reactant combines with Cl to form HCl, and the other part of the reactant becomes joined to the carbonyl group.

Reaction with concentrated ammonia solution

Similar to alcohols, acyl chlorides react readily with concentrated ammonia solution. You can now predict the reaction that occurs: the H of the second reactant combines with Cl to form HCl, and the other part of the reactant becomes joined to the carbonyl group. The NH_2 group joined to the carbonyl group produces a different functional group – amide (not to be confused with amine). The equation for the reaction of ethanoyl chloride with ammonia is as follows:

$$CH_3-COCl + NH_3 \longrightarrow CH_3-CONH_2 + H-Cl$$

Unlike with water and alcohols, a further reaction occurs. The reactant is a base and the product is an acidic gas, so these react together to form ammonium chloride:

$NH_3 + HCl \rightarrow NH_4Cl$

You could write a single equation that combines both of these reactions:

$CH_3COCl + 2NH_3 \rightarrow CH_3CONH_2 + NH_4Cl$

You will learn more about amides later in this book.

Reaction with amines

This reaction is similar to the previous one. A primary amine can be shown as RNH_2, so you can again predict the reaction that occurs: the H of the second reactant combines with Cl to form HCl, and the other part of the reactant (RNH) becomes joined to the carbonyl group. The equation for the reaction of ethanoyl chloride with the simplest primary amine, methylamine is as follows:

$$CH_3-COCl + CH_3NH_2 \longrightarrow CH_3-CONHCH_3 + H-Cl$$

The product of this reaction is a substituted amide, often called an N-substituted amide. The capital N emphasises that there is an alkyl group attached to N, rather than the usual situation in which it is attached to another C atom. You can see from the product name (N-methylethanamide) that N- is used as a locant, in the same way as in names such as 2-methylbutane – it shows where the methyl group is attached.

Secondary amines (represented as R_2NH) react in the same way, although the product will contain two substituted alkyl groups. A sample equation is:

$CH_3COCl + (CH_3)_2NH \rightarrow CH_3CON(CH_3)_2 + HCl$

The organic product is N,N-dimethylethanamide.

You might be able to predict why this type of reaction does not occur with a tertiary amine (R_3N). With three alkyl groups, there is now no H atom to react with Cl to form hydrogen chloride.

Learning tip
Be sure not to confuse amines (RNH_2) with amides ($RCONH_2$).

Questions

1. Write the names of the organic product formed in the four similar reactions of propanoyl chloride with water, methanol, ammonia and methylamine.

2. Write an equation for butanoyl chloride reacting with:
   (a) propan-1-ol
   (b) ethylamine.

17.3 4 Esters

By the end of this section, you should be able to...
- identify the ester functional group
- understand the hydrolysis reactions of esters, in acidic and alkaline solution

Introduction
In previous sections, you have come across esters as the products of reactions of other organic compounds. We will now take a closer look at esters, including their reactions.

Naming esters
This is straightforward, but you need to remember these points:
- As for carboxylic acids and acyl chlorides (but not amides), an ester name contains two words.
- The first word comes from the alkyl group joined to O.
- The second word comes from the alkyl group joined to C.

Table A shows some examples.

Structural formula	Name
$HCOOCH_3$	methyl methanoate
CH_3COOCH_3	methyl ethanoate
$HCOOCH_2CH_3$	ethyl methanoate
$CH_3COOCH_2CH_3$	ethyl ethanoate
$CH_3COOCH_2CH_2CH_3$	propyl ethanoate

table A

Physical properties
Esters are colourless liquids with relatively low melting and boiling temperatures, and are insoluble in water. All of the hydrogen atoms in their molecules are attached to carbon atoms, so hydrogen bonding is not possible.

The generally pleasant smells of esters are largely responsible for the familiar odours of flowers and fruits. Examples include:

pentyl ethanoate — pears
3-methylbutyl ethanoate — bananas
methyl butanoate — apples

Esters are present in perfumes, food flavourings, solvents, anaesthetics and biofuels.

fig A Benzyl ethanoate is responsible for food flavourings that resemble apples and pears.

Hydrolysis of esters
The only reactions of esters that you need to know about are two slightly different types of **hydrolysis**. The term 'hydrolysis' means breaking down (*lysis*) using water (*hydro*). It can be considered as the reverse of esterification. You may remember equations such as this from a previous section:

$$CH_3COOH + CH_3CH_2OH \rightleftharpoons CH_3COOCH_2CH_3 + H_2O$$

This shows the formation of ethyl ethanoate from the corresponding carboxylic acid and alcohol. The reversible arrow indicates that a mixture of ethyl ethanoate and water reacts together to form ethanoic acid and ethanol. This reverse reaction could be described as the hydrolysis of ethyl ethanoate.

Hydrolysis in acidic solution
We partly covered this type of hydrolysis in the previous paragraph. For a given ester, warming it with water will cause hydrolysis. As with esterification, the reaction is slow. A catalyst such as sulfuric acid will speed up hydrolysis, but will not affect the position of equilibrium, so the reaction will not go to completion:

$$CH_3COOCH_2CH_3 + H_2O \rightleftharpoons CH_3COOH + CH_3CH_2OH$$

Hydrolysis in alkaline solution
The disadvantage of using an alkali such as aqueous sodium hydroxide instead of sulfuric acid is that the reaction produces a carboxylate salt instead of a carboxylic acid. However, the advantage easily outweighs this disadvantage – the reaction goes to completion instead of reaching an equilibrium. The equation for methyl propanoate is:

$$CH_3CH_2COOCH_3 + NaOH \rightarrow CH_3CH_2COO^- + Na^+ + CH_3OH$$

Converting the carboxylate salt into a carboxylic acid only needs a dilute acid to be added:

$$CH_3CH_2COO^- + H^+ \rightarrow CH_3CH_2COOH$$

The final organic products are methanol and propanoic acid.

Saponification

The Latin word for soap is *sapo*, so saponification means 'soap-making'. This is a particular example of the alkaline hydrolysis of esters. Soaps and detergents have similar uses, but the main difference is that detergents use organic compounds obtained from crude oil, while soaps use organic compounds obtained from vegetable (and, originally, animal) sources.

Vegetable oils contain large quantities of triglycerides. These are triesters, each of which consists of a large ester molecule that can be hydrolysed to one alcohol and three carboxylic acid molecules. The hydrolysis of a typical triglyceride is shown in **fig B**.

$$CH_2-O-CO-C_{17}H_{35}$$
$$CH-O-CO-C_{17}H_{35} + 3NaOH \longrightarrow CH_2OH-CHOH-CH_2OH + 3C_{17}H_{35}COO^-Na^+$$
$$CH_2-O-CO-C_{17}H_{35}$$

a triglyceride – an ester of long-chain carboxylic acids and a triol

propane-1,2,3-triol (glycerol)

sodium octadecanoate (sodium stearate)

fig B An example of saponification.

In this example, hydrolysis produces an alcohol with three hydroxyl groups (glycerol). Glycerol has many uses, including as a sweetener in foods and toothpastes, and in skin care products. It is now also being used as a component of the liquid in electronic cigarettes.

The other product of this reaction, commonly known as sodium stearate, is a very common ingredient of most soaps.

Learning tip
Acidic hydrolysis of an ester is a reversible reaction, but alkaline hydrolysis reactions go to completion.

Questions

1. What are the names of these esters?
   (a) $CH_3CH_2COOCH_2CH_3$
   (b) $(CH_3)_2CHCOOCH_3$

2. Write equations for the hydrolysis of propyl butanoate in:
   (a) acidic conditions
   (b) alkaline conditions.

Key definition
Hydrolysis is the breaking of a compound by water into two compounds.

17.3 5 Polyesters

By the end of this section, you should be able to...

- understand how polyesters are formed by condensation polymerisation reactions

Addition polymerisation

Examples of addition polymerisation were covered in **Book 1**, so may be useful to remind you of this type of polymerisation. The monomer propene can combine with many thousands of other propene molecules to form a very long chain. This process can be represented as:

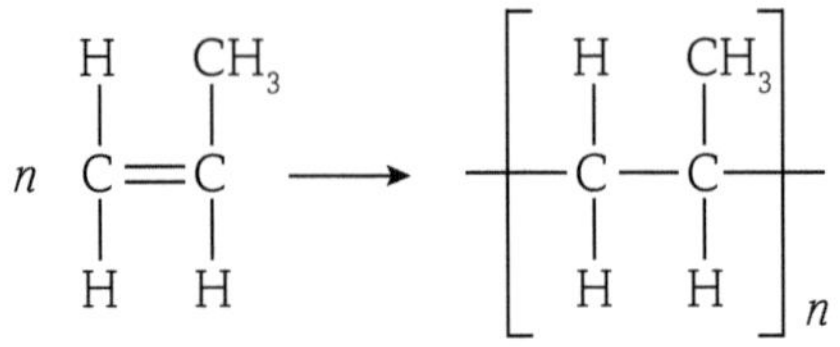

$$n \ \underset{\substack{|\\H}}{\overset{\substack{H\\|}}{C}} = \underset{\substack{|\\H}}{\overset{\substack{CH_3\\|}}{C}} \longrightarrow \left[\underset{\substack{|\\H}}{\overset{\substack{H\\|}}{C}} - \underset{\substack{|\\H}}{\overset{\substack{CH_3\\|}}{C}} \right]_n$$

fig A This equation represents the polymerisation of propene.

One thing to notice is that all of the atoms in the monomer molecules end up in the structure of the polymer.

Condensation polymerisation

There are two main differences between addition polymerisation and **condensation polymerisation**. In condensation polymerisation:

- each time two monomer molecules join together, another small molecule is formed
- usually two different monomers react together.

In many examples of condensation polymerisation, the small molecule formed is water, which is the origin of the term 'condensation'. In other examples, the small molecule formed is hydrogen chloride, but the term 'condensation' is still used to describe those examples.

One type of polymer formed in this way is a polyamide, which we will meet in **Section 17.5.4**.

Polyesters

The example we will use to introduce condensation polymerisation is the formation of polyesters. You will remember that esters are the products of the reaction between alcohols and carboxylic acids. However, once one molecule of an alcohol has reacted with one molecule of a carboxylic acid, the reaction is complete. The only organic product is a slightly larger ester molecule.

For a reaction to produce a polymer, we need two monomers, each with two reactive groups at either end. The alcohol usually involved in this type of reaction, more accurately called a 'diol', has two hydroxyl groups, one at each end:

$$H-O-CH_2-CH_2-O-H$$

One of the most common carboxylic acids used, more accurately a dicarboxylic acid, has the structure below. The common name for this compound is terephthalic acid.

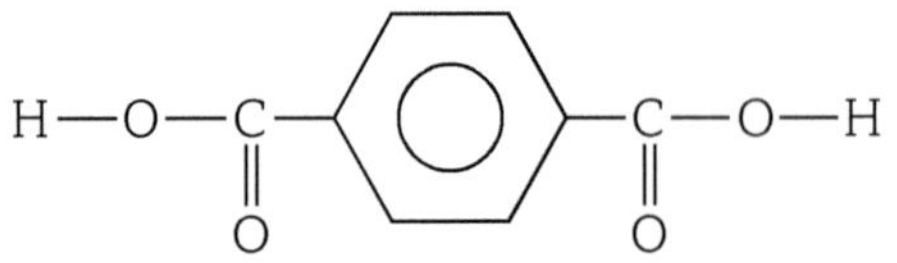

$$H-O-\underset{\substack{\|\\O}}{C}-\underset{}{\bigcirc}-\underset{\substack{\|\\O}}{C}-O-H$$

17.3 Carboxylic acids

Consider how these molecules react together. The OH of the carboxylic acid and the H of the alcohol react together to form the small molecule, water, and the two molecules are linked together by an ester group. This equation shows what happens during the formation of a typical polyester.

$$H-O-\underset{\underset{O}{\|}}{C}-\underset{}{\bigcirc}-\underset{\underset{O}{\|}}{C}-(O-H \quad H)-O-CH_2-CH_2-O-H \rightarrow H-O-\underset{\underset{O}{\|}}{C}-\underset{}{\bigcirc}-\underset{\underset{O}{\|}}{C}-O-CH_2-CH_2-O-H$$

This larger molecule that has formed still has reactive groups at both ends, so it can continue reacting with other molecules in the same way as above, until a very long polymer chain has formed. The structure of the polymer is:

$$\left[\underset{\underset{O}{\|}}{C}-\underset{}{\bigcirc}-\underset{\underset{O}{\|}}{C}-O-CH_2-CH_2-O \right]_n$$

Because of the original name of the acid (terephthalic acid), the polymer was originally known as Terylene.

Polyesters have a very wide range of uses, including soft drink bottles, food packaging, many types of clothing, and duvet fillings.

It is also possible to use a dioyl chloride instead of a dicarboxylic acid, although this is not used as a monomer in industry. The reaction occurs in the same way – the only difference is that the small molecule formed is hydrogen chloride, not water.

Learning tip

When showing how a polyester forms, remember that in the formation of the small molecule the H comes from the diol, and the OH from the dicarboxylic acid (or the Cl from the dioyl chloride).

Questions

1 Why is it not possible to make a polymer by reacting together molecules of $HOOCCH_2COOH$ and CH_3OH?

2 Draw the repeat unit of the polymer formed between molecules of $HOOCCOOH$ and $CH_3CH(OH)CH(OH)CH_3$.

Key definition

Condensation polymerisation refers to the formation of a polymer, usually by the reaction of two different monomers, and in which a small molecule is also formed.

17.4 1 Benzene – a molecule with two models

By the end of this section, you should be able to...

- understand that the bonding in benzene has been represented using the Kekulé and the delocalised model, the latter in terms of overlap of p-orbitals to form pi (π) bonds
- understand that evidence for the delocalised model of the bonding in benzene is provided by data from enthalpy changes of hydrogenation and carbon–carbon bond lengths
- understand why benzene is resistant to bromination, compared with alkenes, in terms of delocalisation of π-bonds in benzene and the localised electron density of the π-bond in alkenes

What are aromatic compounds?

The sections in **Chapter 17.4** deal with compounds containing a benzene ring. You have already met benzene rings in previous sections, but no detail was given about them. For example, in **Section 17.2.3** on carbonyl compounds, there was a reaction involving 2,4-dinitrophenylhydrazine, which contains a benzene ring. In **Section 17.3.5** on carboxylic acids, one of the monomers used to make polyesters was terephthalic acid, which also contains a benzene ring.

Origin of the term

You know that the word '**aromatic**' in everyday language refers to smells, usually pleasant. Some herbs used in cooking are often described as aromatic. The word *smell* is a neutral word, but when you replace it by *aroma* it implies that the smell is pleasant. On the other hand, if you replace it by *odour* it implies that the smell is unpleasant. Think of the aroma of perfume, and the odour of unwashed socks!

fig A Many brightly coloured dyes are made from aromatic compounds.

Aliphatic and aromatic

With a few exceptions, almost all the organic compounds you have met so far in this course could have been described as 'aliphatic', although this is a term we did not use to describe them because there was no point. You can regard aliphatic compounds as all those that are not aromatic!

Benzene

The compound benzene is at the heart of every aromatic compound, so the first thing we must do is understand what is special about benzene.

Benzene can be described as an arene, as can many of its derivatives. You already know that *-ene* indicates the presence of a C=C double bond in a molecule (as in ethene), so you would expect that an arene also contains C=C double bonds. The answer is – yes and no. This obviously needs careful explanation!

Physical properties of benzene

Benzene is a colourless liquid with a boiling temperature of 80 °C, and it is insoluble in water. It is present in crude oil and the fuels obtained from it, so the petrol tank of a car contains some benzene. The one place you will not find it is in a school or college laboratory, because of its toxic nature; in particular, it is a carcinogen (i.e. it can cause cancer). It was first isolated by the English chemist Michael Faraday in 1825. Within a few years, the compound became known as benzene and its molecular formula was established.

Its molecular formula is C_6H_6, which suggests that it is highly unsaturated, because the alkane with six carbon atoms has 14 hydrogen atoms. Finding a structural formula to fit the molecular formula and which was supported by studies of its chemical reactions was a challenge for chemists.

The Kekulé structure

One of the better known stories in the history of chemistry is how the German chemist Friedrich August Kekulé came to suggest the structure of benzene that still bears his name. The story goes like this... Kekulé was dozing in front of a fire. In the flames, he imagined some snake-like molecules dancing, one of which held its own tail and whirled around in the flames. Inspired by this dream about snakes, in 1865 he proposed a cyclic or ring structure for benzene. **Fig B** shows the displayed and skeletal forms of this structure.

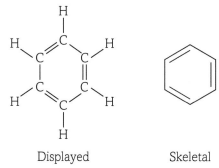

Displayed   Skeletal

fig B Kekulé structure of benzene.

Perhaps Kekulé was also influenced by Josef Loschmidt, who had suggested a ring structure for benzene a few years earlier. Loschmidt's role in developing the structure of benzene is little remembered, although you may remember his name from **Book 1** as the source of the symbol L for the Avogadro constant.

Problems with the Kekulé structure

We still use the original skeletal structure to represent benzene today. However, since it was first used, there has been mounting experimental evidence that does not fit with the structure. Several problems arose in the 19th century, although more became apparent in the 20th century.

Problem 1

If benzene contains three C=C double bonds, it should readily decolourise bromine water in an addition reaction. However, it does not decolourise bromine water. When bromine does react with benzene, a substitution reaction occurs (see **Section 17.4.2**). This evidence suggests that there are no C=C double bonds. **Fig C** shows the results of shaking bromine water with a liquid alkane, benzene and a liquid alkene in separate test tubes.

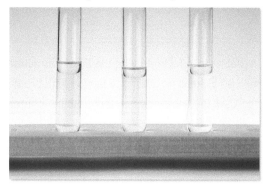

fig C The yellow colour is caused by unreacted bromine.

Problem 2

As more compounds containing the benzene ring were discovered, including those such as dibromobenzene ($C_6H_4Br_2$), another problem appeared. If the Kekulé structure is correct, there should be four isomers with this molecular formula:

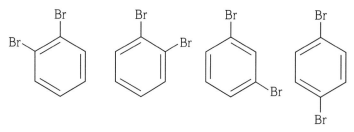

fig D These are the four isomers of $C_6H_4Br_2$, assuming that the Kekulé structure is correct.

However, only three were known to exist. So, the two isomers with bromine on adjacent carbon atoms are identical. It is not the case that adjacent atoms in one isomer are joined by C—C and in the other by C=C. This suggests that the bonds between the carbon atoms in the benzene ring are the same, not different.

Problem 3

When data about the lengths of covalent bonds in molecules became available, there was another problem. **Table A** shows the bond lengths in benzene and cyclohexene (a cyclic alkene with five C—C single bonds and one C=C double bond).

Bond	Bond length/pm
C—C in cyclohexene	154
C=C in cyclohexene	134
C—C in benzene	139

table A

This evidence suggests that the carbon–carbon bonds in benzene are all the same, and perhaps also intermediate in character between C—C and C=C bonds.

Problem 4

When data about enthalpy changes of hydrogenation became available, there was yet another problem. **Table B** shows these enthalpy changes for three relevant compounds.

Compound	Enthalpy change of hydrogenation/kJ mol^{-1}
cyclohexene	−120
cyclohexa-1,4-diene	−239
cyclohexa-1,3,5-triene (theoretical)	−360
benzene (actual)	−208

table B

All of these hydrogenation reactions form the cyclic alkane cyclohexane. This evidence suggests the following.

- The first two values indicate that the enthalpy change for adding 1 mol of H_2 to 1 mol of C=C bonds is around −120 kJ mol^{-1} (the value for cyclohexa-1,4-diene is double that of cyclohexene because there are twice as many C=C bonds).

- The Kekulé structure could be named cyclohexa-1,3,5-triene, for which the value should be treble that of cyclohexene.
- The actual benzene has a value very much lower (152 kJ mol^{-1} lower) than a theoretical structure with three C=C double bonds would have.

The values for cyclohexa-1,3,5-triene (theoretical) and benzene (actual) can be represented on an enthalpy level diagram, as shown below.

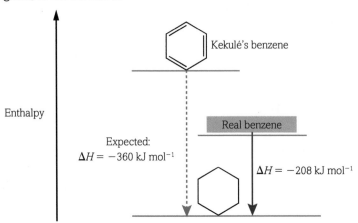

fig E This diagram shows the enthalpy changes for Kekulé's benzene and real benzene.

A new model for benzene

The weaknesses of the Kekulé structure and growing evidence from other areas of chemistry paved the way for a new model of benzene. In **Book 1**, we explained the C=C double bond in ethene in terms of single sigma C—C bonds and pi bonds formed by the sideways overlap of orbitals.

Now consider the situation in benzene after all the sigma (σ) bonds have formed. Each carbon atom has one electron in a p-orbital. You can imagine the formation of three individual π-bonds, but instead imagine the formation of one large π-bond made up of all six electrons. Three are in a doughnut shape above the atoms and three are in another doughnut shape below the atoms. Together these six electrons form a delocalised π-bond, as shown in **Fig F**.

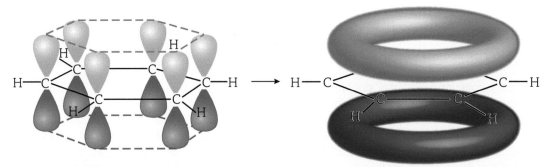

fig F The delocalised structure of benzene forms when the p-orbitals overlap sideways forming a π-electron cloud above and below the plane of the carbon atoms. Each carbon atom contributes one electron to the cloud, giving a delocalised system of six electrons in total.

Problems solved?

Let us see how this new model overcomes the problems associated with the Kekulé structure.

Problem 1   There are no individual C=C bonds, so there is no addition reaction with bromine.

Problem 2   There are three, not four, isomers of $C_6H_4Br_2$ because when the bromine atoms are on adjacent carbon atoms there is no difference in the arrangement of electrons between these atoms.

Problem 3   The carbon–carbon bonds are all the same length because they are identical and are not individual C—C and C=C bonds.

Problem 4   When charge is spread around in a species, there is increased stability, which explains the 152 kJ mol^{-1} greater stability of benzene compared to cyclohexa-1,3,5-triene.

The fact that benzene undergoes a substitution reaction with bromine to form C_6H_5Br, rather than addition to form $C_6H_4Br_2$, can be explained because substitution preserves the stability of the delocalised electrons in the pi bond. The addition reaction would instead produce a compound with two C=C double bonds, and that would lack the extra stability of the product of the substitution reaction.

A new meaning of 'aromatic'

Now we can use the old word '**aromatic**' in a new way – it refers to a hydrocarbon ring containing delocalised electrons. The traditional skeletal formula for benzene can now be replaced by this:

fig G The circle represents the delocalised pi bonds with six electrons in total.

Learning tip

When explaining the bonding in a benzene molecule, be careful to avoid confusion between electrons in p-orbitals and delocalised electrons in π-bonds.

Questions

1. Using the Kekulé structure for benzene, draw the isomers of tribromobenzene, $C_6H_3Br_3$.

2. How many electrons in a benzene molecule are involved in:
   (a) σ-bonding
   (b) π-bonding?

Key definitions

The original meaning of **aromatic** was a description of the smell of certain organic compounds.

The new meaning of **aromatic** is a description of the bonding in a compound – delocalised electrons forming pi (π) bonding in a hydrocarbon ring.

17.4　2　Some reactions of benzene

By the end of this section, you should be able to...

- understand the reactions of benzene with:
  (i) oxygen in air (combustion with a smoky flame)
  (ii) bromine, in the presence of a catalyst
  (iii) a mixture of concentrated nitric and sulfuric acids
  (iv) halogenoalkanes and acyl chlorides with aluminium chloride as catalyst (Friedel–Crafts reaction)

Reactions as a hydrocarbon

In **Section 17.4.1**, hydrogenation was mentioned. This reaction occurs in the same way with benzene as with alkenes. The same is true for combustion.

Hydrogenation

This is done by mixing benzene with hydrogen and heating under pressure with a nickel catalyst. The equation for the reaction can be shown using molecular or skeletal formulae:

$$C_6H_6 + 3H_2 \rightarrow C_6H_{12}$$

or:

[skeletal formula: benzene + 3H₂ → cyclohexane]

Note the plain hexagon (without a circle) used for the skeletal formula for cyclohexane.

Combustion

As with all hydrocarbons, benzene burns in air, although with a smoky flame. This is typical of many compounds with a high carbon-to-hydrogen ratio. The equation for complete combustion is:

$$C_6H_6 + 7\tfrac{1}{2}O_2 \rightarrow 6CO_2 + 3H_2O$$

Reactions as an arene

The most important reactions of benzene are substitution reactions, in which the products are also aromatic. We will consider the mechanisms of these reactions in the next section (**Section 17.4.3**).

> **Learning tip**
>
> Bromination refers to putting bromine into a molecule. This could be by an addition reaction, but in this section it refers only to a substitution reaction.

Bromination

The term 'bromination' is ambiguous. It could refer to the addition reaction between benzene and bromine. This reaction does occur in the presence of ultraviolet radiation, but it is of little importance. In this book, the term refers to a substitution reaction.

Benzene and bromine are heated under reflux, in the presence of a catalyst called a **halogen carrier**. Halogen carriers are usually metal–halogen compounds such as aluminium chloride, aluminium bromide or iron(III) bromide. For bromination, it is enough to add iron filings because they react with some of the bromine to form iron(III) bromide. The products are bromobenzene and hydrogen bromide. The equation for the reaction is:

[benzene + Br₂ → bromobenzene + HBr]

Nitration

This refers to the substitution of a hydrogen atom by a nitro group. The nitro group has the formula NO_2 and should not be confused with the gas nitrogen dioxide. The reaction is carried out by warming benzene with a mixture of concentrated nitric and sulfuric acids. The nitric acid can be considered as the source of the NO_2 group and the sulfuric acid as a catalyst. The products are nitrobenzene and water. The equation for the reaction is:

[benzene + HNO₃ —H₂SO₄→ nitrobenzene + H₂O]

Friedel–Crafts reactions

Reactions with this name were developed by Charles Friedel (a French chemist) and James Crafts (an American chemist). There are several types of reaction bearing their names, but we will consider only two of them: alkylation and acylation. They both have these features in common:

- using a reagent represented by XY, one of the hydrogen atoms in benzene is substituted by Y, and the other product is HX
- a catalyst is needed – often aluminium chloride, although other halogen carriers such as iron(III) chloride and iron(III) bromide also work
- anhydrous conditions are needed because water would react with the catalyst and sometimes also with the organic product.

Friedel–Crafts alkylation reactions

Alkylation means the substitution of one of the hydrogen atoms of benzene by an alkyl group. The reagent is a halogenoalkane, and the products are an alkylbenzene (methylbenzene) and hydrogen chloride. The equation for the reaction with chloromethane is:

$$C_6H_6 + CH_3Cl \xrightarrow{AlCl_3} C_6H_5CH_3 + HCl$$

Friedel–Crafts acylation reactions

Acylation means the substitution of one of the hydrogen atoms in benzene by an acyl group. The reagent is an acyl chloride, and the products are a ketone (phenylethanone) and hydrogen chloride. The equation for the reaction with ethanoyl chloride is:

$$C_6H_6 + CH_3COCl \xrightarrow{AlCl_3} C_6H_5COCH_3 + HCl$$

Naming aromatic compounds

The rules for naming aromatic compounds are similar to those used for aliphatic compounds. Prefixes, locants, hyphens and commas, and multiplying prefixes are used in the same way. In some names, you will see 'phen' or 'phenyl' used to represent the benzene ring when it is attached to another functional group. For example, the organic product of the last reaction is phenylethanone: 'ethan' refers to the two carbon atoms in CH_3CO, 'one' indicates that the compound is a ketone and 'phenyl' refers to the C_6H_5 group. You might expect 'benzyl' to be used rather than 'phenyl', but it is not. The explanation for this is beyond the scope of this book.

fig. A Shoe polish is one of the many commercial products containing aromatic compounds.

Questions

1 Write an equation for the reaction between benzene and 2-bromopropane.

2 What is the name of the organic product formed in the reaction between benzene and butanoyl chloride?

Key definition

A **halogen carrier** is a catalyst that helps to introduce a halogen atom into a benzene ring.

17.4 3 Electrophilic substitution mechanisms

By the end of this section, you should be able to...

- understand the mechanism of the electrophilic substitution reactions of benzene (halogenation, nitration and Friedel–Crafts reactions), including the generation of the electrophile

Electrophilic substitution

Although you have not met this mechanism so far, you have learned the meaning of each word. The reaction between methane and chlorine in the presence of ultraviolet radiation is an example of radical substitution. The reaction between ethene and bromine is an example of electrophilic addition.

Once you understand the meaning of each word, it is not a big step to understand the meaning of electrophilic substitution. Before we consider the mechanisms of the individual reactions, it is worth considering what they all have in common:

- The benzene ring is electron-rich. Although it is electrically neutral, the delocalised electrons in the pi (π) bond above and below the atoms mean that the molecule attracts electrophiles – species attracted to negative charge.

- Using Y^+ to represent the electrophile, as Y^+ approaches the delocalised π-bond, it attracts two of the six electrons, which form a covalent bond with it. This gives an intermediate species with a positive charge (shown by convention in the centre of the hexagon), in which one carbon atom is joined to both Y and H. There are now only four delocalised electrons, and these are represented by two-thirds of a circle.

- This intermediate species lacks the stability of an aromatic compound (because the circle is broken), so in the next step the H leaves as H^+ and the two electrons in the C—H bond join the remaining four to restore the delocalised π-bond (with its relatively high stability).

When drawing these mechanisms, or skeletal formulae, a substituent on the benzene ring can be shown in any of the six positions. It is usually shown at the top of the hexagon or the next position clockwise.

To help you see the similarities and differences in these mechanisms, they are all set out in the same way.

Bromination

Step 1: formation of the electrophile

Bromine reacts with the catalyst to form Br^+:

$$AlCl_3 + Br_2 \rightarrow Br^+ + AlCl_3Br^-$$

Step 2: electrophilic attack

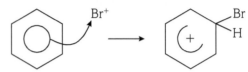

Note that the H atom shown in the product of this step is already present in benzene, although it is not shown. It is shown here because it will be lost in Step 3.

Step 3: formation of the aromatic product

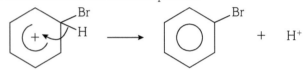

Step 4: formation of the inorganic products

$$AlCl_3Br^- + H^+ \rightarrow AlCl_3 + HBr$$

Nitration

Step 1: formation of the electrophile

Nitric acid and sulfuric acid react to form NO_2^+ (the nitryl ion or nitronium ion):

$$HNO_3 + H_2SO_4 \rightarrow NO_2^+ + HSO_4^- + H_2O$$

Step 2: electrophilic attack

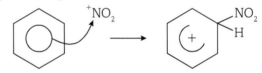

Step 3: formation of the aromatic product

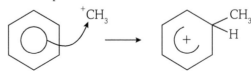

Step 4: formation of the inorganic products

$$HSO_4^- + H^+ \rightarrow H_2SO_4$$

Alkylation

Step 1: formation of the electrophile

Chloromethane reacts with the catalyst to form CH_3^+:

$$AlCl_3 + CH_3Cl \rightarrow CH_3^+ + AlCl_4^-$$

Step 2: electrophilic attack

Step 3: formation of the aromatic product

[benzene ring with CH$_3$ and H, with + charge] → [benzene ring with CH$_3$] + H$^+$

Step 4: formation of the inorganic products

$AlCl_4^- + H^+ \rightarrow AlCl_3 + HCl$

Acylation

Step 1: formation of the electrophile

Ethanoyl chloride reacts with the catalyst to form CH_3CO^+:

$AlCl_3 + CH_3COCl \rightarrow CH_3CO^+ + AlCl_4^-$

Step 2: electrophilic attack

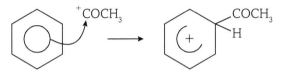

Step 3: formation of the aromatic product

[benzene ring with COCH$_3$ and H, with + charge] → [benzene ring with COCH$_3$] + H$^+$

Step 4: formation of the inorganic products

$AlCl_4^- + H^+ \rightarrow AlCl_3 + HCl$

Learning tip

Note the similarities between all these mechanisms. The first step involves formation of the electrophile, and the last step shows what happens to the replaced H$^+$. The second and third steps show the actual substitution in the benzene ring.

Questions

1. Write an equation for Step 1 in the bromination reaction in which the halogen carrier is iron(III) bromide.

2. Write equations for Steps 2 and 3 in the alkylation reaction using 1-bromopropane as a starting material.

17.4 Phenol

By the end of this section, you should be able to...

- understand the reaction of phenol with bromine water
- understand reasons for the relative ease of bromination of phenol compared with benzene

What is phenol?

Most aromatic compounds have a benzene ring and an aliphatic part. In previous sections of this topic you have covered some important reactions of the benzene ring. You know many of the reactions of aliphatic compounds from earlier chapters and from **Book 1**. In this section, we will look at one more aromatic compound with a familiar aliphatic group, the hydroxyl (OH) group.

Phenol consists of a hydroxyl group joined to a benzene ring. It has some properties similar to an aliphatic alcohol such as ethanol, and you would expect it to undergo electrophilic substitution reactions because it contains a benzene ring. Its structure is:

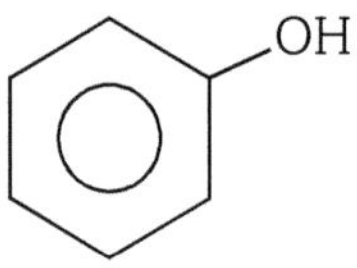

Bromination of phenol

You will remember from **Section 17.4.2** that benzene undergoes bromination, but the reaction needs a catalyst and the reaction mixture has to be heated under reflux. In contrast, the bromination of phenol occurs at room temperature without a catalyst, and works with bromine water. The bromine water is decolourised during the reaction, and the organic product forms as a white precipitate. The equation for the reaction is:

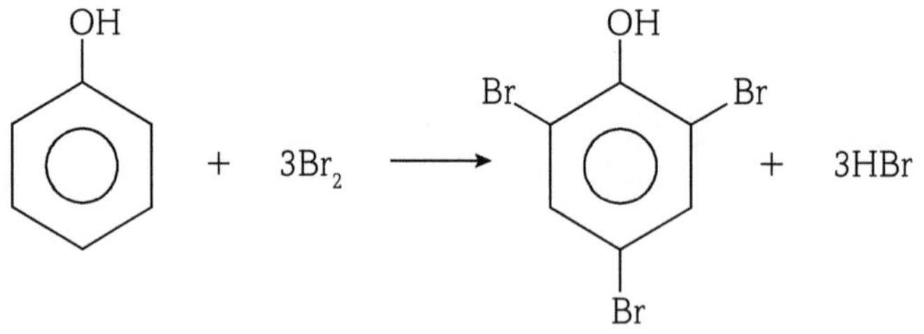

Another obvious difference between benzene and phenol is that the substitution reaction occurs three times, as there are three bromine atoms in the organic product. You can probably see why the name is 2,4,6-tribromophenol. The carbon joined to OH is carbon 1, the other carbons are numbered clockwise, and the three bromine atoms are joined to carbons 2, 4 and 6 in the ring.

It is beyond the scope of this book to try to explain why the reaction involves substitution by three bromine atoms and not any other number. However, we can try to explain why the reaction occurs much more readily with phenol than with benzene:

- The oxygen in the OH group has lone pairs of electrons, and these electrons can merge with the electrons in the delocalised pi (π) bond.

- The electron density above and below the ring of atoms is increased, a process sometimes referred to as 'activation' because the molecule is now much more reactive towards electrophiles.

- Bromine molecules, although originally non-polar, are polarised as they approach the benzene ring. Eventually, the Br—Br bond breaks and the Br^+ electrophile attacks the benzene ring.

Fig A shows how the delocalised electrons in the π-bond change because of the presence of the oxygen.

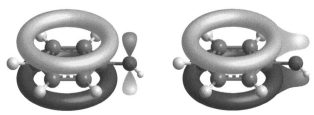

fig A The electrons in the p-orbitals of the oxygen atom become part of the delocalised electron system. This increases the electron density above and below the ring and makes the molecule more attractive to electrophiles such as Br^+.

Uses of phenol and its derivatives

One of the earliest and best known uses of phenol was in antiseptic surgery, especially by the surgeon Joseph Lister. Since then, other antiseptics have been developed from phenol. One of the most familiar of these is the product of the chlorination of phenol – this is 2,4,6-trichlorophenol, better known as the antiseptic TCP. Nowadays, phenol is widely used in many industries, notably in the manufacture of polymers and pharmaceuticals.

Phenolphthalein

You will remember this substance as an indicator in acid–base titrations, and will not be surprised to learn that it contains a benzene ring joined to a hydroxyl group! Although you do not need to know its structure, notice that its colour change involves a change in the structure of phenol.

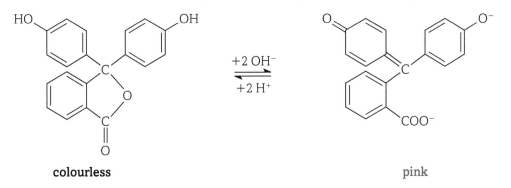

fig B Antiseptics based on phenol are useful for treating a grazed knee.

fig C The phenolphthalein colour change involves a change in the structure of phenol.

The colourless form is a molecule with three benzene rings and two OH groups, while the pink form is an ion with no OH groups and two benzene rings.

Learning tip

Be careful to distinguish between the use of 'benz' and 'phen' in organic names. In general, 'benz' comes after the other groups in the molecule.

Questions

1. Suggest why phenol is a solid at room temperature and is slightly soluble in water.
2. Write an equation for the formation of TCP from phenol.

17.5 1 Amines and their preparations

By the end of this section, you should be able to...

- identify the amine functional group
- understand, in terms of reagents and general reaction conditions, the preparation of primary aliphatic amines:
    (i) from halogenoalkanes
    (ii) by the reduction of nitriles
- understand that aromatic nitro-compounds can be reduced, using tin and concentrated hydrochloric acid, to form amines

Introduction to amines

These nitrogen-containing compounds have some similarities with ammonia. The three bonding pairs of electrons around nitrogen are distributed in a trigonal pyramidal shape. The nitrogen atom has a lone pair of electrons and three bonds to one or more alkyl groups. If there is one alkyl group, the amine is classed as primary, if there are two, then it is secondary, and with three it is tertiary. Fig A shows the structures of the simplest amine in each of these classes.

fig A These are the simplest examples of primary, secondary and tertiary amines.

fig B Some cold and flu remedies contain amines.

In this book we will focus on primary amines. Methylamine is a gas at room temperature, ethylamine has a boiling temperature of 17 °C, and propylamine and butylamine are volatile liquids. This list of names shows the naming system used – the suffix is *-amine* and the usual codes methyl, ethyl, propyl, butyl etc. are used.

We will also meet the simplest aromatic amine, $C_6H_5NH_2$, which is phenylamine.

Amines occur widely in nature and many drugs (both legal and illegal) are amines, including the group of compounds called amphetamines.

Preparation of aliphatic amines

There are two main ways of making primary aliphatic amines, starting from halogenoalkanes or from nitriles.

Preparation from halogenoalkanes

Although you do not need to know the mechanisms of these reactions, a reference to the mechanism will help you understand why certain reaction conditions are needed. The method involves heating a halogenoalkane with ammonia. Because ammonia is a gas, this must be done under pressure and in a sealed container. Alternatively, the halogenoalkane can be mixed with concentrated aqueous ammonia. The equation for the preparation of methylamine is:

$$CH_3Cl + NH_3 \rightarrow CH_3NH_2 + HCl$$

The reaction involves nucleophilic attack by the lone pair of electrons of ammonia on the electron-deficient carbon atom in the halogenoalkane. Notice that the amine formed also has a nitrogen atom containing a lone pair of electrons. This means that it could also act as a nucleophile, competing with ammonia in the attack on the halogenoalkane. This would result in the reaction:

$$CH_3Cl + CH_3NH_2 \rightarrow (CH_3)_2NH + HCl$$

The product of this reaction is a secondary amine, in this case dimethylamine.

To prevent such unwanted side-reactions occurring, or at least reduce the chances of them happening, the ammonia is used in excess, so that it outnumbers the molecules of primary amine formed. Some of the excess ammonia reacts with the acidic hydrogen chloride formed, so a better equation to represent the preparation is:

$$CH_3Cl + 2NH_3 \rightarrow CH_3NH_2 + NH_4Cl$$

Preparation from nitriles

Nitriles can be reduced to primary amines by reduction, using the reducing agent lithium tetrahydridoaluminate, which you may remember from **Section 17.2.2** can also be used to reduce carbonyl compounds. As before, the reactants are mixed in dry ether, to ensure that there is no water that could affect the reaction. The equation for the reduction of ethanenitrile is:

$$CH_3CN + 4[H] \rightarrow CH_3CH_2NH_2$$

[H] represents hydrogen atoms produced by the reagent in a way that you do not need to know.

Preparation of aromatic amines

A specific method is used to prepare aromatic amines, especially phenylamine, which is made by the reduction of nitrobenzene. In this method, the reducing agent is tin mixed with concentrated hydrochloric acid, and the reaction mixture is heated under reflux. The reduction is achieved partly through the oxidation of tin to tin(II) ions and tin(IV) ions, and partly through the hydrogen produced in the reaction between tin and the acid. For simplicity, we will represent the reducing agent by [H]. The equation for the reaction is:

$$C_6H_5NO_2 + 6[H] \rightarrow C_6H_5NH_2 + 2H_2O$$

As with other amines, phenylamine is basic and will react with the acid present to form the phenylammonium ion, but this can easily be converted into phenylamine by adding an alkali such as sodium hydroxide solution:

$$C_6H_5NH_3^+ + OH^- \rightarrow C_6H_5NH_2 + H_2O$$

In the 19th century, phenylamine was called aniline, and was the source of large numbers of dyes used mainly for clothing. It is still used in the manufacture of synthetic indigo, most familiar as the colour of blue denim jeans. Nowadays, much phenylamine is used in the manufacture of polymers and pharmaceuticals.

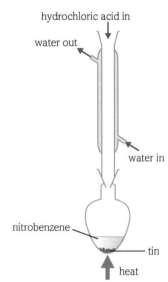

fig C Apparatus suitable for converting nitrobenzene into phenylamine.

Learning tip

Note that in the name used for a nitrile, the code indicating the number of carbon atoms includes the C in the CN group, so $CH_3CH_2CH_2CN$ is butanenitrile, not propanenitrile.

Questions

1. Write an equation for the preparation of propylamine starting from:
   (a) a halogenoalkane
   (b) a nitrile

2. Explain why another organic product would be formed in 1(a) if the two reactants were used in a 1:1 ratio.

17.5 (2) Acid–base reactions of amines

By the end of this section, you should be able to...

- understand the reactions of primary aliphatic amines, using butylamine as an example, with:
  (i) water to form an alkaline solution
  (ii) acids to form salts
- understand reasons for the difference in basicity of ammonia, primary aliphatic and primary aromatic amines given suitable data

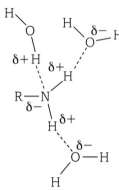

fig A The diagram shows the formation of hydrogen bonds between a primary amine and water.

Reactions with water

The first few members of the homologous series of primary aliphatic amines are completely miscible (capable of mixing) with water, although as the hydrocarbon part of the molecule becomes proportionately larger, the solubility decreases.

They dissolve in water because they can form hydrogen bonds with water molecules (see **fig A**).

Phenylamine is only slightly soluble in water.

As well as dissolving in water, amines also react slightly with water to form alkaline solutions. Compare the equations using methylamine and ammonia:

$$CH_3NH_2 + H_2O \rightleftharpoons CH_3NH_3^+ + OH^-$$

$$NH_3 + H_2O \rightleftharpoons NH_4^+ + OH^-$$

You can see the similarity – this is because the nitrogen atom in both molecules can use its lone pair of electrons to form a dative bond with the hydrogen of a water molecule. The cation in the first equation is the methylammonium ion – think of it as an ammonium ion in which one hydrogen has been replaced by a methyl group. So, how do amines and ammonia compare as bases? What is the position of equilibrium in these reactions?

Comparing basicities

The **basicity** (basic strength) of a base can be quantified using the constant K_a or the constant pK_a. The pK_a of water is 7.00, and any value greater than 7.00 indicates a basicity greater than that of water. Water is equally good as a base and as an acid. **Table A** shows values for ammonia and some amines.

Name	Formula	pK_a
ammonia	NH_3	9.30
methylamine	CH_3NH_2	10.64
ethylamine	$CH_3CH_2NH_2$	10.73
propylamine	$CH_3CH_2CH_2NH_2$	10.84
phenylamine	$C_6H_5NH_2$	4.62

table A

What these values show is that methylamine is a stronger base than ammonia, and that extending the hydrocarbon chain causes further, but smaller, increases in basicity. Phenylamine is a much weaker base than any of the aliphatic amines, and is a weaker base than even water.

Methylamine is a stronger base than ammonia because the methyl group is electron-releasing, and so has an increased electron density on nitrogen compared with ammonia. The ethyl and propyl groups are only slightly more electron-releasing than methyl, so their effects are very similar. **Fig B** shows the situation.

$$CH_3 \rightarrow \overset{\delta-}{NH_2} \quad CH_3CH_2 \rightarrow \overset{\delta-}{NH_2} \quad CH_3CH_2CH_2 \rightarrow \overset{\delta-}{NH_2}$$

fig B The diagram represents the slight increase in electron-releasing effect of alkyl groups, causing a slight increase in electron density on the nitrogen atoms.

The equation for the reaction of butylamine with water is

$$CH_3CH_2CH_2CH_2NH_2 + H_2O \rightleftharpoons CH_3CH_2CH_2CH_2NH_3^+ + OH^-$$

The situation with phenylamine is very different. Think back to the diagram in **Section 17.4.4** showing that in phenol the lone pairs of electrons on oxygen are attracted to the delocalised electrons in the pi bond. A similar thing happens with phenylamine, and this makes the nitrogen less electron-rich and the lone pair of electrons less available for donating to the hydrogen of a water molecule. **Fig C** shows how the delocalised electrons now extend above and below the nitrogen. So, as the original lone pair of electrons on nitrogen is now part of the delocalised system, it is less available for bonding to the hydrogen of a water molecule.

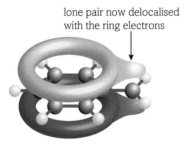

fig C The electrons from the lone pair on the nitrogen have joined with the delocalised electrons on the benzene ring.

Reactions with acids

Even though the amines vary in their basicity, they all react with strong acids to form ionic salts. Sample equations are:

$CH_3NH_2 + HNO_3 \rightarrow CH_3NH_3^+ + NO_3^-$
methylamine → methylammonium nitrate

$CH_3CH_2CH_2CH_2NH_2 + HCl \rightleftharpoons CH_3CH_2CH_2CH_2NH_3^+ + Cl^-$
butylamine → butylammonium chloride

Learning tip

Note the big difference in basicity between aliphatic amines and aromatic amines, and the small differences in basicity between different aliphatic amines.

Questions

1. Write an equation for the reaction of ethylamine with:
   (a) water
   (b) sulfuric acid.

2. Predict, with a reason, whether dimethylamine has a lower or higher basicity than methylamine.

Key definition

The **basicity** of a base is the extent to which it can donate a lone pair of electrons to the hydrogen atom of a water molecule.

17.5 · 3 · Other reactions of amines

By the end of this section, you should be able to...

- understand the reactions of primary aliphatic amines, using butylamine as an example, with:
  (i) ethanoyl chloride
  (ii) halogenoalkanes
  (iii) copper(II) ions to form complex ions

Reactions with ethanoyl chloride

Although you do not need to know the mechanisms of these reactions, the name of the mechanism will help you understand how the reactions occur. The reaction type is **addition–elimination**, which means that the two molecules join together, and then a small molecule is eliminated – in these examples, hydrogen chloride. The organic product contains a new functional group – amide – in which a carbonyl group is next to a NH group. The equation for the reaction between ethanoyl chloride and butylamine is:

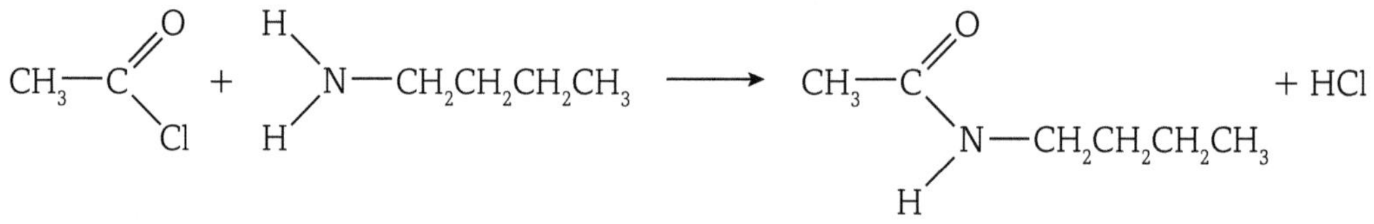

$$CH_3-COCl + H_2N-CH_2CH_2CH_2CH_3 \longrightarrow CH_3-CO-NH-CH_2CH_2CH_2CH_3 + HCl$$

We will meet amides in a later section, but we can introduce the naming system here. The most common amide is ethanamide, and the product of the above reaction can be thought of as the result of replacing one of the hydrogens in the NH_2 group by the butyl group. Its name contains two words – the first is the alkyl group from the amine and the second indicates the number of carbon atoms in the original acyl chloride. As the butyl group is attached via a nitrogen atom, N is used as the locant. The name is N-butylethanamide.

Paracetamol

This is one of the commonest pharmaceuticals used to relieve the symptoms of fever and pain. It is manufactured in a sequence of reactions, one of which is an addition–elimination reaction. Its structure is:

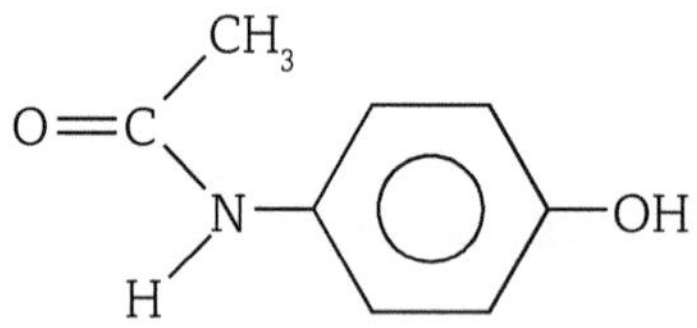

$$O=C(CH_3)-NH-C_6H_4-OH$$

You can see how paracetamol got its name. The 'para' part indicates that the two groups attached to the benzene ring are at opposite ends of the ring. The 'acetam' part comes from the old name for ethanamide, which used to be called acetamide. Finally, the 'ol' part indicates the presence of a hydroxyl group.

Reactions with halogenoalkanes

Although you do not need to know the mechanisms of these reactions, you can see that the two would react together because a halogenoalkane contains an electron-deficient carbon atom and an amine contains an electron-rich nitrogen atom. Using general formulae, the reaction can be represented as:

$$R_1NH_2 + R_2X \rightarrow R_1NHR_2 + HX$$

where R_1 is the alkyl group in the amine and R_2 is the alkyl group in the halogenoalkane. The reaction is another example of substitution. The organic product is a secondary amine and the inorganic product is a hydrogen halide, often hydrogen chloride.

An example equation using butylamine and chloroethane is:

$$CH_3CH_2CH_2CH_2NH_2 + CH_3CH_2Cl \rightarrow CH_3CH_2CH_2CH_2NHCH_2CH_3 + HCl$$

Notice that the organic product also contains an electron-rich nitrogen atom, so it can also react with chloroethane. The equation for this further reaction is:

$$CH_3CH_2CH_2CH_2NHCH_2CH_3 + CH_3CH_2Cl \rightarrow CH_3CH_2CH_2CH_2N(CH_2CH_3)_2 + HCl$$

The organic product of this reaction is a tertiary amine.

Once again, the tertiary amine contains a nitrogen atom with a lone pair of electrons, which can also react with the halogenoalkanes. The equation for this further reaction is:

$$CH_3CH_2CH_2CH_2N(CH_2CH_3)_2 + CH_3CH_2Cl \rightarrow CH_3CH_2CH_2CH_2N^+(CH_2CH_3)_3 + Cl^-$$

Note that the equation is different this time – HCl is not formed because this would require the loss of H from the nitrogen of the organic reactant, which a tertiary amine does not have. The product is an ionic compound related to ammonium chloride ($NH_4^+ + Cl^-$), except that all the hydrogens in the ammonium ion have been replaced by alkyl groups. It is known as a quaternary ammonium salt.

Reaction with halogenoalkanes does not seem to be a good method of preparation, as it is likely that a mixture of similar products will be formed. If there is an excess of the halogenoalkane, then the sequence of all four reactions is more likely to occur. Producing a quaternary ammonium salt in this sequence is what happens in industry. These compounds have many uses, and are usually found in fabric softeners.

Reactions with copper(II) ions

You will remember from **Topic 15** that ammonia can act as a lone pair donor in its reactions with transition metal ions. This is the overall equation for its reaction with hexaaquacopper(II) ions:

$$[Cu(H_2O)_6]^{2+} + 4NH_3 \rightarrow [Cu(NH_3)_4(H_2O)_2]^{2+} + 4H_2O$$

Amines also have a lone pair of electrons on nitrogen, so can take part in similar reactions. The observations are the same as with ammonia – first of all a pale blue precipitate forms, then with excess butylamine, the precipitate dissolves to give a deep blue solution. The equations for these reactions look complicated, but the only difference is the use of $CH_3CH_2CH_2CH_2NH_2$ in place of NH_3.

Formation of the pale blue precipitate:

$$[Cu(H_2O)_6]^{2+} + 2CH_3CH_2CH_2CH_2NH_2 \rightarrow [Cu(H_2O)_4(OH)_2] + 2CH_3CH_2CH_2CH_2NH_3^+$$

Formation of the deep blue solution:

$$[Cu(H_2O)_4(OH)_2] + 4CH_3CH_2CH_2CH_2NH_2 \rightarrow [Cu(CH_3CH_2CH_2CH_2NH_2)_4(H_2O)_2]^{2+} + 2H_2O + 2OH^-$$

Learning tip
Be careful not to confuse amines, ammines and amides, all of which appear in this section.

Questions

1. Write an equation for the reaction between propanoyl chloride and propylamine.

2. Draw the structure of the quaternary ammonium salt formed when ethylamine reacts with 1-bromopropane.

Key definition

An **addition–elimination** reaction occurs when two molecules join together, followed by the loss of a small molecule.

17.5 4 Amides and polyamides

By the end of this section, you should be able to...
- identify the amide functional group
- understand that amides can be prepared from acyl chlorides
- know that the formation of a polyamide is a condensation-polymerisation reaction
- draw the structural formulae of the repeat units of condensation polymers formed by reactions between dicarboxylic acids and diamines

Amides

We met amides in previous sections as the products of reactions. Amides have a functional group consisting of a carbonyl group joined to an amino group.

$$-C\overset{O}{\underset{NH_2}{\diagup}}$$

As with other compounds in which two functional groups share the same carbon, amides have properties that are different from both carbonyl compounds and amines. They are solids (except for methanamide, which is a liquid), and the lower aliphatic amides are soluble in water because they contain two electronegative atoms and polar bonds and so can form hydrogen bonds with water. The carbon atom is very electron-deficient because it is joined to both nitrogen and oxygen.

Preparation of amides

A convenient way to prepare amides in the laboratory is by mixing an acyl chloride with concentrated aqueous ammonia. Although you do not need to know the mechanisms of these reactions, it helps to understand the reaction if we refer to the mechanism. The lone pair of electrons on the nitrogen of the ammonia molecule is strongly attracted to the electron-deficient carbon atom of the acyl chloride. The chlorine of acyl chloride combines with one of the hydrogen atoms of ammonia to form hydrogen chloride, which appears as misty fumes. The equation for the reaction between propanoyl chloride and ammonia is:

$$CH_3CH_2COCl + NH_3 \rightarrow CH_3CH_2CONH_2 + HCl$$

Note that ammonia is basic and that the inorganic product is acidic, so there will be a reaction between the two molecules:

$$NH_3 + HCl \rightarrow NH_4Cl$$

So, an alternative equation for the reaction, combining these equations, is:

$$CH_3CH_2COCl + 2NH_3 \rightarrow CH_3CH_2CONH_2 + NH_4Cl$$

Polyamides

In **Section 17.3.5**, we learned about polyesters as examples of polymers formed by condensation polymerisation reactions. The formation of these polymers needs two monomers – a dicarboxylic acid and a diol. The formation of polyamides also needs two monomers – a dicarboxylic acid and a diamine.

The simplest examples of a dicarboxylic acid and a diamine are $HOOCCOOH$ and $H_2NCH_2NH_2$. You can see how these molecules can react together – one has acidic groups and the other has basic groups, so they can react with each other to form a CONH group with the elimination of H_2O.

Learning tip
In the formation of a polyamide, the small molecule formed is water (if a dicarboxylic acid is used) or hydrogen chloride (if a dioyl chloride is used).

Nylon

Nylon is almost certainly the most familiar example of a polyamide. Both of the monomers used in the production of its most common form contain six carbon atoms. Common examples of these monomers are hexanedioic acid (known as adipic acid in the polymer industry) and hexane-1,6-diamine (also known as 1,6-diaminohexane). What both of these monomers have in common is a number of CH_2 groups with reactive groups at each end of the chain. The formation of this polymer is shown below:

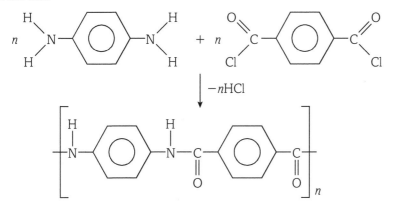

hexanedioic acid   1,6-diaminohexane   nylon 6,6

fig A Balloon fabric is made from nylon.

There are different types of nylon, and this example is described as nylon 6,6 because both of the monomers contain six carbon atoms.

Kevlar®

If benzene rings take the place of the CH_2 groups in the monomers used to make nylon 6,6, and a dioyl chloride is used in place of a dicarboxylic acid, then the monomers and the polymer structure can be shown like this:

fig B Kevlar® in action in a bulletproof vest.

This polymer produced is known as Kevlar®, and has very many uses, the most familiar of which is as body armour (bullet-proof and stab-proof vests). However, the list of applications continues to grow, as a web search will reveal.

Questions

1 Write the overall equation for the formation of ethanamide from an acyl chloride.

2 Write an equation for the formation of Kevlar® starting from a diamine and a dicarboxylic acid.

17.5 5 Amino acids

By the end of this section, you should be able to...

- identify molecules that are amino acids
- draw the structural formulae of the repeat units of condensation polymers formed by reactions between amino acids
- understand the properties of 2-amino acids, including:
  (i) acidity and basicity in solution, as a result of the formation of zwitterions
  (ii) effect of aqueous solutions on plane-polarised monochromatic light

What are amino acids?

The name suggests the presence of an amino group and a carboxylic acid group. Unlike amides, these two groups are separated by a carbon atom and so retain most of their typical properties. Below is the general displayed formula of an amino acid.

When R (the symbol for an alkyl group) is only a hydrogen, the amino acid is the simplest one, glycine. Over 20 amino acids are found in humans. Some of these are synthesised in the body, and others must be included in the diet. Proteins are made from amino acids, and these will be considered in the next section. Although R is normally used to represent an alkyl group, in the case of amino acids it can represent a more complex structure. **Table A** shows information about some important amino acids.

Name	Abbreviation	Structure	Isoelectric point
alanine	ala	$H_2N-CH(CH_3)-COOH$	6.0
cysteine	cys	$H_2N-CH(CH_2-SH)-COOH$	5.1
glutamic acid	glu	$H_2N-CH(CH_2-CH_2-COOH)-COOH$	3.2
glycine	gly	H_2N-CH_2-COOH	6.0
lysine	lys	$H_2N-CH(CH_2-CH_2-CH_2-CH_2-NH_2)-COOH$	9.7

table A

Amino acids are often referred to as 2-amino acids because the NH_2 is attached to the second carbon atom in the chain, counting the C in COOH as carbon 1. Note that most amino acids have one NH_2 group and one COOH group, although some have two of one and one of the other.

Acidic and basic properties

The values of the **isoelectric points** of the amino acids have been included to help you understand their acid–base character. They are all soluble in water, and you can imagine that one of two reactions might occur. The molecule could act as a base and form an alkaline solution in a reaction like this:

$H_2N\text{—}CHR\text{—}COOH + H_2O \rightleftharpoons H_3N^+\text{—}CHR\text{—}COOH + OH^-$

Alternatively, it could act as an acid and form an acidic solution in a reaction like this:

$H_2N\text{—}CHR\text{—}COOH + H_2O \rightleftharpoons H_2N\text{—}CHR\text{—}COO^- + H_3O^+$

There is a third alternative – an H$^+$ ion could transfer from the COOH group to the NH$_2$ group in a reaction like this:

$H_2N\text{—}CHR\text{—}COOH \rightleftharpoons H_3N^+\text{—}CHR\text{—}COO^-$

The product of this reaction is electrically neutral because it has a positive charge and a negative charge that balance each other. Such species are called **zwitterions**. *Zwitter* is the German word for hybrid, meaning a cross between two things and having the characteristics of both.

Now to the isoelectric point. This is the pH at which the zwitterion exists in aqueous solution. A low isoelectric point indicates that the molecule is predominantly acidic (glutamic acid has a value of 3.2), while a high value indicates that it is predominantly basic (lysine has a value of 9.7). You can see that these values are related to the numbers of NH$_2$ and COOH groups in the molecule (glutamic acid has an extra carboxylic acid group and lysine has an extra amino group).

Salt formation

All amino acids can form salts with acids and bases. For example, alanine can react with acids to form this protonated structure:

$$H_3N^+\text{—}CH(CH_3)\text{—}COOH$$

Glutamic acid can react with sodium hydroxide to form three possible salts. This is because there are two COOH groups, so either of them can react, or both can react. One salt has the structure:

$$H_2N\text{—}CH(CH_2\text{—}CH_2\text{—}COOH)\text{—}COO^-Na^+$$

You may have heard of this salt – its name is monosodium glutamate and is often used as a flavour enhancer in food.

Optical activity

Almost all amino acids contain a chiral centre (the C of the CH group) and so are optically active. The exception is glycine, which has a CH$_2$ group instead. The enantiomers rotate the plane of polarisation of polarised light – some enantiomers are dextrorotatory (+) and others are laevorotatory (−). If an amino acid is synthesised in the laboratory, then a racemic mixture is formed.

Learning tip

Practice writing equations for different amino acids to show how they can react as both acids and bases.

Questions

1. Draw the structure of glycine in a solution with:
   (a) a pH of 6.0
   (b) a very low pH
   (c) a very high pH.

2. Draw structures to show how the two enantiomers of cysteine are related to each other.

Key definitions

The **isoelectric point** of an amino acid is the pH of an aqueous solution in which it is neutral.

A **zwitterion** is a molecule containing positive and negative charges but which has no overall charge.

17.5 6 Peptides and proteins

By the end of this section, you should be able to...

- understand that the peptide bond in proteins:
  (i) is formed when amino acids combine by condensation polymerisation
  (ii) can be hydrolysed to form the constituent amino acids, which can be separated by chromatography

What is a peptide?

When two amino acid molecules react together, an acid–base reaction occurs. The OH of the COOH group combines with one of the H atoms of the NH_2 group to form water. This is a condensation reaction, in which the two amino acids are joined together by an amide group (CO—NH). The bond that forms is known as a **peptide bond** and the organic product is a dipeptide.

More than one dipeptide

When two amino acids combine together to form a dipeptide, there are always two possibilities – sometimes more. For example, when glycine reacts with alanine, the OH could be lost from either molecule, as shown below.

alanine + glycine:

$$H_2N-CH(CH_3)-COOH + H_2N-CH_2-COOH \longrightarrow H_2N-CH(CH_3)-CO-NH-CH_2-COOH + H_2O$$

glycine + alanine:

$$H_2N-CH_2-COOH + H_2N-CH(CH_3)-COOH \longrightarrow H_2N-CH_2-CO-NH-CH(CH_3)-COOH + H_2O$$

Tripeptides

When one molecule of each of three different amino acids reacts together to form a tripeptide, the six possibilities can be summarised using the three-letter abbreviations shown in the table in the previous section. They are:

ala–cys–glu   ala–glu–cys   cys–ala–glu   cys–glu–ala   glu–ala–cys   glu–cys–ala

If you work out the possibilities in which two or more molecules of the same amino acid in these three react to form a tripeptide, you will find that there are many more!

Polypeptides and proteins

Most peptides in living things contain many tens of amino acids joined together, and there comes a point where if the chain is very long the polypeptide is described as a protein. The main difference between a long-chain polypeptide and a protein is that proteins have further levels to their structures. These are to do with the way in which the polypeptide chains interact with each other in three dimensions, to give secondary, tertiary and quaternary structures, but these are beyond the scope of this book.

Analysing proteins

Many proteins have very large molar masses and complex structures. In recent decades, more and more proteins have been analysed. As several million different proteins are thought to exist, this work is still continuing but far from complete. **Table A** shows three examples of proteins, the first two of which have vital roles in the human body.

Protein	Where found	Approximate molar mass/g mol^{-1}	Approximate number of amino acids
insulin	pancreas	5700	51
haemoglobin	blood	66 000	574
urease	soya beans	480 000	4500

table A

The first step in discovering the structure of a protein is to find which amino acids are present. The second is to find the order in which they occur in the polypeptide chains.

Hydrolysing proteins

The polypeptide chains in a protein can be broken down into their individual amino acids by prolonged heating with concentrated hydrochloric acid. This breaks the peptide bonds between the amino acids, although because of the strongly acidic conditions, all the amino acids formed will have their NH$_2$ groups protonated as $^+$NH$_3$ groups. You can see this process more easily in the hydrolysis of a dipeptide. As an example, we will use the dipeptide formed between alanine and glycine, referred to earlier in this section. The equation for this hydrolysis is:

$$H_2N-CH(CH_3)-CO-HN-CH_2-COOH + H_2O + 2H^+ \longrightarrow H_3\overset{+}{N}-CH(CH_3)-COOH + H_3\overset{+}{N}-CH_2-COOH$$

You can see that this hydrolysis reaction is the reverse of the original dipeptide formation, except that the two NH$_2$ groups are protonated, which explains the inclusion of the two H$^+$ ions in the equation.

Using chromatography

The different types of chromatography will be covered in later sections, but you already know about simple chromatography from your previous study of chemistry. A mixture of amino acids produced by hydrolysis of a protein can be spotted onto chromatography paper. Using a suitable solvent, the individual amino acids will rise to different heights during the experiment. As amino acids are colourless, the chromatogram can be sprayed with a developing agent so that the positions of the amino acids can be seen. You may read about the use of ninhydrin as a developing agent. Although this works well, it is not normally used now because of its toxic nature.

However, once the positions of the amino acids have been established, their R_f values can be calculated and so the individual amino acids can be identified. Working out how they are all joined together is rather more complicated and beyond the scope of this book.

Insulin is a protein with the molecular formula $C_{256}H_{381}N_{65}O_{79}S_6$. It is made up of a total of 51 amino acids, including 17 different ones. Its structure is shown in abbreviated form in **fig A**.

17.5 6

```
phe           gly
 |             |
val           ile
 |             |
asp           val
 |             |
gln           glu
 |             |
his           gln
 |             |
leu           cys
 |             |
cys—S—S—cys
 |             |
gly           thr
 |             |
ser           ser
 |             |
his           val
 |             |
leu           cys
 |             |
val           ser
 |             |
glu           leu
 |             |
ala           tyr
 |             |
leu           gln
 |             |
tyr           leu
 |             |
leu           glu
 |             |
val           asp
 |             |
cys—S         tyr
 |      \      |
gly      S—cys
 |             |
gly           asp
 |
arg
 |
gly
 |
phe
 |
phe
 |
tyr
 |
thr
 |
pro
 |
lys
 |
thr
```

fig A This diagram shows the sequence of amino acids in insulin from a sheep. You can see that there are actually two separate sequences of amino acids in which cysteine molecules are joined by disulfide bonds.

Learning tip

Two possible dipeptides can form between two different amino acids. If one or both of the amino acids has more than one NH_2 or COOH group, then more possible dipeptides can be formed.

Questions

1. Write the structures of the two dipeptides formed when alanine and cysteine react together.

2. Write the structures of the two amino acids formed when this dipeptide is hydrolysed by concentrated hydrochloric acid:

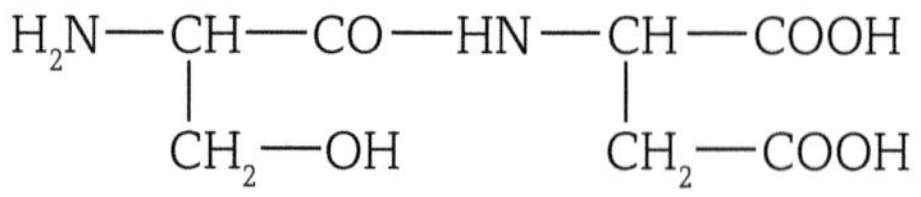

$$H_2N-CH-CO-HN-CH-COOH$$
$$\quad\quad\quad |\quad\quad\quad\quad\quad\quad\quad |$$
$$\quad\quad\quad CH_2-OH\quad\quad\quad CH_2-COOH$$

Key definition

A **peptide bond** is the bond formed by a condensation reaction between the carbonyl group of one amino acid and the amino group of another amino acid.

17.6 1 Principles of organic synthesis

By the end of this section, you should be able to...

- plan reaction schemes, of up to four steps, to form both familiar and unfamiliar compounds
- understand methods of increasing the length of the carbon chain in a molecule by the use of magnesium to form Grignard reagents and the reactions of the latter with carbon dioxide and with carbonyl compounds in dry ether

What is organic synthesis?

The term 'synthesis' (the plural is 'syntheses') refers to making something new from what already exists. In organic chemistry, it usually means using a familiar compound to make an unfamiliar compound. You have already learned many examples of converting one compound into another – for example, converting a halogenoalkane into an alcohol. In this section, we will consider converting one compound into another compound that cannot be achieved in one step, but needs two or more steps. To keep things relatively simple, in this book the number of steps will be limited to four.

The ability to plan a reaction scheme to achieve this synthesis is not easily learned. It is not like recalling a simple mathematical expression and putting the numbers into a calculator to obtain the answer. Instead, it relies on a thorough knowledge of a range of chemical reactions and the ability to select the ones that are appropriate to the problem.

Here is a simple example of a two-step synthesis – how to convert bromoethane to ethyl propanoate. You could start by considering the reactions of bromoethane, and then consider reactions in which ethyl propanoate is a product. You could consider the question as a puzzle in which you need to find the identity of X in this sequence:

bromoethane → X → ethyl propanoate

If you know a product of a reaction of bromoethane that is the same as a compound that can be converted into ethyl propanoate, then you only need to include the reagents and conditions to achieve each of these two steps. When the synthesis involves three steps, it is more difficult because you need to identify two intermediate compounds in the synthesis. If there are four steps, then we have reached the limit of what is reasonable to expect at this point in your study of chemistry!

In the example above, X is ethanol, and the steps are:

Step 1: bromoethane → ethanol
reflux with aqueous potassium hydroxide (hydrolysis)

Step 2: ethanol → ethyl propanoate
heat with propanoic acid using an acid catalyst (esterification).

Preparing for planning a synthesis

One thing that will help you learn how to plan a synthesis is to put together an outline of all the reactions you need to know. You could leave out reactions that are less useful in synthesis, such as combustion, polymerisation and chemical tests. You could try to produce a diagram (like a spider diagram) that would include all the reactions, but that would probably be too complicated to easily use. It might be better to produce a series of smaller ones. For example, **fig A** shows reactions based on alcohols.

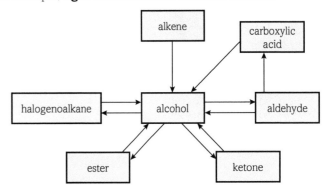

fig A Reactions of alcohols.

Where the arrows point in both directions, this means that you need to know both conversions. Note that reactions from **Book 1** are included. If you copy this onto a sheet of paper, you can add the formulae of reagents and any special conditions on the arrows. You could use different colours to emphasise different features.

Extending a carbon chain

It is possible that one of the steps in your synthesis will involve extending an existing carbon chain by one or more carbon atoms. You have already learned three ways of doing this:

1 Reacting a halogenoalkane with a cyanide ion forms a nitrile with one more carbon atom than the halogenoalkane.
2 The addition of hydrogen cyanide to a carbonyl compound.
3 The alkylation of benzene, which introduces an alkyl group into a benzene ring.

A different way, which is more versatile, involves the use of Grignard reagents.

Grignard reagents

Victor Grignard was a French chemist who devised a new way to synthesise organic compounds. His method involved the use of organometallic compounds containing magnesium, which are now known as Grignard reagents. They are made by heating under reflux the chosen halogenoalkane with magnesium in a solvent of dry ether. Haloarenes such as bromobenzene are also used. Bromoalkanes are used in preference to other halogenoalkanes.

Grignard reagents contain magnesium covalently bonded to both the alkyl group and the halogen. A general equation for the formation of a Grignard reagent is:

$$R\text{—}Br + Mg \rightarrow R\text{—}Mg\text{—}Br$$

Grignard reagents react with water, so they are both made and used in a solvent of dry ether.

Grignard reactions

Once a Grignard reagent has been made, it can then be converted into a range of organic compounds, depending on the reagent used. In this book we will consider only two of these reactions – with carbon dioxide and with carbonyl compounds. They can be summarised as follows.

1 with carbon dioxide $RMgBr \rightarrow RCOOH$ carboxylic acid
2a with methanal $RMgBr \rightarrow RCH_2OH$ primary alcohol
2b with an aldehyde R'CHO $RMgBr \rightarrow RR'CHOH$ secondary alcohol
2c with a ketone R'COR'' $RMgBr \rightarrow RR'R''COH$ tertiary alcohol

R, R' and R'' represent different alkyl groups, although they could be the same. Note that both carbon dioxide and methanal increase the carbon chain by one carbon atom. Both aldehydes and ketones produce a branch, of variable length, on the existing carbon chain.

Having made the appropriate Grignard reagent, the second reagent (carbon dioxide or the carbonyl compound) is added. After the reaction is complete, a dilute acid is added to obtain the desired organic product.

fig B Victor Grignard won the Nobel prize in chemistry for his work in organic synthesis.

Examples of Grignard reactions

Formation of 3-methylbutanoic acid

$$CH_3\text{—}CH\text{—}CH_2\text{—}COOH$$
$$|$$
$$CH_3$$

This is an example of reaction 1, described above. The COOH part comes from carbon dioxide, so the starting compound must supply the $(CH_3)_2CHCH_2$ part, which means using $(CH_3)_2CHCH_2Br$, i.e. 1-bromo-2-methylpropane. The equations for the reactions that occur in this example are:

Step 1: formation of the Grignard reagent

$$(CH_3)_2CHCH_2Br + Mg \rightarrow (CH_3)_2CHCH_2MgBr$$

Step 2: reaction with chosen reagent

$$(CH_3)_2CHCH_2MgBr + CO_2 \rightarrow (CH_3)_2CHCH_2COOMgBr$$

Step 3: hydrolysis using a dilute acid

$$(CH_3)_2CHCH_2COOMgBr + H_2O \rightarrow (CH_3)_2CHCH_2COOH + Mg(OH)Br$$

The inorganic product of Step 3 will react with the dilute acid. You do not need to know details of the mechanisms.

Formation of propan-1-ol

This and the other examples are abbreviated as reaction schemes, rather than being shown as three steps.

Propan-1-ol, $CH_3CH_2CH_2OH$, is a primary alcohol. So, methanal is the source of the CH_2OH part, and the starting compound is bromoethane, which is the source of the CH_3CH_2 part.

$$CH_3CH_2Br \xrightarrow{Mg/ether} CH_3CH_2MgBr \xrightarrow[\text{then } H_2O/H^+]{HCHO} CH_3CH_2CH_2OH$$

Formation of pentan-2-ol

Pentan-2-ol, $CH_3CH(OH)CH_2CH_2CH_3$, is a secondary alcohol, so ethanal is the source of the $CH_3CH(OH)$ part, and the starting compound is 1-bromopropane, which is the source of the $CH_3CH_2CH_2$ part.

$$CH_3CH_2CH_2Br \xrightarrow{Mg/ether} CH_3CH_2CH_2MgBr \xrightarrow[\text{then } H_2O/H^+]{CH_3CHO} CH_3CH(OH)CH_2CH_2CH_3$$

Formation of 2-methylpropan-2-ol

2-Methylpropan-2-ol, $(CH_3)_3COH$ is a tertiary alcohol, so propanone is the source of the $(CH_3)_2COH$ part, and the starting compound is bromomethane, which is the source of the CH_3 part.

$$CH_3Br \xrightarrow{Mg/ether} CH_3MgBr \xrightarrow[\text{then } H_2O/H^+]{CH_3COCH_3} (CH_3)_3COH$$

Learning tip

Practise choosing reagents for a Grignard synthesis. First, identifying the product type (acid or alcohol) tells you the type of compound to react with the Grignard reagent. Then, work out how many carbon atoms are needed in the halogenoalkane used to form the Grignard reagent. See how you get on with the questions in this section!

Questions

1. Draw the structure of the organic product obtained in a Grignard synthesis starting from 1-bromo-2,2-dimethylpropane and propanal.

2. Draw the structures of the two compounds needed in a Grignard synthesis of 3,3-dimethylhexan-4-ol.

17.6 2 Hazards, risks and control measures

By the end of this section, you should be able to...

- select and justify suitable practical procedures for carrying out reactions involving compounds with functional groups included in the specification, including identifying appropriate control measures to reduce risk, based on data about hazards

Safety in chemistry laboratories

Incidents that cause harm to people are rare in school and college chemistry laboratories. One of the reasons for this is that all laboratories must consider the hazards of carrying out chemistry experiments and use safe methods of working.

This applies to all chemistry experiments, whether you are given the instructions for doing them, or whether you plan your own. When you plan an organic synthesis, you must consider the hazards associated with the reactants, the substance you are synthesising, and with any intermediate compounds formed.

Hazards, risks and control measures

The hazards of chemical substances relate to the inherent properties of each substance and the way in which it will be used. Most people would consider that water (H_2O) is completely safe and has no hazards. In most situations this is the case. However, consider a beaker of boiling water. The steam coming from the boiling water, and the boiling water itself, could both cause harmful effects if they came into contact with your skin.

Now consider a substance that most people would consider dangerous. You probably used dilute hydrochloric acid years ago in experiments involving marble chips. You will have been told to wear eye protection when using the acid, because of the harm it could do if it got into your eyes.

The **hazard** exists because hydrochloric acid is corrosive.

The **risk** is that hydrochloric acid may get into your eyes and cause harm.

The control measure is to wear eye protection to reduce the risk of the acid getting into your eyes.

This simple example should help you to understand the difference between hazard and risk – these two words are often thought, wrongly, to have the same meaning.

Hazard warning symbols

A long time ago, bottles containing certain chemicals were labelled with the word POISON. This early attempt to prevent harm to laboratory workers was well intentioned, but some people may have thought that bottles without such a label contained harmless substances.

More recently, symbols (sometimes called pictograms) have been used as labels for bottles to identify the actual hazard of the substance inside. The actual symbols have changed over the years, and you may still see older ones used, especially those in squares with an orange background, such as these:

The symbols in current use are red diamond shapes. **Table A** shows some of the more common ones, along with a simple description of their meanings.

Symbol	Meaning	
	Health hazard	includes warnings on skin rashes, eye damage and ingestion
	Corrosive	can cause skin burns and permanent eye damage
	Flammable	can catch fire if heated or comes into contact with a flame
	Acute toxicity	can cause life-threatening effects, even in small quantities

table A

In some cases, the substance may have more than one symbol, especially when it is an aqueous solution. For example, you are likely to use hydrochloric acid in three different concentrations:

- when used in a titration, it will have a concentration of approximately 0.1 mol dm^{-3}
- as a general laboratory reagent, it will probably have a concentration of approximately 1 or 2 mol dm^{-3}
- a bottle labelled 'concentrated hydrochloric acid' may have a concentration of more than 10 mol dm^{-3}.

These very different concentrations have different risks.

There are several other hazard symbols, including those that you are more likely to find in biology or physics laboratories.

Control measures

It is the responsibility of whoever is in charge of the laboratory to identify risks and hazards, and to prescribe appropriate control measures. This responsibility is shared with the student doing the experiment, who should follow the guidance. There are several organisations that provide advice and support to schools and colleges, and it is likely that the health and safety information you see in your laboratory has come from one of these sources.

As a student, you may be asked to plan an experiment or a synthesis, including identifying the hazards and control measures needed, so you need to be familiar with this information.

fig A You will also find examples of hazards, risks and control measures in your home.

Apparatus

You also need to consider risks associated with apparatus. For example, mercury thermometers are often replaced by spirit thermometers or digital thermometers, which have much lower risks. Heating may be done using an electrical heating mantle instead of a Bunsen burner. You may use glass apparatus with ground glass joints, as these are safer to set up than those that use corks or bungs. You also have to consider where to support apparatus with a clamp and stand.

Learning tip

Become familiar with the hazards of the substances you use in an experiment and consider how any risks of using them may be reduced.

Questions

1. Suggest some control measures you should use in an experiment to convert bromoethane into ethanol.

2. Use your chemical knowledge to suggest why spilling sodium hydroxide solution onto your skin may cause more harm than hydrochloric acid of the same concentration.

Key definitions

A **hazard** is a property of a substance that could cause harm to a user.

A **risk** is the possible effect that a substance may cause to a user, and this will depend on factors such as concentration and apparatus. The level of risk is controlled using control measures.

17.6 3 Practical techniques in organic chemistry – Part 1

By the end of this section, you should be able to...

- understand the following techniques used in the preparation and purification of organic compounds: refluxing, simple distillation, steam distillation, fractional distillation

Different practical techniques

In the preparation and purification of any organic compound, there are likely to be several techniques used. Each preparation or synthesis will use one or more of the techniques in this section and the next. Most of them have been covered in **Book 1**, but all are included in these sections for completeness.

Heating under reflux

Some reactions involving organic compounds are slow at room temperature, so the obvious thing to do is heat the reaction mixture. However, as many organic compounds are volatile, there is a risk that they will escape from the reaction mixture during the heating process. The normal way to prevent this happening is to heat the reaction mixture in a flask fitted with a reflux condenser. This technique is also known as refluxing. All the vapours rising from the reaction mixture during heating enter the condenser and change back into liquids and return to the flask so that the unreacted compounds can react.

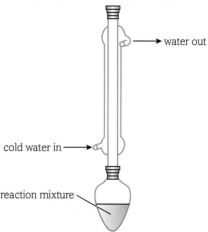

fig A Apparatus for heating under reflux.

The flask can be heated using hot water or oil in a beaker heated by a Bunsen burner or by using an electric mantle. The use of anti-bumping granules in the reaction mixture helps to make the boiling smooth. A gentle flow of cold water enters at the bottom of the condenser.

Methods of separation

Simple distillation

This is a separation technique used to obtain a liquid product from a reaction mixture that has a boiling temperature much lower than the other substances in the reaction mixture.

Distillation of an impure liquid involves heating it in a flask connected to a condenser. The liquid with the lowest boiling temperature evaporates or boils off first and passes into the condenser first, and so can be collected in the receiver separately from another liquid that may evaporate later. The purpose of the thermometer is to monitor the temperature of the vapour as it passes into the condenser. If the temperature remains steady, this is an indication that one compound is being distilled. If after a while the temperature begins to rise, this indicates that a different compound is being distilled.

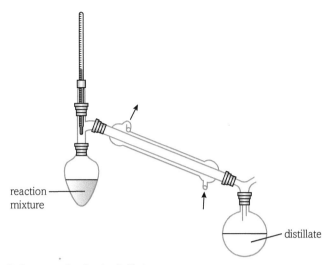

fig B Apparatus for simple distillation.

The advantages of using simple distillation, rather than fractional distillation, are that it is easier to set up and is quicker. The disadvantage is that it does not separate the liquids as well as fractional distillation. It should only be used if the boiling temperature of the liquid being purified is very different from the other liquids in the mixture, ideally a difference of more than 25 °C.

Steam distillation

Steam distillation is a technique used to separate an insoluble liquid from an aqueous solution. It involves passing steam into a reaction mixture that contains an aqueous solution and a liquid that forms a separate layer. The agitation of the liquid caused by the steam bubbling through the mixture ensures that both the insoluble liquid and the aqueous solution are on the surface of the mixture and so can form part of the liquid that evaporates. The advantage of this process is that the insoluble liquid is removed from the reaction mixture at a temperature below its normal boiling temperature.

An example is phenylamine, a liquid with a boiling temperature of 184 °C. During steam distillation of a reaction mixture containing phenylamine, a liquid containing phenylamine and water distils at a temperature of 98 °C. An advantage of doing this is that the temperature at which phenylamine distils is much lower than its boiling temperature, so there is less chance of decomposition.

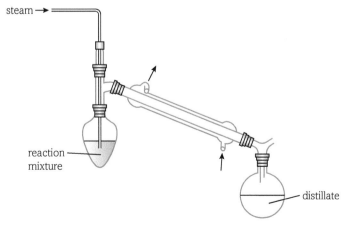

fig C Apparatus for steam distillation.

Fractional distillation

Fractional distillation uses the same apparatus as simple distillation, but with a fractionating column between the heating flask and the still head. The column is usually filled with glass beads or pieces of broken glass, which act as surfaces on which the vapour leaving the column can condense, and then be evaporated again as more hot vapour passes up the column. Effectively, the vapour undergoes several repeated distillations as it passes up the column, which provides a better separation.

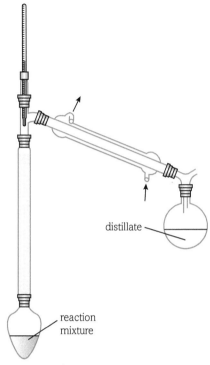

fig D Apparatus for fractional distillation.

Fractional distillation takes longer than simple distillation. It is best used when the difference in boiling temperatures is small, and when there are several compounds to be separated from a mixture.

Learning tip

Remember that there are three types of distillation, and that you need to consider which one to use in a particular case.

Questions

1. Explain briefly why the condenser is vertical in refluxing but nearly horizontal in the three methods of distillation.

2. Suggest, with a reason, which fractionating column provides the better separation – very small glass beads or larger pieces of glass.

17.6 4 Practical techniques in organic chemistry – Part 2

By the end of this section, you should be able to...
- understand the following techniques used in the preparation and purification of organic compounds: washing, drying, solvent extraction, recrystallisation, melting temperature determination, boiling temperature determination

More methods of separation

Solvent extraction
As the name suggests, this method involves using a solvent to remove the desired organic product from the other substances in the reaction mixture. Several solvents can be used, but the choice depends mainly on these features:

- The solvent added should be immiscible (i.e. does not form a mixture) with the solvent containing the desired organic product.
- The desired organic product should be much more soluble in the added solvent than in the reaction mixture.

The method is this:

- Place the reaction mixture in a separating funnel, and then add the chosen solvent – it should form a separate layer.
- Place the stopper in the neck of the funnel and gently agitate the contents of the funnel for a short while (put a finger on the stopper, invert, open the tap, agitate in a circular motion, close the tap and return the funnel to its normal position).
- Allow the contents to settle into two layers.
- Remove the stopper and open the tap to allow the lower layer to drain into a flask, and then do the same to allow the upper layer to drain into a separate flask.

fig A Apparatus for separating immiscible liquids.

If a suitable solvent is used and the method is followed correctly, most of the desired organic product will have moved into the added solvent. It is better to use the solvent in small portions rather than in a single volume (e.g. four portions of 25 cm³ rather than one portion of 100 cm³) because this is more efficient. Using more portions of solvent, but with the same total volume, removes more of the desired organic product.

The desired organic product has been removed from the reaction mixture, but is now mixed with the added solvent. Therefore, simple distillation or fractional distillation must now be used to separate the desired organic product from the solvent used.

Washing
You might assume that washing is a technique used to remove impurities from a solid. Although this is correct, it is also used to remove impurities from a liquid. Washing could involve the use of water or an organic solvent. Whichever liquid is used, it must be chosen carefully so as to dissolve the impurities but as little as possible of the substance being purified.

An impure solid is stirred in some of the solvent, then the mixture filtered. If the solid is already in a filter funnel, the solvent could be added on top of the solid.

A liquid would be mixed with a solvent chosen so that it will dissolve little, if any, of the liquid to be purified. The mixture is then shaken in a separating funnel. After allowing the two liquid layers to separate, the tap is opened to allow each layer to drain into a separate container.

Drying
No special technique is needed to dry an organic solid – it just needs to be left in a warm place or in a desiccator with a suitable drying agent.

fig B Drying an organic solid in a desiccator

Many organic liquids are prepared using inorganic reagents, which are often used in aqueous solution. A liquid organic product may partially, or even completely, dissolve in water. Therefore, water may be an impurity that needs to be removed by a drying agent. Obviously, one important feature of a drying agent is that it does not react with the organic liquid.

Several drying agents are available, but the most common ones are anhydrous metal salts – usually calcium chloride, magnesium sulfate and sodium sulfate. The property that these compounds have in common is that they form hydrated salts. So, when they come into contact with water in an organic liquid, they absorb the water as water of crystallisation.

The drying agent is added to the organic liquid and the mixture is swirled or shaken, and then left for a period of time. Before use, a drying agent is powdery, but after absorbing water it looks more crystalline. If a bit more drying agent is added and remains powdery, then this is an indication that the liquid is dry. The liquid also goes from cloudy to clear when water is removed.

The drying agent is removed either by decantation (pouring the organic liquid off the solid drying agent) or by filtration.

Filtration

It is likely that after preparation, an organic solid will need to be filtered at some stage. This always happens as part of recrystallisation, so the two pieces of apparatus (Buchner and Hirsch funnels) that are normally used will be considered under the next heading. The use of a vacuum pump means that these methods are described as filtration under reduced pressure.

Recrystallisation

When an organic solid has been prepared, it is likely to need purification. A traditional way of removing impurities is the technique of recrystallisation. The principle behind this technique is that a solid compound is dissolved in a suitable solvent that can dissolve all or most of any impurities but very little of the compound being purified. The steps used in a typical purification are:

- Add the impure solid to a conical flask.
- Add some of the chosen solvent and warm until the mixture nears the boiling temperature of the mixture.
- If there is still some undissolved solid, add further solvent and warm until the mixture boils again.
- Continue adding further solvent and heating until all of the soluble solid has dissolved.
- If insoluble impurities are present, then hot filtration could be done using fluted filter paper in a heated funnel.
- Allow the liquid to cool until crystals of the organic solid have formed.
- More crystals can be obtained by cooling the solution below room temperature in an ice bath.
- The mixture is then filtered to remove soluble impurities using a Buchner funnel or a Hirsch funnel, and then dried in a desiccator or oven.

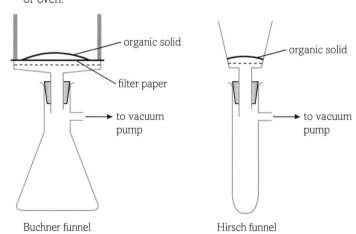

fig C Apparatus for filtration under reduced pressure.

Learning tip
Study the information about recrystallisation carefully. Many students do not understand why there might be two filtrations needed. Some throw away the substance they have just purified!

Testing for purity

Having prepared an organic compound, a simple test can be performed to give an indication of whether the compound is pure. If the compound is a solid, then its melting temperature can be measured. If it is a liquid, then its boiling temperature can be measured.

Determination of melting temperature

For solids, impurities reduce the melting temperature. If you measure the melting temperature of your organic compound, then you can compare it with the known value of the pure compound. This will enable you to estimate how pure it is.

The traditional way to measure a melting temperature of a solid is to place some of the solid in a small capillary tube attached to the bulb of a thermometer and then place the assembly in a liquid that has a boiling temperature above that of the melting temperature of the solid.

In practice, this apparatus is often replaced by various electrical devices.

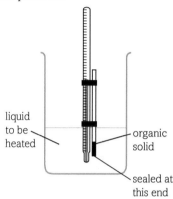

fig D Apparatus for determination of melting temperature.

Determination of boiling temperature

For liquids, impurities increase the boiling temperature.

The boiling temperatures of pure organic compounds have been carefully measured and are widely available in data books and online. If you measure the boiling temperature of your organic compound, you can compare the value you obtain with the known value of the pure compound and then estimate how pure the compound is.

The apparatus used depends on the volume of liquid available, and whether it is toxic or flammable. The apparatus used for simple distillation can be used in most cases.

A word of caution. This test may not be conclusive because you may not be able to measure the boiling temperature of your organic compound accurately enough – your thermometer might read too low or too high. So, even if your measured boiling temperature exactly matches the one in the data book or online, you may wrongly assume that your compound is pure.

It is also worth remembering that different organic compounds can, by coincidence, have the same boiling temperature. For example, both 1-chloropentane and 2-methylpropan-1-ol boil at 108 °C.

Questions

1. Explain why the solvent extraction method will not work if the solvent dissolves the impurities.

2. Explain why, in recrystallisation, insoluble impurities are removed from a hot solution, not a cooled one.

17.6 5 Simple chromatography

By the end of this section, you should be able to...

- understand that chromatography separates components of a mixture between a mobile phase and a stationary phase
- calculate R_f values from one-way chromatograms

Paper chromatography

You may be familiar with paper chromatography from your early experiences of chemistry. This is a suitable place to remind you of this technique and to explain some of the principles that relate to most types of chromatography. The word 'chromatography' comes from the Greek for colour (*chroma*) and writing (*graphein*) because mixtures of coloured substances were used in early experiments. However, the technique can also be used with colourless substances because these can be detected in other ways without considering their colours. Chromatography is used:

- to separate a mixture into its individual components
- to identify the components of a mixture by considering how far they have travelled up the paper.

If you have synthesised an organic compound, chromatography is a good way to check whether it is pure and, if not, what impurities are present.

Apparatus

All forms of paper chromatography need the following:

- A container (usually glass) with a lid. The lid is there to prevent evaporation of the solvent. The container could be a beaker or rectangular tank.
- Paper. In introductory experiments you may have used filter paper or kitchen towel, but specially formulated chromatography paper gives better results. There also needs to be a method of supporting the paper in the container.
- A solvent. In some cases, water works well, but often the solvent is a mixture of organic compounds, and is chosen to fit the characteristics of the components of the mixture.

Whichever variation is used, the mixture is spotted onto the paper a short distance from one edge. Often, spots of known substances are spotted separately at the same level as the mixture. A small volume of the chosen solvent is added to the container, and then the paper is inserted and suitably supported. The lid is replaced and the apparatus left until the solvent front has reached almost to the top of the paper. The paper is then removed, the position of the solvent front marked, and the paper left to dry. The resulting dried paper is known as a chromatogram. **Fig A** shows typical apparatus used to set up a chromatography experiment and the resulting chromatogram.

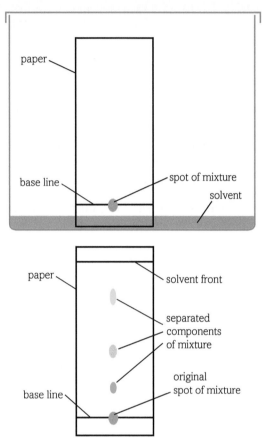

fig A The chromatogram is shown in a rather idealised way – the spots are often elongated and their centres are difficult to identify.

How does chromatography work?

All types of chromatography depend on the use of a **stationary phase** and a **mobile phase**. Each component in the mixture is attracted to both phases, but more strongly to one than the other.

- A component that is strongly attracted to the stationary phase but weakly attracted to the mobile phase will not travel very far up the paper.
- A component that is weakly attracted to the stationary phase but strongly attracted to the mobile phase will travel a long way up the paper.

In paper chromatography, the stationary phase is the water trapped in the fibres of the chromatography paper, and the mobile phase is the solvent.

We have used colours to illustrate the different substances in the mixture, but most organic compounds are colourless. Chromatography can still be used, but the components on the chromatogram must be visualised using ultraviolet radiation or by spraying with a chemical reagent that will react with the component to form a coloured product.

224

Thin layer chromatography

The apparatus and method used in thin layer chromatography (TLC) is very similar to those of paper chromatography. The only difference is that instead of paper, a sheet of glass or plastic coated in a thin layer of a solid such as silica or alumina is used.

fig B This is a TLC plate after a chromatography experiment.

Calculating R_f values

The chromatogram is analysed and the distance travelled by the solvent (from the base line to the solvent front) is measured. For each component in the mixture, the distance it has travelled (starting from the base line) is measured.

The R_f value is calculated using the expression:

$$R_f = \frac{\text{distance travelled by component}}{\text{distance travelled by solvent}}$$

R_f values have no units because they are a ratio of two distances. The letters in the term refer to 'retardation' (sometimes 'retention') and 'factor'. Each component has a characteristic R_f value, so in theory different components can be identified by reference to a table of known R_f values. Unfortunately, R_f values depend on the solvent used and other factors. In practice, paper and thin layer chromatography are usually carried out with known substances included in the same experiment, so that a comparison can be made.

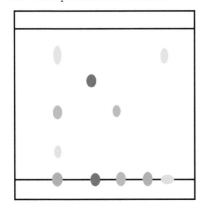

fig C Chromatogram.

This chromatogram suggests that the mixture on the left contains three different substances. Two of these three are probably the green and blue substances on the right. The third substance (yellow) cannot be identified by this experiment.

Learning tip

Remember, when calculating R_f values, the distance moved by the solvent is measured from the base line to the solvent front.

Column chromatography

Column chromatography uses the same principles as thin layer chromatography. The stationary phase is alumina or silica packed into a tube (a burette will do) and soaked in a solvent. The mixture is placed on top of the stationary phase and more solvent (the mobile phase) added on top. When the tap is opened, the solvent drips through the tip and the components of the mixture begin to move down the tube and separate.

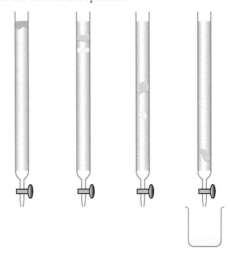

fig D Separation of a mixture by column chromatography.

More solvent is added at the top, and eventually one component (yellow in this example) leaves the column and can be collected in a container. If the experiment is continued, the blue component can be collected in the same way.

The advantage of column chromatography is that much larger quantities of material can be separated than with paper chromatography.

Questions

1. Suggest why, in paper chromatography, the non-polar substance hexane has a high R_f value.

2. Calculate the R_f value of component X in this diagram.

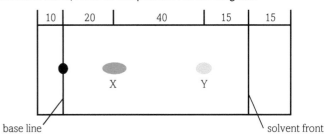

Key definitions

The **stationary phase** in paper chromatography is the liquid or solid that does not move.

The **mobile phase** is the liquid that moves through the stationary phase and transports the components.

17.7 1 Traditional methods of analysis

By the end of this section, you should be able to...

- deduce the empirical formulae, molecular formulae and structural formulae of compounds from data obtained from combustion analysis, elemental percentage composition and characteristic reactions of functional groups

Introduction

Chapter 17.6 is all about organic synthesis – planning how to convert one organic compound into another, including common techniques used in preparation and purification. In this section we will look at methods of analysis – different ways of finding the structure of an organic compound.

Modern methods, such as infrared spectroscopy, are increasingly used. In many situations two or more methods are used together. In this section we will first consider the traditional methods of analysis, many of which were introduced in **Book 1**.

Determining empirical and molecular formulae

Qualitative analysis

The purpose of qualitative analysis is to find out which elements are present in a compound. We will take as our example compound X, which is found to contain only carbon, hydrogen and oxygen.

Quantitative analysis

One method of finding the percentage composition by mass of a compound is combustion analysis. A known mass of the compound is burned, and the masses of carbon dioxide and water formed are found by measuring the increase in mass of U-tubes containing the absorbers (shown in **fig A**).

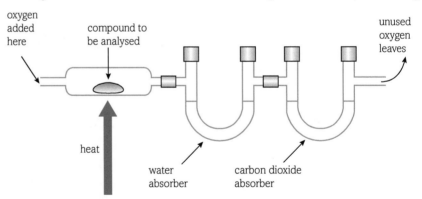

fig A Apparatus for combustion analysis.

Table A shows the masses for the combustion of compound X.

Substance	Mass/g
compound X	1.98
water formed	1.44
carbon dioxide formed	3.53

table A

The first step is to calculate the masses of carbon and hydrogen in the carbon dioxide and water. The relative molecular mass (M_r) of carbon dioxide is 44.0, but because the relative atomic mass (A_r) of carbon is 12.0, the proportion of carbon in carbon dioxide is always $\frac{12.0}{44.0}$. Similarly, the proportion of hydrogen in water is always $\frac{(2 \times 1.0)}{18.0}$.

All of the carbon in the carbon dioxide comes from the carbon in the organic compound. Similarly, all of the hydrogen in the water comes from the hydrogen in the organic compound. So, in this example:

$$\text{mass of carbon} = \frac{3.53 \times 12.0}{44.0} = 0.963\,g$$

$$\text{and mass of hydrogen} = \frac{1.44 \times 2.0}{18.0} = 0.160\,g$$

These two masses add up to 1.123 g. However, the original mass of the organic compound was 1.98 g, so the difference must be the mass of oxygen present in the organic compound (1.98 − 1.123 = 0.857 g).

Calculation of empirical formula

This is done by dividing the mass of each element by its relative atomic mass (there is no need to convert the masses to percentages). **Table B** is the calculation table.

	C	H	O
mass of element/g	0.963	0.160	0.857
A_r	12.0	1.0	16.0
division by A_r	0.0803	0.160	0.0536
ratio	1.5	3	1
whole number ratio	3	6	2

table B

The empirical formula of X is therefore $C_3H_6O_2$.

Learning tip

Remember that you can do empirical formula calculations starting from the masses of the elements present, or from the percentage composition by mass. Also, if the supplied percentages do not add up to 100, then you should assume that the remainder is oxygen.

Determination of molecular formula

To deduce the molecular formula of X from its empirical formula, an additional piece of information is needed – the molar mass of X. Let us assume that this is found by experiment to be 74.0 g mol^{-1}. We can now compare the 'molar mass' of the empirical formula (74.0) with the actual molar mass of X. In this example they are both the same, so we can deduce that the molecular formula of X is $C_3H_6O_2$ – the same as the empirical formula.

Determining structural formulae

For a compound (X) with a molecular formula of $C_3H_6O_2$ there are several possibilities for the structure. Using traditional methods, we can obtain information from chemical tests to help decide which functional groups are present. Although it is not a good idea to jump to an obvious conclusion, one way for the atoms in X to be arranged is as CH_3—CH_2—COOH. This suggestion could be confirmed, or not, by a simple chemical test – the addition of sodium hydrogencarbonate. If effervescence is observed, this evidence would support the suggestion that X is CH_3—CH_2—COOH. If there is no effervescence, then X cannot have this structure. It would be sensible to test X for the presence of other functional groups before making a decision about its structure. **Table C** shows the tests done on samples of X, and the results.

	Test	Result with X
1	add sodium hydrogencarbonate	no change
2	add 2,4-dinitrophenylhydrazine solution	orange-yellow precipitate
3	add Tollens' reagent and warm	no change
4	heat with acidified potassium dichromate(VI)	colour change from orange to green
5	test product from test 4 with Tollens' reagent and warm	no change
6	test product from test 4 with sodium hydrogencarbonate	effervescence

table C

The conclusions that can be made from the results of these tests are:

1. X does not have a COOH group
2. X has a carbonyl (C=O) group
3. X does not have an aldehyde (CHO) group
4. X contains a group that can be oxidised – as this is not an aldehyde group, it could be an alcohol group
5. the oxidation product of test 5 is not an aldehyde, but could be a ketone or a carboxylic acid
6. the oxidation product of test 5 is a carboxylic acid, so X could contain a primary alcohol group.

A structure that is consistent with the results of these tests is:

$$CH_3—CO—CH_2—OH$$

Limitations of traditional methods of analysis

Even with this very simple example, you can see how difficult it is to be sure of the structure of X, whose molecules contain only 11 atoms. With compounds whose molecules contain many more atoms than this, you can see that traditional methods of analysis become less suitable.

We will consider the use of more modern methods in the following sections.

Questions

1. A compound, S, has the percentage composition by mass C = 40.00, H = 6.67 and O = 53.33. Its M_r is 60.0 and it causes effervescence with sodium hydrogencarbonate. Suggest its identity.

2. Another compound, T, has the same percentage composition and M_r value as S. It does not cause effervescence with sodium hydrogencarbonate. Suggest its identity.

17.7 2 Determining structures using mass spectra

By the end of this section, you should be able to...

- use data from mass spectra to:
  (i) suggest possible structures of a simple organic compound given relative molecular masses, accurate to four decimal places
  (ii) calculate the accurate relative molecular mass of a compound, given relative atomic masses to four decimal places, and therefore identify a compound

Mass spectrometry so far

In **Book 1**, you learned about two aspects of mass spectrometry. They were:

- determining the relative atomic masses of elements from isotopic abundances
- using fragmentation patterns to determine the structures of organic compounds.

In this section, we will learn about another use – determining the molecular formula (although not always the structure or identity) of an organic compound from a precise relative molecular mass obtained from **high resolution mass spectrometry** (HRMS).

High resolution mass spectrometry

Without going into technical detail, there are mass spectrometers that can provide a value for a relative molecular mass to four or more decimal places. Values with this degree of precision sometimes enable a compound to be positively identified from the relative molecular mass alone, without the need for any other information. Consider compounds with $M_r = 58$ (or 58.0). There are several possible structures, including those in **table A**:

Formula	Structure	Accurate M_r
C_4H_{10}	$CH_3CH_2CH_2CH_3$ or $(CH_3)_3CH$	58.0780
C_3H_6O	CH_3COCH_3 or CH_3CH_2CHO	58.0417
$C_2H_6N_2$	$H_2NCH=CHNH_2$	58.0530

table A

fig A HRMS equipment looks very little different from other mass spectrometry equipment.

Once an accurate M_r has been obtained from HRMS, it is possible to decide whether the compound has the molecular formula C_4H_{10} or C_3H_6O or $C_2H_6N_2$. Note that if the accurate M_r is found to be 58.0417, it is still not possible to decide whether the compound is propanone or propanal because they both have the same accurate M_r value.

Learning tip

HRMS gives information only about molecular formulae, not about structure. Structures can be deduced from the fragmentation patterns in mass spectrometry.

Calculating the accurate relative molecular mass

A variation on this method involves you doing some very straightforward but tedious calculations. You know that relative atomic masses are relative to the average mass of a carbon atom (^{12}C) taken as exactly 12, which for our purposes means 12.0000. **Table B** shows the accurate atomic masses of some common elements:

Element	Symbol	Accurate atomic mass
hydrogen	H	1.0078
carbon	C	12.0000
nitrogen	N	14.0031
oxygen	O	15.9949

table B

You will need to use information from **table B** in the following examples.

WORKED EXAMPLE 1

A compound is found to have an M_r value of 84.0573. Which of these structures does the compound have?

$$CH_2=CHCH_2CH_2CH_2CH_3 \text{ or } CH_3CH=CHCH_2CHO$$

Answer
The molecular formula of $CH_2=CHCH_2CH_2CH_2CH_3$ is C_6H_{12} and its molecular formula is calculated as follows:

$$(6 \times 12.0000) + (12 \times 1.0078) = 84.0936$$

The molecular formula of $CH_3CH=CHCH_2CHO$ is C_5H_8O and its molecular formula is calculated as follows:

$$(5 \times 12.0000) + (8 \times 1.0078) + 15.9949 = 84.0573$$

This means that the compound is $CH_3CH=CHCH_2CHO$.

WORKED EXAMPLE 2

Three compounds have these structures:

A $CH_3CH_2CH_2CH=CH_2$
B $CH_3CH(NH_2)CN$
C $CH_2=CHCOCH_3$

In a high resolution mass spectrometer compound Y is found to have an M_r value of 70.0423.
What is the identity of Y?

Answer

The molecular formula of A is C_5H_{10} and its M_r value is:

$(5 \times 12.0000) + (10 \times 1.0078) = 70.0780$

The molecular formula of B is $C_3H_6N_2$ and its M_r value is:

$(3 \times 12.0000) + (6 \times 1.0078) + (2 \times 14.0031) = 70.0530$

The molecular formula of C is C_4H_6O and its M_r value is:

$(4 \times 12.0000) + (6 \times 1.0078) + 15.9949 = 70.0417$

Although none of the calculated M_r values exactly matches the accurate M_r value obtained from the high resolution mass spectrometer, the value is much closer to C than to A or B, so compound Y is C.

WORKED EXAMPLE 3

The high resolution mass spectrum shown in **fig B** has two major peaks.

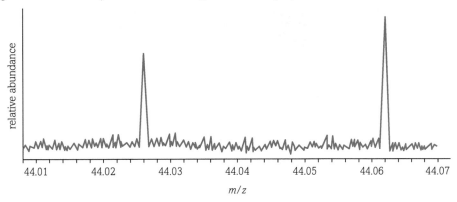

fig B A high resolution mass spectrum.

Suggest which of these compounds are responsible for the mass spectrum:

P C_3H_8
Q CH_3CHO
R $O=C=O$

Answer

M_r of P = $(3 \times 12.0000) + (8 \times 1.0078) = 44.0624$

M_r of Q = $(2 \times 12.0000) + (4 \times 1.0078) + 15.9949 = 44.0261$

M_r of R = $12.0000 + (2 \times 15.9949) = 43.9898$

The two peaks correspond to P and Q, but not to R, so propane and ethanal are present, but not carbon dioxide.

Key definition

High resolution mass spectrometry (HRMS) is a type of mass spectrometry that can produce M_r values with several decimal places, usually four or more.

Questions

1. Two compounds have the structures $CH_3CH_2CH_2CH_2NH_2$ and $HN=CHCOOH$. In a high resolution mass spectrum, one of these compounds has $M_r = 73.0812$. Explain how you can decide which compound it is. You do not need to show any working.

2. A compound is thought to be either 1,2-diaminoethane or ethanoic acid. Its accurate M_r value is 60.0213. Which compound is it? Show your working.

17.7 3 Chromatography – HPLC and GC

By the end of this section, you should be able to...

- recognise that high performance liquid chromatography, HPLC, and gas chromatography, GC:
  (i) are types of column chromatography
  (ii) separate substances because of different retention times in the column

High performance liquid chromatography

High performance liquid chromatography (HPLC) is a refinement of column chromatography, described in **Section 17.6.5**. The main differences are:

- The solvent is forced through a metal tube under high pressure, rather than being allowed to pass through by gravity.
- The particle size of the stationary phase is much smaller, which leads to better separation of the components.
- The sample is injected into the column.
- The components are detected after passing through the column, usually by their absorption of ultraviolet radiation.
- The whole process is automated and the results are quickly available on a computer display.

Fig A shows a typical setup.

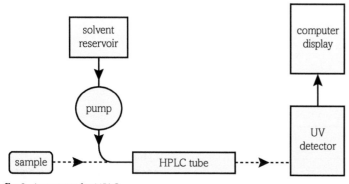

fig A Apparatus for HPLC.

As each component reaches the detector, a signal is displayed on the computer screen. Just as retention factor (R_f) values can be obtained from paper and thin layer chromatography, in HPLC values of retention time are obtained. The retention time of a component is the time taken from injection to detection, and these values can be important in identifying the components. However, retention times depend on several variables, including:

- the nature of the solvent
- the pressure used
- the temperature inside the column.

Gas chromatography

Gas chromatography (GC) is another refinement of column chromatography. The main differences are:

- The metal tube can be several metres long and is coiled to save space.
- The stationary phase is a solid or liquid coated on the inside of the tube.
- The mobile phase is an inert carrier gas (often nitrogen or helium).
- The sample is injected into the column, as in HPLC.
- The components passing through the column are detected.
- The whole process is automated and the results are quickly available on a computer display.

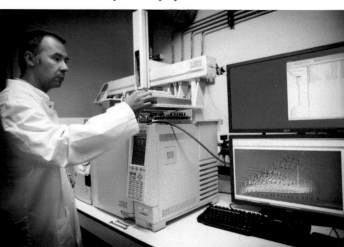

fig B The complete GC equipment takes up very little room.

Fig C shows a typical setup.

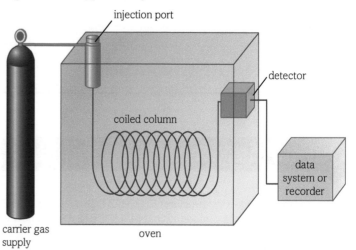

fig C Apparatus for GC.

230

After the sample is injected, the components vaporise and move through the coiled tube with the carrier gas. They move at different speeds, depending on how strongly they are attracted to the stationary phase. Those with weaker attractions move more quickly and have shorter retention times. **Fig D** shows the separation occurring in the coiled tube.

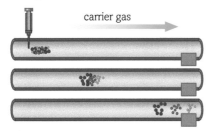

fig D A gas chromatograph separating three components.

The relative concentrations of the different components are often displayed on a graph like that in **fig E**.

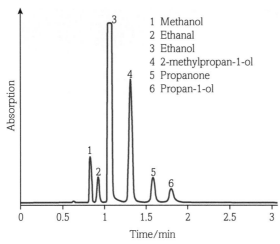

1 Methanol
2 Ethanal
3 Ethanol
4 2-methylpropan-1-ol
5 Propanone
6 Propan-1-ol

fig E The areas under the peaks represent the relative concentrations of the components.

Learning tip

Traditional column chromatography (using a burette, for example) has one advantage over HPLC and GC – reasonable amounts of the components can be collected. In GC the components are often detected in a flame, so they are destroyed during the detection process.

Questions

1. How does the movement of the mobile phase in HPLC differ from that in traditional column chromatography?

2. Suggest reasons why retention times for a given substance may be different when obtained from different gas chromatograms.

17.7 4 Chromatography and mass spectrometry

By the end of this section, you should be able to...

- recognise that high performance liquid chromatography, HPLC, and gas chromatography, GC may be used in conjunction with mass spectroscopy, in applications such as forensics or drugs testing in sport

Limitations of HPLC and GC

The two types of chromatography described in **Section 17.7.3** are very useful for separating small quantities of components in a mixture, but they are not very good at positively identifying them. This is partly because of the difficulty in controlling all of the variables (such as solvent, pressure and temperature) and partly because different substances may have the same retention times. Components are often identified by reference to a database of the retention times of known substances. However, if a component has a retention time for which there is no reference, then these techniques provide no useful information.

Here are two areas in which the results of these chromatography methods have to be exactly correct:

- in providing forensic evidence (for use in a court of law)
- in detecting banned drugs in sportsmen and women and racehorses.

They are used in other areas, including analysis of pollutants in the environment, detecting explosives in airport baggage, and in space probes on other planets.

Because HPLC and GC do not give results that are beyond doubt, they are often combined with a different technique that you learned about in **Book 1**. This is mass spectrometry (MS), a technique that cannot separate a mixture but which can provide information about the structures of compounds. The combinations of these techniques are often abbreviated to HPLC-MS and GC-MS.

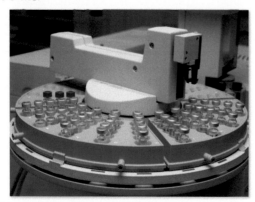

fig A GC-MS is widely used for forensic purposes.

GC-MS

As the two combined techniques are very similar, we will focus only on GC-MS. **Fig B** shows the stages in a typical setup.

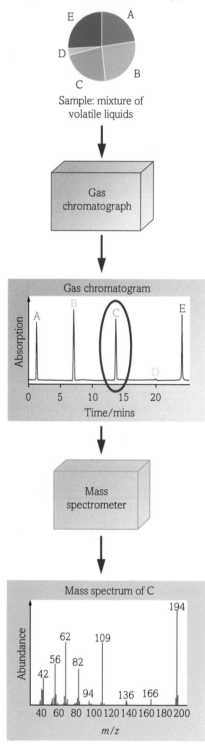

fig B Stages in GC-MS.

Imagine a sample containing a mixture of five different substances, A–E.

Stage 1: The mixture is injected into a gas chromatograph.

Stage 2: Each component has a different retention time, so emerges at a different time.

Stage 3: One at a time, each component enters the mass spectrometer.

Stage 4: Component C has its mass spectrum displayed.

Stage 5: The m/z values and relative abundances of components are compared with a database of known substances.

Stage 6: When a match is found, component C has been positively identified.

Very small traces of substances, for example in a human hair or a paint fleck, can be accurately identified by this method.

Problems with drug testing

Although the use of GC-MS for detecting small amounts of banned drugs is usually accepted as producing reliable results, sometimes a problem arises. One example involves the anabolic steroid called nandrolone. Anabolic steroids have legitimate uses in medicine, especially in the treatment of anaemia, osteoporosis and some forms of cancer. Its effects include muscle growth, increased red blood cell production and bone density.

Mainly because of their effects in promoting muscle growth, sporting authorities have prohibited the use of anabolic steroids in sportsmen and women, including athletes. When someone takes nandrolone, it is converted in the human body into a similar compound, 19-norandrosterone, which is then excreted in urine. Participants may be routinely tested for the presence of 19-norandrosterone in their urine, and the International Olympic Committee has set a limit of 2 micrograms (µg) per cm³ of urine, which is only 0.000 000 002 g per dm³. A participant with a concentration higher than this has 'tested positive'. The skeletal formulae of these two related compounds are:

nandrolone

19-norandrosterone

In the 1998 Olympic Games, over 600 participants were tested for nandrolone, and none tested positive. In the years that followed, some very high-profile athletes in a range of sports have been banned from competition because of testing positive for nandrolone. However, many scientists believe that the number of positive tests is higher than expected. The suspicion is that nandrolone may be an ingredient in some nutritional supplements that participants are allowed to take. Another theory is that a combination of permitted dietary supplements and exercise may increase the concentration of nandrolone in the human body.

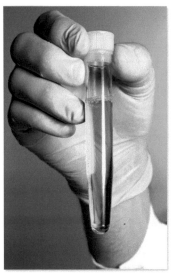

fig D A urine sample from an athlete about to be tested for illegal drugs.

Learning tip

In quoting very small concentrations of drugs, remember that a microgram (µg) is one millionth of a gram, 1×10^{-6} g.

Questions

1. What is the main function of each technique in HPLC-MS?

2. Which of the two compounds nandrolone and 19-norandrosterone would you expect to react with:

    (a) bromine water

    (b) 2,4-dinitrophenylhydrazine?

17.7 5 Principles of NMR spectroscopy

By the end of this section, you should be able to...
- understand the key aspects of NMR spectroscopy
- understand how ^{13}C and ^{1}H are detected by NMR

What is NMR?

You learned about two major techniques used in analysing organic compounds in **Book 1**. These are mass spectrometry and infrared (IR) spectroscopy. In **Section 17.7.2** you learned more about mass spectrometry (high resolution mass spectrometry), but there will be no more discussion of infrared spectroscopy.

This section and the following sections in **Topic 17** are about a third major technique – **nuclear magnetic resonance spectroscopy** (NMR). This is important for determining the structure of organic compounds. However, some people worry about the term 'nuclear' because of its use in other contexts – nuclear power stations and nuclear waste. Every major hospital has a department with an NMR scanner, although they are normally described as MRI (magnetic resonance imaging) scanners. Dropping the word 'nuclear' in hospital use is sensible, because the technique has no connection with radioactive substances, but it does have a connection with nuclei in atoms.

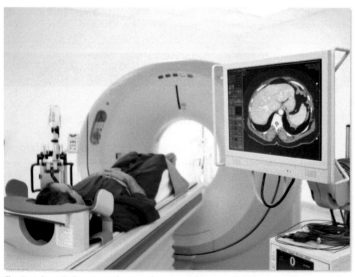

fig A A hospital patient about to have an MRI scan. MRI is a development of NMR spectroscopy.

A very much simplified explanation

NMR is much more difficult to fully understand than MS or IR spectroscopy, but fortunately our emphasis will be on interpreting the spectra produced, and not on fully understanding how it works. Even so, it will help if you read a bit of the theory behind it. **Table A** shows some of the key points.

Point	Comment
Nucleons (protons and neutrons) have spin.	You know that electrons in atoms that pair up in orbitals have opposite spins. Well, protons in the nucleus also have opposite spins.
Nuclei have either an even or an odd number of nucleons.	In nuclei with an even number of nucleons, these spins cancel out. In nuclei with an odd number of nucleons, these spins cancel out, except for the odd one, which leaves the nucleus with residual spin.
Residual spin causes a tiny magnetic field.	Such nuclei can be imagined as tiny magnets.
These nuclei are affected by an external magnetic field.	Nuclei can be imagined as either lining up or opposing the external magnetic field.
There is a difference in energy between these two different states of a nucleus.	As the spins of nuclei 'flip' between lining up and opposing the external magnetic field the nuclei can absorb electromagnetic radiation.
The nuclei of two adjacent (neighbouring) atoms in a molecule influence each other, and electrons also have an effect.	This means that electromagnetic radiation is absorbed differently by different atoms in a molecule, and these are detected as different 'signals' at different frequencies.

table A

Key aspects of NMR

There are several aspects of NMR that you do need to know about, so read on!

NMR spectrum

Just as the output of a mass spectrometer and of an infrared spectrometer produces a spectrum – a kind of graph – so each compound tested using this method produces a characteristic NMR spectrum. The vertical axis is sometimes labelled 'absorption' (for 'absorption of radio frequency energy') and has no units, although the label is often omitted. The horizontal axis is labelled chemical shift (or sometimes with the δ symbol – the same symbol as used for partial charges) and has the units 'ppm' (parts per million). The scale usually starts with a zero value on the right, with the values increasing to the left.

234

Which atoms are detected by NMR?

For an atom to show up in an NMR spectrum, it must have an odd number of nucleons (protons and neutrons). That rules out carbon atoms, or at least ^{12}C atoms, as they each have three pairs of protons and three pairs of neutrons. However, in naturally-occurring carbon, about 1.1% of the carbon atoms are of the isotope ^{13}C, and they have an odd number of neutrons (seven). So, carbon atoms in molecules can be detected because some of them are ^{13}C atoms. Hydrogen atoms are also detected – at least the ^{1}H isotopes that make up 99.985% of all hydrogen atoms. In the context of NMR, ^{1}H hydrogen atoms are often referred to as protons, so an alternative name for the process is 'proton NMR'.

We often refer to these two atoms (^{13}C and ^{1}H) as 'producing a signal'. In this book we will consider only these two atoms, although NMR can be successfully used for other atoms, including ^{15}N and ^{31}P.

Suitable solvents

The organic compound being analysed has to be dissolved in a solvent, but most solvent molecules contain carbon and hydrogen atoms. This means that they will produce signals that interfere with those produced by the compound being analysed. The commonest solvent used has the formula $CDCl_3$. The symbol D stands for deuterium, which is the ^{2}H isotope of hydrogen. Deuterium atoms contain a proton and a neutron, and their spins cancel out. This solvent produces no signals that could interfere with the signals from hydrogen atoms. It produces only one signal that can interfere with the signals from carbon atoms, and this can easily be removed from the spectrum.

TMS

Carbon and hydrogen atoms in different chemical environments will produce different signals. Therefore, there needs to be a reference standard that will give a prominent signal. The compound used for this purpose is tetramethylsilane (usually abbreviated to TMS). Its structure is shown below:

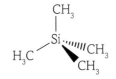

fig B The largest part of NMR equipment is often the large vacuum flask containing liquid helium, needed to cool the magnets to below 4 K.

You can see that it resembles a methane molecule, but with silicon instead of carbon and methyl groups instead of hydrogen atoms. SiH_4 is called silane, by analogy with methane, so perhaps you can understand the origin of the name tetramethylsilane. The molecule is suitable because it contains 12 hydrogen atoms, all joined in the same way in a symmetrical arrangement, so it produces a single strong signal that is easy to identify. It is also chemically unreactive, so will not react with most organic compounds.

Chemical shift

Each ^{13}C and ^{1}H atom in a compound produces a signal at a characteristic **chemical shift** (δ value), depending on the other atoms it is joined to. The reference standard, TMS, is arbitrarily assigned a chemical shift of zero ($\delta = 0.0$ ppm), and all other atoms have values related to this. Chemical shift can be thought of as the resonant frequency of an atom. In **Sections 17.7.6** and **17.7.7** you will learn to use a data booklet containing values (or more often, ranges) of chemical shift to identify specific atoms in molecules. Here are two examples:

carbon in a ketone group —C—**C**O—C— $\delta = 210$ ppm
carbon in an alcohol   —**C**—OH $\delta = 60$ ppm

Note that these values refer only to the carbon atoms shown in bold, and not to any other carbon atoms or to any hydrogen atoms in the molecule.

Learning tip

Do not worry if you think you do not understand the details of NMR theory – by far the most important thing is for you to understand its application.

Questions

1. Some organic compounds contain these atoms: ^{16}O, ^{19}F, ^{32}S. Explain which of these are not suitable for use in NMR spectroscopy, and why.

2. Suggest why water (H_2O) is less suitable than TMS as a reference in NMR spectroscopy.

Key definitions

Nuclear magnetic resonance spectroscopy is a technique used to find the structures of organic compounds. It depends on the ability of nuclei to resonate in a magnetic field.

The **chemical shift** of a proton (or group of protons) is a number (in the units ppm) that indicates its behaviour in a magnetic field relative to tetramethylsilane. It can be used to identify the chemical environment of the carbon atoms or of the hydrogen atoms (protons) attached to it.

17.7 6 ¹³C NMR spectroscopy

By the end of this section, you should be able to...

- understand that ¹³C NMR spectroscopy provides information about the positions of ¹³C atoms in a molecule
- use data from ¹³C NMR spectroscopy to:
  (i) predict the different environments for carbon atoms present in a molecule given values of chemical shift, δ
  (ii) justify the number of peaks present in a ¹³C NMR spectrum because of carbon atoms in different environments

What does a ¹³C NMR spectrum show?

The actual spectrum of a compound is usually shown as a horizontal line just above the horizontal scale, but with vertical lines showing the signals produced by the ¹³C carbon atoms in the molecule. These vertical lines are usually referred to as signals or peaks.

- The number of vertical lines tells you the number of different **chemical environments** of carbon atoms in the molecule, and not necessarily the total number of carbon atoms.
- The positions of the vertical lines on the horizontal scale tell you the chemical shifts of each carbon atom and with reference to a list of chemical shifts, allow you to deduce the chemical environments.

The two isomers of propanol can be used to illustrate these points.

Propan-1-ol has the structure CH_3—CH_2—CH_2—OH. Each molecule contains three carbon atoms, all in different chemical environments. You might think that the two carbon atoms in the CH_2 groups are in the same chemical environment, but they are not. The second one is joined to CH_3 and CH_2, but the third one is joined to CH_2 and OH, which means they are in different chemical environments. There should therefore be three peaks in the spectrum.

Propan-2-ol has the structure CH_3—$CH(OH)$—CH_3. Each molecule contains three carbon atoms, but those in the two CH_3 groups are in the same chemical environment. This is because each CH_3 group is joined to $CH(OH)CH_3$. This means that there should be two peaks in the spectrum. **Fig A** shows the ¹³C spectra of these isomers.

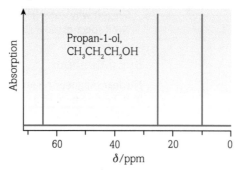 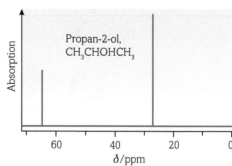

fig A ¹³C NMR spectra of propan-1-ol and propan-2-ol.

The spectra confirm the number of peaks expected, but show two other features.

- The less important feature is that in propan-1-ol the three peaks are of equal height, but in propan-2-ol the two peaks have different heights. The greater height indicates the greater number of carbon atoms causing that peak.
- The more important one is that the peaks have different chemical shifts.

Interpreting chemical shifts

To interpret chemical shifts, you need a table of values and corresponding carbon atoms or a chart. The chart in **fig B** will be provided in an examination, and you need to become familiar with using it.

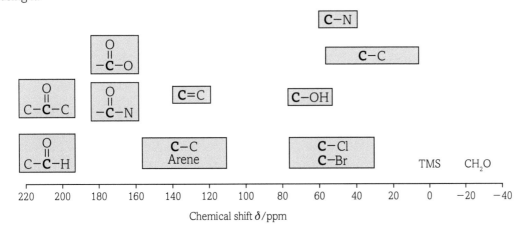

fig B ^{13}C NMR chemical shift ranges.

We will use this chart to analyse the two spectra for propanol.

Propan-1-ol:

In the spectrum for propan-1-ol the peaks appear at 65, 25 and 10 ppm. The only relevant ranges in the chart are C—C (the range is approximately 60–0 ppm) and C—OH (the range is approximately 75–55 ppm).

This means that the peak at 65 ppm corresponds to C—OH, so this is the carbon in the CH_2OH group. The peaks at 25 and 10 ppm correspond to C—C, so these are the carbons in the CH_3 and CH_2 groups. Notice that it is not possible to decide whether the peak at 25 ppm is due to the CH_3 or the CH_2 group.

Propan-2-ol:

In the spectrum for propan-2-ol the peaks appear at 65 and 27 ppm. The only relevant ranges in the chart are again C—C and C—OH.

This means that the peak at 65 ppm corresponds to C—OH, so this is the carbon in the CHOH group. The peak at 27 ppm corresponds to C—C, so this is due to the carbons in the two CH_3 groups. The peak at 27 ppm is bigger than the one at 65 ppm because it is caused by two carbon atoms, not one.

Fig C shows the interpretation of these two spectra.

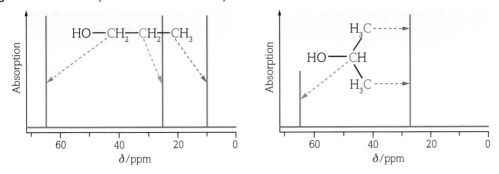

fig C Interpreting the ^{13}C NMR spectra of propan-1-ol and propan-2-ol.

Practice makes perfect!

The way to check how well you understand how to use ^{13}C NMR spectra is to try lots of examples. In this section we will continue with some examples of varying types.

WORKED EXAMPLE 1

Predicting the number of peaks

You need to consider each carbon atom and whether it is uniquely positioned in the molecule, or whether there is one or more carbon atoms identically positioned in the molecule.

For each of these compounds, predict the number of peaks due to the carbon atoms:

A $CH_3-CH_2-CH_2-CH_3$
B $(CH_3)_2-CH-CH_2-CH_3$
C $(CH_3)_4C$

The answers are in **table A**.

Compound	Number of peaks	Comments
A	2	One due to both CH_3 groups (these are both joined to $CH_2CH_2CH_3$ so are identical). One due to both CH_2 groups (these are both joined to CH_3 and CH_2CH_3 so are identical).
B	4	One due to the two left-hand CH_3 groups (these are identically joined to the rest of the molecule) – this should be bigger than the other two peaks because it is due to two carbon atoms. One due to the CH group. One due to the CH_2 group. One due to the right-hand CH_3 group (this is not identical to the other CH_3 groups because it is joined to CH_2 and they are joined to CH).
C	2	One due to the four CH_3 groups, which are all identically joined to the central C atom – this should be much bigger than the other peak as it is due to four carbon atoms. One due to the central C atom.

table A

WORKED EXAMPLE 2

Predicting the appearance of a spectrum

You need to consider all of the carbon atoms, as in example 1, but this time also use the chart in **fig B** when considering the other atoms joined to them.

For each of these compounds, predict the number of peaks and chemical shift ranges due to the carbon atoms:

A $CH_3-CH_2-CH_2-NH_2$
B $(CH_3)_2CH-COOH$
C $(CH_3)_3N$

The answers are in **table B**.

Compound	Number of peaks	Chemical shift ranges	Comments
A	3	2 peaks in range 60–0 1 peak in range 63–35	Due to CH_3 and first CH_2 group. Due to CH_2 joined to NH_2.
B	3	1 peak in range 60–0 1 peak in range 60–0 1 peak in range 182–165	Due to the two CH_3 groups. Due to the CH group. Due to the COOH group.
C	1	1 peak in range 63–35	Due to the three CH_3 groups, which are all identical.

table B

WORKED EXAMPLE 3

Interpreting a spectrum

A compound has the molecular formula C_3H_6O and its ^{13}C NMR spectrum is shown in **fig D**.

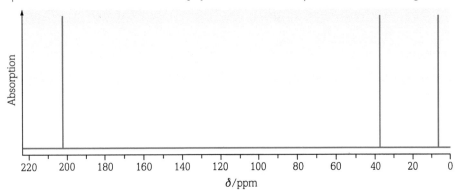

fig D Which compound has this ^{13}C NMR spectrum?

Use the spectrum and the chemical shift chart to explain whether the compound's structure is

$CH_2=CH-CH_2OH$ or CH_3-CH_2-CHO.

Answer

The peak at δ = 203 ppm is due to a carbon in an aldehyde or ketone

The peaks at δ = 37 and 6 ppm are due to a carbon joined to another carbon by a single bond

This interpretation fits CH_3-CH_2-CHO but not $CH_2=CH-CH_2OH$

If the structure were $CH_2=CH-CH_2OH$, there would have been a peak in the range δ = 140–115 ppm corresponding to C=C, and a peak in the range δ = 75–55 ppm corresponding to C–OH.

Learning tip

When you need to decide how many different chemical environments there are for the carbon atoms in a molecule, it is a good idea to write the structure with the carbon atoms separated, e.g. $CH_3-CH(OH)-CH_3$

Question

1. Work out the number of different chemical environments for the carbon atoms in the first six straight-chain alkanes.

Key definition

Chemical environments of carbon atoms in a molecule are related to whether the carbon atoms are identically, or differently, positioned within a molecule.

17.7 7 ¹H NMR spectroscopy

By the end of this section, you should be able to...

- understand that high resolution proton NMR provides information about the positions of ¹H atoms in a molecule
- use data from high resolution ¹H NMR spectroscopy to:
  (i) predict the different types of proton present in a molecule given values of chemical shift, δ
  (ii) relate relative peak areas, or ratio numbers of protons, to the relative numbers of ¹H atoms in different environments

Low resolution ¹H NMR spectroscopy

¹H NMR spectroscopy comes in two forms – low resolution and high resolution. The low resolution form is no longer used, but if we start with this, it will help you to understand the high resolution form more easily.

As with ^{13}C NMR spectroscopy, you may see a peak at $\delta = 0.0$ ppm. This is due to the reference standard tetramethylsilane (TMS) and should be ignored. The chemical shifts are relative to TMS.

What does a low resolution ¹H NMR spectrum show?

The actual spectrum of a compound is usually shown as a horizontal line just above the horizontal scale, but with **peaks** produced by the ¹H hydrogen atoms in the molecule. The horizontal scale often starts from $\delta = 12.0$ ppm on the left to 0.0 ppm on the right. Quite often, only the part of the scale where the peaks appear is shown.

- The number of peaks tells you the number of different chemical environments of the hydrogen atoms in the molecule, and not the total number of hydrogen atoms.
- The positions of the peaks on the horizontal scale tell you the chemical shifts of each hydrogen atom, and with reference to a list or chart of chemical shifts, these allow you to deduce the types of chemical environment.
- The areas under the peaks represent the relative numbers of hydrogen atoms in each environment.
- Sometimes a number is added at the side of each peak to show the relative area under the peak.

Another way to represent the relative areas is by an **integration trace**. This is a horizontal line that becomes higher as it passes each peak. The increases in height can be measured – they represent the relative areas under the peaks.

The two isomers of propanol can be used to illustrate these points.

Propan-1-ol has the structure CH_3–CH_2–CH_2–OH. There are eight hydrogen atoms, in four different chemical environments. You might think that the two pairs of hydrogen atoms in the CH_2 groups are in the same chemical environment, but they are not. The first CH_2 group is joined to CH_3 and CH_2, but the second one is joined to CH_2 and OH, and these are different chemical environments. This means that there should be four peaks in the spectrum. **Fig A** shows the low resolution ¹H NMR spectrum of this isomer with an integration trace.

Propan-2-ol has the structure CH_3–CH(OH)–CH_3. There are eight hydrogen atoms, but those in the two CH_3 groups are in the same chemical environments. This is because each CH_3 group is joined to CH(OH)CH_3. The hydrogen atoms in the CH group and in the OH group are in unique chemical environments. This means that there should be three peaks in the spectrum. **Fig B** shows the low resolution ¹H NMR spectrum of this isomer with the ratio of peak areas indicated by circled numbers.

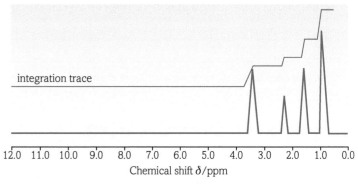

fig A The low resolution ¹H NMR spectrum of propan-1-ol.

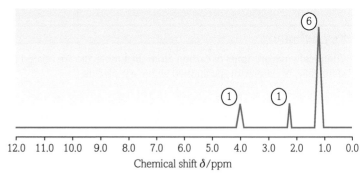

fig B The low resolution ¹H NMR spectrum of propan-2-ol.

Interpreting chemical shifts

To interpret chemical shifts, you need a table of reference values with corresponding hydrogen atoms or a chart. **Fig C** shows a simplified version of the chart that will be provided in an examination, and you need to become familiar with using it.

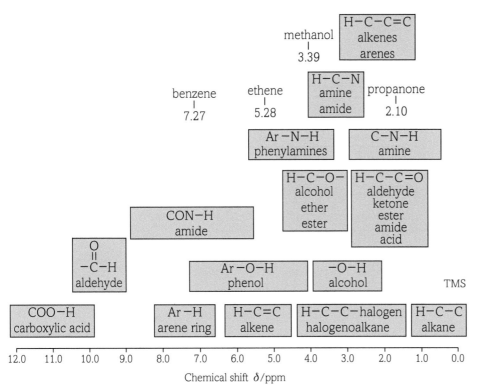

fig C ¹H NMR chemical shift ranges.

For two of the chemical shift ranges in this chart, the compounds included show trends in chemical shift values, as shown in this **table A**.

Type of compound	Trend
halogenoalkane	$R_2CHF > R_2CHCl > R_2CHBr > R_2CHI$
alkane	$R_3CH > R_2CH_2 > RCH_3$

table A

Butanal and butanone

We will use the chart to interpret the low resolution ¹H NMR spectra of these isomers with the molecular formula of C_4H_8O. **Fig D** shows the spectrum for butanal.

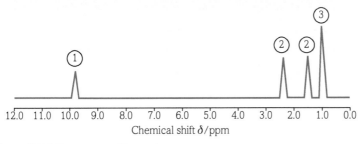

fig D The low resolution ¹H NMR spectrum of butanal.

The peak at $\delta = 9.8$ ppm is within the aldehyde range, so must be due to the H in the CHO group.

The peak at $\delta = 2.4$ ppm is within the range for H—C—C=O, so must be due to the two H atoms in the CH_2 group joined to the CHO group.

The peak at $\delta = 1.6$ ppm is within the range for H—C—C (alkane), so must be due to the two H atoms in the CH_2 group not directly joined to the CHO group. Note that although the spectrum is for an aldehyde, the shift for an alkane is appropriate because this CH_2 group is part of the $CH_3CH_2CH_2$ alkyl group.

The peak at $\delta = 1.0$ ppm is within the range for H—C—C (alkane), so must be due to the three H atoms in the CH_3 group.

Now for the spectrum of butanone, shown in **Fig E**.

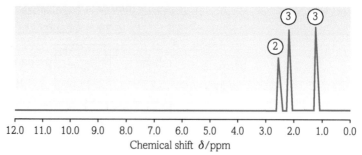

fig E The low resolution ^{1}H NMR spectrum of butanone.

The peak at $\delta = 2.5$ ppm is within the range for H—C—C=O, so must be due to the two H atoms in the CH$_2$ group joined to the CO group.

The peak at $\delta = 2.1$ ppm is within the range for H—C—C=O, so must be due to the three H atoms in the CH$_3$ group joined to the CO group.

The peak at $\delta = 1.1$ ppm is within the range for H—C—C (alkane), so must be due to the three H atoms in the CH$_3$ group not directly joined to the CO group.

Learning tip

Low resolution ^{1}H NMR spectra indicate three things:
- The number of peaks represents the number of different chemical environments of the hydrogen atoms (protons).
- The areas under the peaks indicate the relative numbers of protons in each chemical environment.
- The chemical shift of a peak can indicate the functional group responsible (e.g. aldehyde or ketone).

Questions

1. Predict the number of peaks in the ^{1}H NMR spectrum of each of the isomers of C$_5$H$_{12}$.

2. Why does an integration trace not always indicate the actual numbers of hydrogen atoms in a molecule?

3. How would you easily decide from a ^{1}H NMR spectrum whether a compound is ethylamine (CH$_3$CH$_2$NH$_2$) or ethanamide (CH$_3$CONH$_2$)?

Key definitions

A **peak** in a ^{1}H NMR spectrum shows the presence of hydrogen atoms (protons) in a specific chemical environment.

An **integration trace** shows the relative numbers of equivalent protons (i.e. in the same chemical environment).

17.7 8 Splitting patterns in ¹H NMR spectroscopy

By the end of this section, you should be able to...

- use data from high resolution ¹H NMR spectroscopy to:
  (i) deduce the splitting patterns of adjacent, non-equivalent protons using the $n+1$ rule and hence suggest the possible structures for a molecule
  (ii) predict the chemical shifts and splitting patterns of the ¹H atoms in a given molecule

High resolution ¹H NMR spectroscopy

In **Section 17.7.7** we introduced NMR by starting with the low resolution form, although only the high resolution form is actually used now. Both forms have two features in common:

- Peaks at different chemical shift values help to identify the different chemical environments of the hydrogen atoms. If the environments are identical, then the protons are often described as being equivalent (i.e. they are **equivalent protons**).
- The relative peak area (i.e. the ratio of peak areas) helps to decide the numbers of hydrogen atoms responsible.

A high resolution ¹H NMR spectrum has one extra feature not seen in a low resolution spectrum – the peaks may have a **splitting pattern**. This means that the peak is split into a group of smaller peaks (sub-peaks) grouped very closely together. When considering these split peaks:

- you can take the chemical shift to be at the centre of the group of sub-peaks
- the relative peak areas are shown in the same way as in a low resolution spectrum – by an integration trace or by showing numbers next to the peaks
- remember that the numbers indicate the ratio of the numbers of protons – they may happen to be the actual numbers of protons, but this is not always the case.

What causes the splitting of peaks?

Although the most important thing is for you to be able to interpret the splitting patterns, it will help if you understand how the splitting is caused. The fundamental point to be aware of is that hydrogen atoms (we will call them protons from now on) joined to a carbon atom have an influence on the protons on adjacent or neighbouring carbon atoms. Consider a molecule of butanone, in which the protons are labelled with letters:

$$\overset{a}{CH_3}-\underset{\underset{O}{\|}}{C}-\overset{b}{CH_2}-\overset{c}{CH_3}$$

fig A The three different chemical environments for the protons in butanone.

The protons labelled a are not influenced by any other protons because there are none on the adjacent carbon atom. The protons labelled b are influenced by the protons labelled c and vice versa. These influences are what cause splitting of the peaks (sometimes referred to as spin–spin coupling), usually considered in terms of the $n+1$ rule. Very simply, if a carbon atom has n protons, then the peaks on adjacent carbon atoms are split into $n+1$ sub-peaks. The split peaks are sometimes referred to as **multiplets**, but more often by a term indicating the number of sub-peaks.

If the peak is not split, it is called a singlet.

If the peak is split into two sub-peaks, it is called a doublet.

If the peak is split into three sub-peaks, it is called a triplet.

If the peak is split into four sub-peaks, it is called a quartet.

The sub-peaks in a split peak have distinctive shapes, due to splitting into different sub-peak areas, as shown in **table A**.

n	$n+1$	Multiplet	Ratio of peak areas
0	1	singlet	1
1	2	doublet	1:1
2	3	triplet	1:2:1
3	4	quartet	1:3:3:1

table A

Learning tip

The splitting pattern of a group of protons indicates nothing about the group of protons being split, but relates only to the number of protons on an adjacent atom.

Understanding the spectrum of butanone

We can apply the information provided above to explain the spectrum of butanone in **table B**.

Protons	Multiplet	Explanation
a	singlet	There are no protons on the adjacent carbon atom (C=O), so there is no splitting.
b	quartet	There are three protons on the adjacent CH_3 group, so applying the $n+1$ rule, the peak for the two protons is split into a quartet.
c	triplet	There are two protons on the adjacent CH_2 group, so applying the $n+1$ rule, the peak for the three protons is split into a triplet.

table B

Two more points to remember

- We have said that protons cause splitting of peaks on adjacent atoms. While this is generally true, it does not apply in cases where the molecule is completely symmetrical. For example, there is splitting of peaks for 1,1-dichloroethane, $CHCl_2$—CH_3. The single proton splits the peak for the CH_3 group into a doublet, and the three protons in the CH_3 group split the CH peak into a quartet, following the $n+1$ rule. However, in 1,2-dichloroethane, CH_2Cl—CH_2Cl, neither of the CH_2 groups affects the protons of the other CH_2 group because all of the protons are equivalent (the CH_2 groups are in the same chemical environment).

- Now consider 1-chloropropane, CH_3—CH_2—CH_2—Cl. This time, the two CH_2 groups are in different chemical environments (so the protons are not equivalent). One is joined to CH_3 and CH_2, and the other is joined to CH_2 and Cl, so there will be splitting of peaks by the CH_2 groups. However, this example has been included to make a different point. The first CH_2 group would be influenced by the CH_3 group on its left and by the CH_2 group on its right. Using the $n+1$ rule, you would predict that the CH_3 and the CH_2 group acting together would cause splitting into six sub-peaks – this splitting pattern would be described as a sextet. In other compounds, a quintet would be formed by four protons on adjacent carbon atoms – these could be two CH_2 groups, or one CH_3 group and one proton.

Practice makes perfect – again!

The way to check how well you understand how to use ^1H NMR spectra is to try lots of examples. We will use three examples, all of which require you to use the chart of chemical shifts in the previous section.

WORKED EXAMPLE 1

Predicting the appearance of a spectrum (chemical shifts and splitting patterns)
What would you predict about the appearance of the ^1H NMR spectrum of methyl propanoate ($CH_3CH_2COOCH_3$)?

Answer
The CH_3 in the ethyl group is part of an alkyl group and is joined only to a hydrocarbon (CH_2) group, so it should have a peak in the range δ = 1.9–0.1 ppm, which is split into a triplet because of the two protons in the adjacent CH_2 group.

The CH_2 in the ethyl group is part of an alkyl group and is joined to a hydrocarbon (CH_3) group but also to the C of the COO (ester) group, so it should have a peak in the range δ = 3.0–1.8 ppm, which is split into a quartet because of the three protons in the adjacent CH_3 group.

The CH_3 in the methyl group is joined only to an oxygen atom in the COO group, so it should have a peak in the range δ = 4.3–3.0 ppm, which is a singlet. This peak is not split because there are no protons on the adjacent atom.

Fig B shows the actual spectrum of methyl propanoate, so you can check whether the predictions are correct.

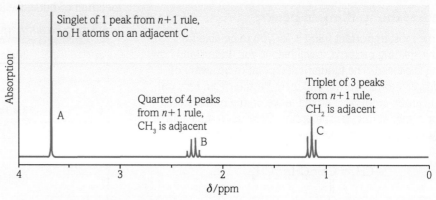

fig B The ^1H NMR spectrum of methyl propanoate.

WORKED EXAMPLE 2

Predicting the structure of a compound from its spectrum

Fig C shows the ^{1}H NMR spectrum of a compound with the molecular formula $C_4H_8O_2$.

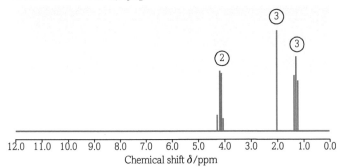

fig C The ^{1}H NMR spectrum of a compound with molecular formula $C_4H_8O_2$.

What is its structure?

Remember that the spectrum shows three features:
- relative peak areas
- chemical shifts
- splitting patterns

all of which need to be considered together.

Answer

The peak on the left is a quartet at $\delta = 4.1$ ppm representing two protons. This suggests a CH_2 group next to a CH_3 group and an O in an alcohol, ether or ester, so $O-CH_2-CH_3$.

The peak in the middle is a singlet at $\delta = 2.0$ ppm representing three protons. This suggests a CH_3 group joined to C=O in an aldehyde, ketone, ester, amide or acid, so CH_3-CO-.

The peak on the right is a triplet at $\delta = 1.3$ ppm representing three protons. This suggests a CH_3 group next to a CH_2 group, so $-CH_2-CH_3$.

The final step is to put all these interpretations together. This is often easy to do but difficult to explain. The only structure that fits the interpretations is $CH_3-CO-O-CH_2-CH_3$, which is the ester ethyl ethanoate.

WORKED EXAMPLE 3

Another prediction of the structure of a compound from its spectrum

Fig D shows the ^{1}H NMR spectrum of a compound with the molecular formula $C_4H_8O_2$ (the same molecular formula as in Example 2).

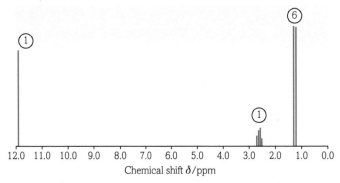

fig D The ^{1}H NMR spectrum of a compound with the formula $C_4H_8O_2$.

What is its structure?

Answer

The peak on the left is a singlet at $\delta = 11.9$ ppm representing one proton. This suggests a proton in a COOH group, so COOH.

The peak in the middle is a quartet (or multiplet) at $\delta = 2.6$ ppm representing one proton. This suggests a CH group joined to a CH_3 group, so CH_3-CH-.

The peak on the right is a doublet at $\delta = 1.2$ ppm representing six protons. This suggests two CH_3 groups next to a CH group, so $(CH_3)_2-CH-$.

The final step is to put all these interpretations together. Again, this is easy to do but difficult to explain. The only structure that fits the interpretations is $(CH_3)_2-CH-COOH$ - this is the acid methylpropanoic acid.

Questions

1. A compound has the molecular formula C_4H_8O. Its ^{1}H NMR spectrum contains only three singlets, with chemical shift values of 4.0, 3.4 and 2.1 ppm. The integration trace shows the ratio 2 : 3 : 3. What is its structure?

2. A compound has the molecular formula $C_5H_{10}O_2$. Its ^{1}H NMR spectrum contains a singlet with a chemical shift of 2.3 ppm, a quartet with a chemical shift of 1.5 ppm, and a doublet with a chemical shift of 1.2 ppm. The integration trace shows the ratio 3 : 1 : 6. What is its structure?

Key definitions

Equivalent protons are hydrogen atoms in the same chemical environment.

A **splitting pattern** is the appearance of a peak as a small number of small sub-peaks very close to each other.

Multiplets are the different splitting patterns observed (singlets, doublets, triplets or quartets).

THINKING BIGGER

MOLECULES OF FRAGRANCE AND TASTE

Behind the flavours and scents that we experience every day there is often a complex mix of organic molecules. The following extract is from *Chemistry World*, the print and online magazine of the Royal Society of Chemistry.

THE SWEET SCENT OF SUCCESS

Chart toppers

To make a fine fragrance, new molecules are mixed with old favourites. But which fragrance compounds are most valued by industry – a kind of fragrance top 10? It's a difficult question to answer. Gautier has no doubt about the value of Hedione (methyl dihydrojasmonate), used in almost all fine fragrances and Firmenich's top seller in terms of volume. The compound was discovered at Firmenich in the early 1960s as an analogue of methyl jasmonate, a key component of jasmine oil. Said to give a warm floral-jasmine note, Hedione has been used in perfumes for over 40 years, first starring in Dior's *Eau Sauvage* in 1966.

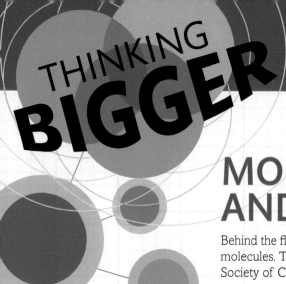

fig A Skeletal structure of Hedione.

fig B α- and β-damascone are fruity components of rose oil.

'The synthesis of Hedione gives four stereoisomers,' explains Gautier. 'Only one of them truly contributes to the odour – (+)-cis-methyl dihydrojasmonate – while the others are thought to modulate other fragrances. Over the years we've been able to synthesise new versions of Hedione which have a much higher quantity of the right isomer.'

Gautier also has a soft spot for the rose ketones: damascenone, and alpha and beta damascone. Firmenich is famed for their discovery and synthesis in the 1960s during a quest to identify the characteristic smell of Bulgarian rose oil. But batch purity problems meant that damascenone was not released commercially until 1982. Then came alpha-damascone (rose-apple note) and beta-damascone (blackcurrant-plum note). The rose ketones broke new ground in the 1980s, giving female perfumes such as Dior's *Poison* their unusual and distinctive fragrance. Today, beta-damascenone and beta-damascone remain two of the most important fragrance ingredients.

fig C The molecular isomers α-damascone and β-damascone.

Chemists are forever improving on the rose ketones, which are still trendsetters. One of Kraft's recent and 'very successful' captives is Pomarose, a 'cut-open' seco-damascone. Pomarose introduced the dried-fruit character of *Poison* to the male market and had its debut in DKNY's male fragrance *Be delicious Men*. 'It has a very specific note of cooked apple, rose, and dried plums. It's very diffusive so it gives a lot of bloom to a fragrance,' enthuses Kraft. Givaudan's process development team had to try 19 different synthesis routes before it managed to produce Pomarose. 'Initially we thought we couldn't produce it. It was a crazy idea that originated from a 1 per cent impurity for which different structures were proposed based on the NMR spectrum. After we had discovered the correct structure we synthesised the "wrong proposal" for fun but it turned out to be so powerful,' recalls Kraft. Interestingly, it wasn't the impurity itself that became Pomarose, but one of the alternative structures.

Where else will I encounter these themes?

Thinking Bigger

Let us start by considering the nature of the writing in the article.

1. What techniques does the author use to make the articles interesting to a wider target audience?

Try to highlight some specific examples from the source in your answer.

Now we will look at the chemistry in detail. Some of these questions will link to topics earlier in this book or **Book 1**, so you may need to combine concepts from different areas of chemistry to work out the answers.

2. a. Deduce the molecular formula of Hedione from the skeletal structure given in **fig A**.
   b. Identify the two chiral centres in the molecule Hedione and explain how four isomers are possible.
   c. Write an equation for the hydrolysis reaction of Hedione and 2 mol dm^{-3} sodium hydroxide solution and draw structures of the products.
3. The mass spectrum below shows the molecular isomers α-damascone and β-damascone both found in rose oil.

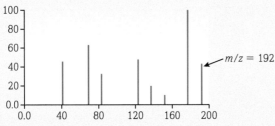

   a. Using the structures in **fig C**, identify which of the two is able to form optical isomers and explain why the other isomer is not.
   b. Suggest what species is responsible for the peak at m/z = 177. Can you suggest why this ion fragment is so abundant in the spectrum?

The skills that you have developed throughout your course will be valuable whether or not you continue your study of chemistry beyond A level. Whenever you read a scientific article, remember to consider the source, think critically and, above all, think like a scientist!

Activity

Many different molecules are now used in adding or augmenting both the flavour and aroma of foods. Choose one of the following groups of flavour molecules and write a report for your classmates on their use as flavouring and smell additives:

(a) Thiazoles
(b) Pyrazines
(c) Ketones and aldehydes
(d) Furanones and lactones

Your report should include the following:

Molecular structures
Some examples of their usage
Any information on how they are either extracted or synthesised.

● From *The Sweet Scent of Success* by Emma Davies, *Chemistry World*, February 2009.

Did you know?

A surprising ingredient for luxury fragrances is ambergris, a yellowish-grey waxy substance that forms in the digestive tract of sperm whales. Despite initially having a strong faecal smell, the substance develops an unusual earthy scent as it ages. Although it has largely been replaced by synthetic alternatives, it is still a sought after substance. In 2013, a dog walker in Morecombe found a 3 kg lump of ambergris on the beach, a find it was estimated to be worth £100 000.

17 Exam-style questions

1. The structures of three organic compounds, **A**, **B** and **C** are shown.

   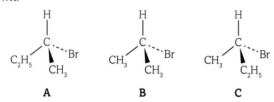

   (a) Explain which two of these compounds are enantiomers. [2]
   (b) Explain why a mixture containing equal numbers of these enantiomer molecules cannot be distinguished from the other compound of the three in a polarimeter. [2]
   (c) Compound **A** reacts with aqueous sodium hydroxide.

   Explain how measurements of the optical activity of compound **A** and the organic product of the reaction can be used to confirm whether the reaction occurs by an S_N1 mechanism or an S_N2 mechanism. [6]

   [Total: 10]

2. The structures of two carbonyl compounds are:

   H—C(H)(H)—C(H)(H)—C(H)(H)—C(H)(=O)   H—C(H)(H)—C(H)(H)—C(=O)—C(H)(H)—H

   (a) State why both compounds are soluble in water. [1]
   (b) (i) Name the mechanism of the reaction between each of the compounds and hydrogen cyanide. [1]
       (ii) Both compounds react with 2,4-dinitrophenylhydrazine.

       Describe how the organic product of each of these reactions could be used to identify the carbonyl compounds. [2]
   (c) Give the structure of the organic product of the reaction of each compound with lithium tetrahydridoaluminate, followed by water. [2]
   (d) Only one of the two compounds reacts on warming with Tollens' reagent.

   Write an equation for the reaction that occurs, using [O] to represent Tollens' reagent, and state the observations that will be made when a positive result is obtained. [2]

   [Total: 8]

3. (a) The structures of three alcohols are:

   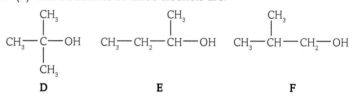

   D   E   F

   Explain which of these alcohols reacts on warming with acidified potassium dichromate(VI) to form a carboxylic acid. [3]
   (b) Draw the displayed formula of the nitrile that can be hydrolysed to the carboxylic acid with the formula $(CH_3)_2CHCH_2COOH$. [1]
   (c) The diagram summarises a sequence of two reactions.

   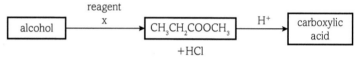

   (i) Deduce the name of the alcohol. [1]
   (ii) Explain how the homologous series to which reagent X belongs can be deduced from the diagram. [2]
   (iii) Deduce the structural formula of the carboxylic acid. [1]
   (iv) The yield of carboxylic acid in the second step can be improved by using a different reagent before the acid in a two-step process.

   Write an equation for each of these steps. [2]

   [Total: 10]

4. Addition and substitution reactions are common types of reaction in organic chemistry.

   (a) (i) Bromine reacts with both propene and benzene.

       Write an equation for each of these reactions. [2]
       (ii) State the type of mechanism in each of these reactions. [2]
   (b) The reagent used in the nitration of benzene is concentrated nitric acid.

       (i) The reaction is usually carried out using a catalyst of concentrated sulfuric acid.

       Write equations to show how the sulfuric acid acts as a catalyst. [2]
       (ii) Draw the mechanism for the reaction with benzene. [3]
   (c) Explain why bromine reacts with phenol at room temperature, and without the need for a catalyst. [3]

   [Total: 12]

5 The structures of two amino acids are:

   H₂N—CH₂—COOH   H₂N—CH—COOH
                              |
                              CH₂SH

        glycine   cysteine

   (a) The isoelectric points of these amino acids are:

       glycine 6.0   cysteine 5.1

       Draw the structure of each amino acid in a solution with
       a pH of 5.5. [2]
   (b) These amino acids react to form form two dipeptides.
       Draw the structure of each dipeptide. [2]
   (c) Describe how you would identify the amino acids in a
       tripeptide using paper chromatography. [6]
       [Total: 10]

6 The compound 1-bromopropane is used as the starting
   material in some syntheses involving Grignard reagents.
   (a) Write an equation for the reaction between
       1-bromopropane and magnesium in dry ether to form a
       Grignard reagent. [1]
   (b) Identify the compound needed to convert this Grignard
       reagent into a primary alcohol and state the name of the
       primary alcohol. [2]
   (c) Identify the compound needed to convert this Grignard
       reagent into a secondary alcohol with molecules containing
       five carbon atoms and state the name of the secondary
       alcohol. [2]
   (d) Identify the compound needed to convert this Grignard
       reagent into a tertiary alcohol with molecules containing
       seven carbon atoms and state the name of the tertiary
       alcohol. [2]
       [Total: 7]

7 The structures of two compounds are:

              CH₃   CH₃
              | |
       CH₃—C—CH₃   CH₂—CH—CH₃
              | |
              OH   OH

   (a) Explain why infrared spectroscopy is not used to distinguish
       these two compounds. [2]
   (b) Explain how the number of peaks in the ¹³C NMR spectra
       of these compounds can be used to identify which
       spectrum is obtained from each compound. [4]
   (c) One of these compounds has a ¹H NMR spectrum that
       shows only two singlets.
       Explain which compound can be identified from this
       information. [3]
   (d) A simplified mass spectrum of each of the compounds is
       shown below.

   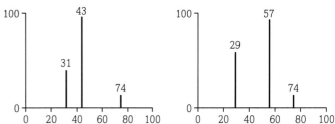

   Explain how this information can be used to identify each
   compound. [6]
   [Total: 15]

Maths skills

In order to be able to develop your skills, knowledge and understanding in chemistry, you will need to have developed your mathematical skills in a number of key areas. This section gives more explanation and examples of some key mathematical concepts you need to understand. Further examples relevant to your A level chemistry studies are given throughout the book and in Book 1.

Using logarithms
Calculating logarithms

Many formulae in science and mathematics involve powers. Consider the equation:

$$10^x = 62$$

The value of x obviously lies between 1 and 2, but how can you find a precise answer? The term logarithm means 'index' or 'power'. Logarithms enable you to solve such equations. You can take the logarithm to base 10 of each side of the equation using the log button of a calculator.

WORKED EXAMPLE

$10^x = 62$

$\log_{10}(10^x) = \log_{10}(62)$

$x = 1.792392...$

You can calculate logarithms using any number as the base by using the $\log_x(y)$ button.

WORKED EXAMPLE

$2^x = 7$

$\log_2(2^x) = \log_2(7)$

$x = 2.807355...$

Many equations relating to the natural world involve powers of e. These are called exponentials. The logarithm to base e is referred to as the natural logarithm and written ln.

Using logarithmic plots

An earthquake measuring 8.0 on the Richter scale is much more than twice as powerful as an earthquake measuring 4.0 on the Richter scale, because the units used for measuring earthquakes are logarithmic. Logarithmic scales in charts and graphs can accommodate enormous increases or decreases in one variable as another variable changes.

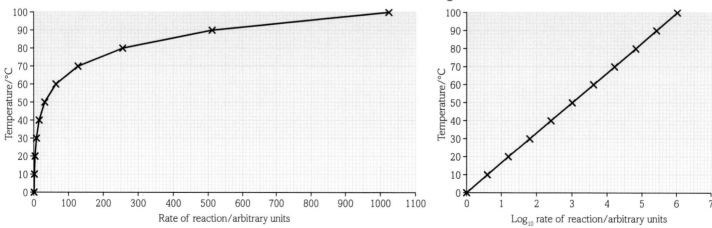

fig A Logarithmic scales are useful when representing a very large range of values, such as in the case of rates of reaction.

Graphs

Understanding that y = mx + c represents a linear relationship

Two variables have a linear relationship if they increase at a constant rate in relation to one another. If you plot a graph with one such variable on the x-axis and the other on the y-axis, you get a straight line.

Any linear relationship can be represented by the equation $y = mx + c$, where the gradient of the line is m and the value at which the line crosses the y-axis is c. An example of a linear relationship is the relationship between degrees Celsius and degrees Fahrenheit, which can be represented by the equation $F = 9/5C + 32$, where C is temperature in degrees Celsius and F is temperature in degrees Fahrenheit.

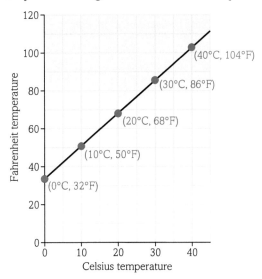

fig B Linear relationship between Fahrenheit and Celsius temperature measurements.

Using the gradient as a measure of a rate of change

Sir Isaac Newton drew tangents to curves on graphs to find the rates of change of the variables as part of his journey towards discovering calculus – a fascinating branch of mathematics. He stated that the gradient of a curve at a given point is exactly equal to the gradient of the tangent to the curve at that point.

To find the gradient at a point on a curve:

1. Use a ruler to draw a tangent to the curve at that point.
2. Calculate the gradient of the tangent using the equation for a linear relationship. This gradient is equal to the gradient of the curve at the point of the tangent.
3. Include the unit with your answer.

Applying your skills

You will often find that you need to use more than one maths technique to answer a question. In this section, you will look at two example questions and consider which maths skills are required and how to apply them.

WORKED EXAMPLE 1

The pH of an aqueous solution is related to the hydrogen ion concentration by the following equation:

$$pH = -\lg[H^+]$$

where hydrogen ion concentration, $[H^+]$, is measured in $mol\ dm^{-3}$.

Calculate the pH of an aqueous solution of $0.001\ mol\ dm^{-3}$ HCl. Assume the acid is completely dissociated. Give your answer to two decimal places.

$$pH = -\lg[H^+]$$

$$pH = -\lg(0.001)$$

$$\lg(0.001) = -3$$

$$pH = 3.00$$

WORKED EXAMPLE 2

A scientist observes an interesting by-product forming during a chemical reaction:

reactant 1 + reactant 2 → desired product + by-product

She decides to investigate how the concentration of this by-product changes during the reaction by plotting a graph of concentration against time. She also plots the change in concentration of the desired product and reactant 1.

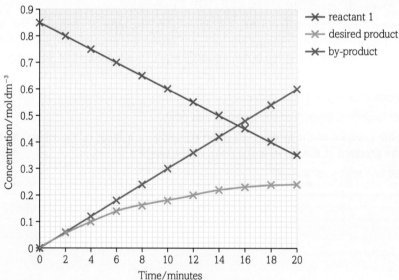

fig C Concentration of reactant and products over the course of a reaction.

(a) What was the concentration of reactant 1 at the start of the reaction?

(b) Calculate the rate of change of concentration of the by-product after 10 minutes. Give your answer in $mol\ dm^{-3}\ s^{-1}$.

(c) Calculate the rate of change of concentration of the desired product after 10 minutes.

(a) To answer this question, you need to find the value for the concentration of reactant 1 when time is equal to zero. That is, you need to find the intercept with the y-axis of the line for the concentration of reactant 1. From the graph, you can see that this is $0.85\ mol\ dm^{-3}$.

(b) In order to calculate the rate of change of by-product concentration during the reaction, you need to find the gradient of its line. The line is straight, meaning that there is a linear relationship between the concentration and time. To find the rate of change, you just have to find the gradient of the line by dividing the change in concentration of by-product by the time taken for the change. Since the line is straight, you can choose any two values.

Take care to use the correct units. Concentration is in mol dm^{-3} but time is in minutes, so you must convert time to seconds to get the correct answer.

concentration after 8 minutes = 0.24 mol dm^{-3}

concentration after 12 minutes = 0.36 mol dm^{-3}

change of concentration = 0.12 mol dm^{-3}

time taken for change = 4 × 60 = 240 s

$$\text{rate of change} = \frac{\text{change in concentration of by-product}}{\text{time taken for change}}$$

$$= \frac{0.12 \text{ mol dm}^{-3}}{240 \text{ s}} = 0.0005 \text{ mol dm}^{-3} \text{ s}^{-1}$$

(c) To find the rate of change of desired product is a little trickier because the line is curved. To do this, you need to draw a tangent to the curve then calculate the gradient of the tangent. A simple way to do this is to draw a triangle with the tangent as its hypotenuse, centred on the 10-minute point as asked (**fig D**).

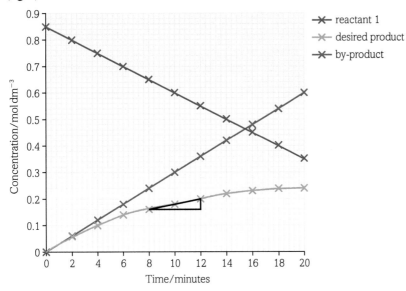

fig D Finding the rate of change of concentration of the desired product by drawing a tangent.

change in concentration between 8 and 12 minutes = 0.04 mol dm^{-3}

time taken for change = 4 × 60 = 240 s

$$\text{rate of change} = \frac{\text{change in concentration of side-product}}{\text{time taken for change}}$$

$$= 0.04 \text{ mol dm}^{-3}/240 \text{ s} = 0.0001666... = 0.0002 \text{ mol dm}^{-3} \text{ s}^{-1}$$

Preparing for your exams

Introduction

It is important to be familiar with the format of the Edexcel A level Chemistry examinations so that you can ensure you are well prepared. Here are some key features:

- You will sit three exam papers, each covering content from both years of your course. The third paper will include synoptic questions that may draw on two or more different topics and questions on conceptual and theoretical understanding of experimental methods.
- You will also have your competency in key practical skills assessed by your teacher in order to gain the Science Practical Endorsement. The endorsement will not contribute to your overall grade but the result (pass or fail) will be recorded on your certificate.

The table below gives details of the three A level exam papers.

A level exam papers

Paper	Paper 1: Advanced Inorganic and Physical Chemistry	Paper 2: Advanced Organic and Physical Chemistry	Paper 3: General and Practical Principles in Chemistry
Topics covered	Topics 1–5 Topic 8 Topics 10–15	Topics 2–3 Topics 5–7 Topic 9 Topic 16–19	Topics 1–19
% of the A level qualification	30%	30%	40%
Length of exam	1 hour 45 minutes	1 hour 45 minutes	2 hours 30 minutes
Marks available	90 marks	90 marks	120 marks
Question types	multiple-choice short open open-response calculation extended writing	multiple-choice short open open-response calculation extended writing	short open open-response calculation extended writing synoptic
Experimental methods?	No	No	Yes
Mathematics	A minimum of 20% of the marks across all three papers will be awarded for mathematics at Level 2 or above.		
Science Practical Endorsement	Assessed by teacher throughout course. Does not count towards A level grade but result (pass or fail) will be reported on A level certificate.		

Exam strategy

Arrive equipped

Make sure you have all of the correct equipment needed for your exam. As a minimum you should take:

- pen (black ink or ball-point pen)
- pencil (HB)
- ruler (ideally 30 cm)
- rubber (make sure it's clean and doesn't smudge the pencil marks or rip the paper)
- calculator (scientific).

Ensure your answers can be read

Your handwriting does not have to be perfect but the examiner must be able to read it! When you're in a hurry it's easy to write key words that are difficult to decipher.

Plan your time

Note how many marks are available on the paper and how many minutes you have to complete it. This will give you an idea of how long to spend on each question. Be sure to leave some time at the end of the exam for checking answers. A rough guide of a minute a mark is a good start, but short answers and multiple choice questions may be quicker. Longer answers might require more time.

Understand the question

Always read the question carefully and spend a few moments working out what you are being asked to do. The command word used will give you an indication of what is required in your answer.

Be scientific and accurate, even when writing longer answers. Use the technical terms you've been taught.

Always show your working for any calculations. Marks may be available for individual steps, not just for the final answer. Also, even if you make a calculation error, you may be awarded marks for applying the correct technique.

Plan your answer

In questions marked with an *, marks will be awarded for your ability to structure your answer logically showing how the points that you make are related or follow on from each other where appropriate. Read the question fully and carefully (at least twice!) before beginning your answer.

Make the most of graphs and diagrams

Diagrams and sketch graphs can earn marks – often more easily and quickly than written explanations – but they will only earn marks if they are carefully drawn.

- If you are asked to read a graph, pay attention to the labels and numbers on the x and y axes. Remember that each axis is a number line.
- If asked to draw or sketch a graph, always ensure you use a sensible scale and label both axes with quantities and units. If plotting a graph, use a pencil and draw small crosses or dots for the points.
- Diagrams must always be neat, clear and fully labelled.

Check your answers

For open-response and extended writing questions, check the number of marks that are available. If three marks are available, have you made three distinct points?

For calculations, read through each stage of your working. Substituting your final answer into the original question can be a simple way of checking that the final answer is correct. Another simple strategy is to consider whether the answer seems sensible. Pay particular attention to using the correct units.

Sample answers with comments

Question type: multiple choice

Which substance has non-polar molecules?

☐ **A** *hydrogen fluoride*

☐ **B** *water*

☐ **C** *carbon dioxide*

☐ **D** *ammonia* [1]

This question is about bond polarity. Electronegative atoms have the ability to draw the electrons in a covalent bond towards them, so that the electron density is unevenly distributed between the two atoms. This results in a polar bond or dipole being set up. All of these molecules contain bonds between atoms with very different electronegativity and hence polar bonds, however, since carbon dioxide is symmetrical overall, then the molecule itself is non-polar.

Question analysis

- Multiple choice questions may require simple recall, as in this case, but sometimes a calculation or some other form of analysis will be required.
- In multiple choice questions you are given the correct answer along with three incorrect answers (called distractors). You need to select the correct answer and put a cross in the box of the letter next to it.
- The three distractors supplied will feature the answers that you are likely to arrive at if you make typical or common errors. For this reason multiple choice questions aren't as easy as you might at first think. If possible try to answer the question before you look at any of the answers.
- If you change your mind, put a line through the box with the incorrect answer (⊠) and then mark the box for your new answer with a cross (☒).
- If you have any time left at the end of the paper, multiple choice questions should be put high on your list of priority for checking answers.

Average student answer

☒ **D** *ammonia*

The student has selected the wrong answer. They haven't remembered that symmetrical molecules can still contain bonds between very electronegative atoms and be non-polar, and just had a guess, or they have not realised that the lone pair of electrons on the nitrogen atom means that ammonia is not symmetrical.

Verdict

This is an incorrect answer because:

- The student has shown a lack of understanding of this topic.

Question type: short open

Complete the electronic configuration for the copper atom Cu and the Cu^{2+} ion.

Cu: $1s^2\ 2s^2\ 2p^6$

Cu^{2+} $1s^2\ 2s^2\ 2p^6$ [2]

You need to know the electronic configuration of the first row transition metal elements. Generally these follow a pattern but there are a couple of exceptions and copper is one of them. The points you need to remember are:

- Copper atoms have a full 3d sub shell and only one 4s electron as this gives a more stable electron arrangement than if the 4s orbital contained two electrons as is the case for the majority of the transition metals.
- For all the transition metals the 4s electrons are lost before the 3d electrons.

256

Preparing for your exams

Question analysis
- Short open questions usually require simple short answers, often one word. Generally, they will be simple recall of the chemistry you have been taught.
- The command word will tell you what you need to do. Here the command word is complete so you simply need to complete the electronic configurations in the space provided.

Average student answer

Cu: $1s^2\ 2s^2\ 2p^6\ 3s^2\ 3p^6\ 3d^9\ 4s^2$

Cu^{2+}: $1s^2\ 2s^2\ 2p^6\ 3s^2\ 3p^6\ 3d^9$

The student has the electronic configuration of the Cu^{2+} ion correct, however, they have clearly forgotten that copper atoms have 10 electrons in the 3d sub shell and only one 4s electron.

Verdict
This is a poor answer because:
- The student has forgotten the electronic configuration of the copper atom.

Question type: open response

Chlorine and bromine are elements in Group 7 of the Periodic Table. Both elements exist in a number of different oxidation numbers and therefore are involved in many redox reactions.

Chlorine dioxide reacts with cold, dilute aqueous sodium hydroxide. The equation for the reaction is:

$2ClO_2(aq) + 2OH^-(aq) \rightarrow ClO_2^-(aq) + ClO_3^-(aq) + H_2O(l)$

Using oxidation numbers, explain why the chlorine in ClO_2 has undergone disproportionation. [3]

Disproportionation is the simultaneous oxidation and reduction of an element in the same reaction. In order to show that disproportionation has occurred, you need to calculate the oxidation state of chlorine in the reacting ClO_2 and the products ClO_2^- and ClO_3^-. You are looking to see that the oxidation state increases for one of the products and decreases for the other.

Question analysis
- The command word in this question is explain. This requires a justification/exemplification of a point, which in this case is that chlorine has undergone disproportionation. In some cases of this type of question, a mathematical explanation could also be used.

Average student answer

In ClO_2 chlorine has an oxidation state of +4. When it reacts its oxidation state changes to +3 in ClO_2^- and +5 in ClO_3^-. This means its oxidation state has both increased and decreased in the same reaction or in other words it has been reduced and oxidised at the same time.

This is an excellent answer and scores all three available marks. The student has calculated the oxidation states so that they can show if disproportionation has occurred or not.

Verdict
This is an excellent answer because:
- The answer is logically structured.
- The student has calculated all three oxidation states correctly and shown that they understand the term disproportionation.

Question type: extended writing

*The equilibrium reaction between carbon dioxide and water is important for maintaining the pH of the body by buffering.

The equation for the reaction is:

$$CO_2 + H_2O \rightleftharpoons HCO_3^- + H^+$$

Explain what is meant by a buffer solution and explain how a buffer solution works by referring to this equation. [6]

Question analysis

- In questions marked with an asterisk (*), marks will be awarded for your ability to structure your answers logically showing how the points that you make are related or follow on from each other where appropriate.
- Four marks are available for making valid points in response to the question. In mark schemes, these are referred to as 'indicative marking points'. To gain all four marks, your answer needs to include six indicative marking points. The remaining two marks are awarded for structuring your answer well, with clear lines of reasoning and linkages between points.
- It is vital to plan out your answer before you write it down. There is always space given on an exam paper to do this so just jot down the points that you want to make before you answer the question in the space provided. This will help to ensure that your answer is coherent and logical and that you don't end up contradicting yourself. However, once you have written your answer go back and cross these notes out so that it is clear they do not form part of the answer.

> It helps with extended writing questions to think about the number of marks available and how they might be distributed. It is reasonable to assume that the marks will be divided equally between the two parts of the question; three marks for explaining what is meant by the term buffer solution and three marks for using the reaction given to explain how a buffer solution works. It is vital that you use this particular reaction to explain how a buffer solution works or you will not be able to access a number of the marks. The command word in this question is explain. Remember, this requires a justification or exemplification of a point.

Average student answer

A buffer solution is a solution that resists changes in pH when a small amount of acid or alkali are added. So if an acid was added then the pH wouldn't change very much and if an alkali was added then the pH wouldn't change very much. Buffers can do this because they contain something that can react with any acid added and they contain another chemical that can react with any alkali added so effectively these are removed and that means that the pH doesn't change very much.

> The student starts to answer this question well and does give a good explanation of what a buffer solution is and gains all three marks for this part of the question. However, rather than then using the equation given in the question to explain in precise terms how a buffer solution works, they simply repeat the information they have already stated. If the question asks you to explain something by referring to a particular reaction you must do that. In this particular case the student has gained none of the three marks available for the second part of this question. In order to gain those marks they should have explained that this reaction works as a buffer because the equilibrium moves to the left-hand side when acid is added as the acid reacts with the HCO_3^- ion and is removed. A reasonable amount of acid can be added before the buffer solution fails because there is a large concentration of HCO_3^- to 'mop up' any hydrogen ions added.

Verdict

This is a poor answer because:

- The student has not used the equation given in the question to explain how a buffer solution works. This will limit the number of marks they are able to obtain.

Question type: calculation

A phosphate buffer system operates in human cells.

The reaction involved is:

$$H_2PO_4^- \rightleftharpoons H^+ + HPO_4^{2-}$$

Calculate the pH of a 0.10 mol dm^{-3} solution of $H_2PO_4^-$

$K_a = 6.2 \times 10^{-8}$ mol dm^{-3} [2]

> For weak acids the position of equilibrium is over to the left-hand side and so the hydrogen ion concentration cannot be assumed to be the same as the original acid concentration as is the case with strong acids. To calculate the pH of a weak acid you need to first write the expression for the equilibrium constant and then rearrange that to work out the hydrogen ion concentration. Remember you can assume that the concentration of $H_2PO_4^-$ is 0.1 mol dm^{-3}. The expression pH = $-\log [H^+]$ can then be used to calculate the pH.

Question analysis

- The command word here is calculate. This means that you need to obtain a numerical answer to the question, showing relevant working. If the answer has a unit, this must be included.
- Always have a go at calculation questions. You may get some small part correct that will gain credit.
- The important thing with calculations is to show your working clearly and fully. The correct answer on the line will gain all the available marks. However, an incorrect answer can gain all but one of the available marks if your working is shown and is correct. Show the calculation that you are performing at each stage and not just the result. When you have finished, look at your result and see if it is sensible.
- Take an approved calculator into every exam and make sure that you know how to use it!

Average student answer

$K_a = [H^+]^2/0.1 = 6.2 \times 10^{-8}$

$[H^+] = \sqrt{(6.2 \times 10^{-8} \times 0.1)} = 7.87 \times 10^{-5}$

pH = $-\log [H^+] = 4.1$

> The student has correctly used the equilibrium expression for the reaction to calculate the H^+ concentration. For this they obtained the first of the two available marks. They then went on to use the expression pH = $-\log [H^+]$ to calculate the pH and gain the second of the two available marks.

Verdict

This is a good answer because:

- the student has completed all steps correctly and has shown their working.

Question type: synoptic

A student carrying out a titration wants to identify the piece of apparatus that contributes most to measurement uncertainties in the experiment.

The percentage measurement uncertainty is marked on the pipette as ±0.06 cm^3, on the volumetric flask as ±0.2 cm^3, and on the burette as ±0.05 cm^3. The average titre was 25.50 cm^3.

Deduce the most significant source of measurement uncertainty in this procedure. [4]

> Paper 3 predominantly contains questions that examine your knowledge of practical chemistry. However, it also includes synoptic questions that can draw on knowledge from any part of the specification. This question is expecting that you can calculate the percentage uncertainty associated with the apparatus used in a titration experiment. These are calculations that you should complete when using titration apparatus.

Question analysis

- A synoptic question is one that addresses ideas from different areas of chemistry in one context. In your answer, you need to use the different ideas and show how they combine to explain the chemistry in the context of the question.
- Questions in Paper 3 may draw on any of the topics in this specification. This question is assessing understanding of experimental methods.

Average student answer

Percentage uncertainty of pipette = 0.06 × 100/25 = 0.24 %

Percentage uncertainty of volumetric flask = 0.2 × 100/250 = 0.08 %

Percentage uncertainty of burette = 0.05 × 100/25.50 = 0.2 %

The pipette has the greatest percentage uncertainty.

> The student has calculated the percentage uncertainty correctly for each piece of apparatus, however, they have forgotten that the percentage for the burette needs to be doubled because this reading is taken twice and therefore the error will be doubled. This has lost them one mark. This answer scores three out of the four available marks.

Verdict

This is a good answer because:

- The student has used the correct formula to calculate percentage uncertainty. There is only one error, and that is not doubling the percentage for the burette. This is only penalised once.

Glossary

An **addition-elimination** reaction occurs when two molecules join together, followed by the loss of a small molecule.

Adsorption is the process that occurs when reactants form weak bonds with a solid catalyst.

An **amphoteric** substance is one that can act both as an acid and as a base.

Amphoteric behaviour refers to the ability of a species to react with both acids and bases.

An **analyser** is a material that allows plane-polarised light to pass through it.

The original meaning of **aromatic** was a description of the smell of certain organic compounds.

The new meaning of **aromatic** is a description of the bonding in a compound – delocalised electrons forming pi (π) bonding in a hydrocarbon ring.

Asymmetric refers to a carbon atom in a molecule that is joined to four different atoms or groups.

Autocatalysis occurs when a reaction product acts as a catalyst for the reaction.

The **basicity** of a base is the extent to which it can donate a lone pair of electrons to the hydrogen atom of a water molecule.

A **bidentate ligand** is one that forms two dative bonds with a metal ion.

A mechanism described as **bimolecular** has two species reacting in the rate-determining step.

A **buffer solution** is a solution that minimises the change in pH when a small amount of either acid or base is added.

Chemical environments of carbon atoms in a molecule are related to whether the carbon atoms are identically, or differently, positioned within a molecule.

The **chemical shift** of a proton (or group of protons) is a number (in the units ppm) that indicates its behaviour in a magnetic field relative to tetramethylsilane. It can be used to identify the chemical environment of the carbon atoms or of the hydrogen atoms (protons) attached to it.

Chiral refers to an atom in a molecule that allows it to exist as non-superimposable forms. It can also refer to the molecule itself.

Complementary colours are colours opposite each other on a colour wheel.

A **complex** is a species containing a metal ion joined to ligands.

A **complex ion** is a complex with an overall positive or negative charge.

Condensation polymerisation refers to the formation of a polymer, usually by the reaction of two different monomers, and in which a small molecule is also formed.

When a base accepts a proton, the species formed is the **conjugate acid** of the base.

When an acid donates a proton, the species formed is the **conjugate base** of the acid.

A **conjugate acid-base pair** consists of either a base and its conjugate acid or an acid and its conjugate base.

The **coordination number** is the number of dative bonds in the complex.

Derivatives are compounds formed from other compounds, especially when the properties of the derivatives can be used to identify the original compound.

Desorption is the process that occurs when products leave the surface of a solid catalyst.

The **electromagnetic spectrum** is the range of all wavelengths and frequencies of all the types of radiation.

The **electromotive force (emf)** is the standard electrode potential of a half-cell (measured under standard conditions of 298 K, 100 kPa pressure and concentrations of 1 mol dm^{-3}) connected to a standard hydrogen electrode.

Enantiomers are isomers that are related as object and mirror image.

The **enthalpy change of hydration**, $\Delta_{hyd}H$, is the enthalpy change when one mole of an ion in its gaseous state is completely hydrated by water.

The **enthalpy change of solution**, $\Delta_{sol}H$, is the enthalpy change when one mole of an ionic solid dissolves in water to form an infinitely dilute solution.

Entropy is a property of matter that is associated with the degree of disorder, or degree of randomness, of the particles.

Equivalent protons are hydrogen atoms in the same chemical environment.

The **half-life** of a reaction is the time taken for the concentration of the reactant to fall to one-half of its initial value.

A **halogen carrier** is a catalyst that helps to introduce a halogen atom into a benzene ring.

A **hazard** is a property of a substance that could cause harm to a user.

High resolution mass spectrometry (HRMS) is a type of mass spectrometry that can produce M_r values with several decimal places, usually four or more.

A **homogeneous catalyst** is one that is in the same phase as the reactants.

Hydrolysis is the breaking of a compound by water into two compounds.

The **instantaneous reaction rate** is the gradient of a tangent drawn to the line of the graph of concentration against time. The instantaneous rate varies as the reaction proceeds (except for a zero order reaction).

An **integration trace** shows the relative numbers of equivalent protons (i.e. in the same chemical environment).

The **isoelectric point** of an amino acid is the pH of an aqueous solution in which it is neutral.

K_w = [H⁺(aq)][OH⁻(aq)]

A **ligand** is a species that uses a lone pair of electrons to form a dative bond with a metal ion.

The **mobile phase** is the liquid that moves through the stationary phase and transports the components.

A **monodentate ligand** is one that forms one dative bond with a metal ion.

A **multidentate ligand** is one that forms several dative bonds with a metal ion.

Multiplets are the different splitting patterns observed (singlets, doublets, triplets or quartets).

Nuclear magnetic resonance spectroscopy is a technique used to find the structures of organic compounds. It depends on the ability of nuclei to resonate in a magnetic field.

Nucleophilic addition is a type of mechanism in which a molecule containing two atoms or groups is added across a polar double bond (usually C=O), and the attacking species in the first step is a nucleophile.

A substance shows **optical activity** if it rotates the plane of polarisation of plane-polarised light.

The **order** of a reactant species is the power to which the concentration of the species is raised in the rate equation.

The **overall order** of a reaction is the sum of all the individual orders.

The **partial pressure** of a gas in a mixture of gases is the pressure that the gas would exert if it alone occupied the volume of the mixture.

A **peak** in a ¹H NMR spectrum shows the presence of hydrogen atoms (protons) in a specific chemical environment.

A **peptide bond** is the bond formed by a condensation reaction between the carbonyl group of one amino acid and the amino group of another amino acid.

The **pH** of an aqueous solution is defined as the reciprocal of the logarithm to the base 10 of the hydrogen ion concentration measured in moles per cubic decimetre pH = –lg [H⁺]. This definition is difficult to remember, so either of the two equations given on page 27 can be used to define pH.

pK_a = –lg K_a

Plane-polarised light is monochromatic light that has oscillations in only one plane.

A **polarimeter** is the apparatus used to measure the angle of rotation caused by a substance.

A **polariser** is a material that converts unpolarised light into plane-polarised light.

A base is a **proton acceptor**.

An acid is a **proton donor**.

A **racemic mixture** is an equimolar mixture of two enantiomers that has no optical activity.

The **rate-determining step** of a reaction is the slowest step in the mechanism for the reaction.

The **rate equation** is an equation expressing the mathematical relationship between the rate of reaction and the concentrations of the reactants.

The (overall) **rate of reaction** is the change in concentration of a species divided by the time it takes for the change to occur. All reaction rates are positive.

A **risk** is the possible effect that a substance may cause to a user, and this will depend on factors such as concentration and apparatus. The level of risk is controlled using control measures.

Six-fold coordination refers to complexes in which there are six ligands forming coordinate bonds with the transition metal ion.

A **splitting pattern** is the appearance of a peak as a small number of small sub-peaks very close to each other.

A **spontaneous process** is one that takes place without continuous intervention from us.

A **square planar** shape contains a central atom or ion surrounded by four atoms or ligands in the same plane and with bond angles of 90°.

The **standard enthalpy change of atomisation** of an element is the enthalpy change measured at a stated temperature, usually 298 K, and 100 kPa when one mole of gaseous atoms is formed from an element in its standard state.

The **stationary phase** in paper chromatography is the liquid or solid that does not move.

A **transition metal** is an element that forms one or more stable ions with incompletely filled d-orbitals.

A mechanism described as **unimolecular** has one species reacting in the rate-determining step.

Unpolarised light has oscillations in all planes at right angles to the direction of travel.

A **zwitterion** is a molecule containing positive and negative charges but which has no overall charge.

The Periodic Table of Elements

Group	(1)	(2)		(3)	(4)	(5)	(6)	(7)	(8)	(9)	(10)	(11)	(12)	(3)	(4)	(5)	(6)	(7)	(0)
	1	2		3	4	5	6	7	8	9	10	11	12	13	14	15	16	17	18
Period 1	1 **H** hydrogen 1.0																		2 **He** helium 4.0
Period 2	3 **Li** lithium 6.9	4 **Be** beryllium 9.0												5 **B** boron 10.8	6 **C** carbon 12.0	7 **N** nitrogen 14.0	8 **O** oxygen 16.0	9 **F** fluorine 19.0	10 **Ne** neon 20.2
Period 3	11 **Na** sodium 23.0	12 **Mg** magnesium 24.3												13 **Al** aluminium 27.0	14 **Si** silicon 28.1	15 **P** phosphorus 31.0	16 **S** sulfur 32.1	17 **Cl** chlorine 35.5	18 **Ar** argon 39.9
Period 4	19 **K** potassium 39.1	20 **Ca** calcium 40.1		21 **Sc** scandium 45.0	22 **Ti** titanium 47.9	23 **V** vanadium 50.9	24 **Cr** chromium 52.0	25 **Mn** manganese 54.9	26 **Fe** iron 55.8	27 **Co** cobalt 58.9	28 **Ni** nickel 58.7	29 **Cu** copper 63.5	30 **Zn** zinc 65.4	31 **Ga** gallium 69.7	32 **Ge** germanium 72.6	33 **As** arsenic 74.9	34 **Se** selenium 79.0	35 **Br** bromine 79.9	36 **Kr** krypton 83.8
Period 5	37 **Rb** rubidium 85.5	38 **Sr** strontium 87.6		39 **Y** yttrium 88.9	40 **Zr** zirconium 91.2	41 **Nb** niobium 92.9	42 **Mo** molybdenum 95.9	43 **Tc** technetium (98)	44 **Ru** ruthenium 101.1	45 **Rh** rhodium 102.9	46 **Pd** palladium 106.4	47 **Ag** silver 107.9	48 **Cd** cadmium 112.4	49 **In** indium 114.8	50 **Sn** tin 118.7	51 **Sb** antimony 121.8	52 **Te** tellurium 127.6	53 **I** iodine 126.9	54 **Xe** xenon 131.3
Period 6	55 **Cs** caesium 132.9	56 **Ba** barium 137.3		57 **La*** lanthanum 138.9	72 **Hf** hafnium 178.5	73 **Ta** tantalum 180.9	74 **W** tungsten 183.8	75 **Re** rhenium 186.2	76 **Os** osmium 190.2	77 **Ir** iridium 192.2	78 **Pt** platinum 195.1	79 **Au** gold 197.0	80 **Hg** mercury 200.6	81 **Tl** thallium 204.4	82 **Pb** lead 207.2	83 **Bi** bismuth 209.0	84 **Po** polonium (209)	85 **At** astatine (210)	86 **Rn** radon (222)
Period 7	87 **Fr** francium (223)	88 **Ra** radium (226)		89 **Ac*** actinium (227)	104 **Rf** rutherfordium (267)	105 **Db** dubnium (268)	106 **Sg** seaborgium (269)	107 **Bh** bohrium (270)	108 **Hs** hassium (269)	109 **Mt** meitnerium (278)	110 **Ds** darmstadtium (281)	111 **Rg** roentgenium (280)	112 **Cn** copernicium (285)		114 **Fl** flerovium (289)		116 **Lv** livermorium (293)		

Key
atomic (proton) number
atomic symbol
name
relative atomic mass

58 **Ce** cerium 140.1	59 **Pr** praseodymium 140.9	60 **Nd** neodymium 144.2	61 **Pm** promethium (145)	62 **Sm** samarium 150.4	63 **Eu** europium 152.0	64 **Gd** gadolinium 157.3	65 **Tb** terbium 158.9	66 **Dy** dysprosium 162.5	67 **Ho** holmium 164.9	68 **Er** erbium 167.3	69 **Tm** thulium 168.9	70 **Yb** ytterbium 173.1	71 **Lu** lutetium 175.0
90 **Th** thorium 232.0	91 **Pa** protactinium 231.0	92 **U** uranium 238.0	93 **Np** neptunium (237)	94 **Pu** plutonium (244)	95 **Am** americium (243)	96 **Cm** curium (247)	97 **Bk** berkelium (247)	98 **Cf** californium (251)	99 **Es** einsteinium (252)	100 **Fm** fermium (257)	101 **Md** mendelevium (256)	102 **No** nobelium (259)	103 **Lr** lawrencium (262)

Index

acid–base indicators 36–7
acid-base reactions 122–3, 124, 125, 126–7
    amines 204–5
    amino acids 211
acid–base titrations 35–8, 43–4
acid dissociation constant K_a 28–9, 33
    determining 34
    from a pH titration curve 43–4
acidic hydrolysis 184, 189
acidic solutions 33
acids
    Brønsted–Lowry 24–5
    dibasic 24, 28–9
    monobasic 24–5
    reactions with amines 205
    strength of 76–7
    see also strong acids; weak acids
activation energy 157
activity effect 27
acyl chlorides 185, 186–7
acylation reactions 197, 199
addition–elimination 206
addition polymerisation 190
adsorption 132, 158
air pollution 133
alanine 210, 211, 212
alcohols
    acyl chloride reaction 186
    carboxylic acid reactions 190–1
    reactions 179, 215
aldehydes 176
    oxidation 178–9
aliphatic amines
    preparation 202–3
    reactions 204–7
aliquots 143
alkaline hydrolysis 154–5, 184, 189
alkaline-manganese batteries 102–3
alkaline solutions 33
alkalis, complex reactions 124, 125, 126–7
alkenes, bonding 176
alkylation reactions 197, 198–9
aluminium oxide 52–3
amides 187, 208
amines 202
    acid-base reactions 204–5
    acyl chloride reactions 187
    copper(II) ion reactions 207
    ethanoyl chloride reactions 206
    halogenoalkane reactions 206–7
    preparation 202–3
    see also butylamine
amino acids 210–11
    chromatography 213–14
    forming peptides 212
ammonia 41
    acyl chloride reactions 187
    basicity 204
    manufacture 158

ammonium ion 41
amphiprotic substances 25
amphoteric substances 25, 123
anabolic steroids 233
antiseptics 201
apparatus
    paper chromatography 224
    risks with 219
aqueous solutions
    pH measurements 32–3
    strong bases 30–1
aromatic amines, preparation 203
aromatic compounds 192, 195
    nomenclature 197
Arrhenius equation 160
asymmetric carbon atom 171
autocatalysis 99, 135, 159

barium sulfate 75
base dissociation constant K_b 33
bases
    base dissociation constant K_b 33
    Brønsted–Lowry 24–5
    nitric acid as a 25
    see also acid-base entries; strong bases; weak bases
basicity 204
batteries 96, 102–3
Benedict's solution 178
benzene
    delocalised structure 194–5
    electrophilic substitution 198–9
    Kekulé structure 193–4
    physical properties 192
    reactions 196–7
benzoic acid 29, 34
bidentate ligands 118–19
bimolecular mechanism 174
blood plasma 41-2
pH 22
boiling temperatures
    carbonyl compounds 181
    carboxylic acids 183
    determination of 223
    organic compounds 177
bonding
    carbonyl compounds 176
    carboxylic acids 182
    covalent 57–8
    dative 110
    delocalised π-bonds 194–5, 198, 200–1
    hydrogen 177, 183
Born–Haber cycles 53–4, 56
bromination
    benzene 193, 196, 198
    phenol 200–1
bromophenol blue 37
bromothymol blue 37, 38
Brønsted–Lowry theory 24–6

buffer solutions 39–42
    calculating the pH 39–40
    definition 39
    and pH curves 43–4
butanal 241–2
butanone 241–2, 243
butylamine 205, 206–7

^{13}C NMR spectroscopy 235, 236–9
calcium carbonate 75–6
calcium fluoride 52–3
calcium nitrate 74–5
calcium oxide 54
cancer treatment 116–17
car batteries 96
carbocation 154, 162
carbon chain, extending 215–16
carbon dioxide 18
    atmospheric 46, 47
    in the body 42
    in Grignard reactions 216
carbon monoxide 121, 133
carbonic acid 41–2
carbonyl compounds
    bonding 176
    in Grignard reactions 216–17
    nomenclature 176
    nucleophilic addition 180–1
    physical properties 176–7
    redox reactions 178–9
carboxylic acids
    alcohol reactions 190–1
    bonding 182
    nomenclature 182
    physical properties 183
    preparation 184
    reactions 184–5
catalysts/catalysis
    autocatalysis 99, 135, 159
    and equilibrium constant 17
    Friedel–Crafts reaction 164–5
    heterogeneous 132–3, 158–9
    homogeneous 134–5, 157–8
catalytic converters 133
cell diagrams 90–1
CFCs 158
chain isomers 170
champagne, bubbles 18–19
charge density 57, 61
chemical environments 236
chemical kinetics see rate of reaction
chemical reactions 50
    change of state 67
    number of moles in 67
    reasons for occurrence 62
    thermodynamics 71, 73
    see also rate of reaction; reaction mechanisms

Index

chemical shift 235
    interpreting in ^{13}C NMR 237
    interpreting in ^{1}H NMR 240–1, 244
    ranges 238, 241
chirality 170–1
chlorine radicals 158
chloroethanoic acid 29, 77
chromate(VI) ions 127–8
chromatogram 224, 225
chromatography 213
    amino acids 213–14
    column 225
    gas 230–1, 232
    GC-MS 232–3
    HPLC 230, 232
    paper 224
    thin layer 225
chromium 109, 126–9
    complexes 126–7, 129
cis-platin 116–17
cis-trans isomers 116
cobalt 109
    complexes 124–5
colorimetry 143
colour 112–13
    change in 122–3, 124–5
    chromium compounds 126–7, 129
    octopus blood 136–7
    vanadium compounds 129
colour wheel 112
column chromatography 225
combustion
    analysis 226
    benzene 196
complementary colours 112
complexes
    chromium 126–7, 129
    cobalt and iron 124–5
    colour in 112–13
    common shapes of 114–15
    naming 111
    square planar 116–17
    stability of 120
    symbols and equations 110
concentration
    hydrochloric acid 218
    hydrogen ion 27–9
    and K_c 16–17
    rate–concentration graph 150–1
concentration–time graph 148–9, 159
condensation polymerisation 190–1
conductometric titrations 38
conjugate acid–base pairs 24–5
    buffers made from 39–41
Contact process 132–3
continuous monitoring method 148–9
control measures 218–19
coordinate bonds 110
coordination number 110
    change in 123
copper 109
    disproportionation 94
    sulfuric acid reaction 93
copper(II) ions 100–1, 113, 122–3
    amine reactions 207
covalent bonding 57–8

cyclohexene 193–4
cysteine 210, 214

damascone 246–7
dative bonds 110
delocalised π-bonds 194–5, 198, 200–1
denticity 118
derivatives 181
desorption 132, 158
1,2-diaminoethane 118–19
dibasic acids 24
    pH 28–9
dibromobenzene 193
dichloroethanoic acid 29
dichromate(VI) ions 127–8
dilution effect 33, 59
2,4-dinitrophenylhydrazine (2,4-DNPH) 180–1
dipeptides 212–13
disproportionation reactions 94
dissociation 25–6
    base dissociation constant K_b 33
    water 30
    see also acid dissociation constant K_a
distillation 220–1
doublets 243–5
drug testing 233
drying agents 222

E numbers 19
EDTA 119–20
electrical conductivity 144
electrochemical cells 90–1
    flow of electrons 92
    half-cells 85–9
electrochemical series 88–9
electrode potential see standard electrode potential
electromagnetic spectrum 112
electromotive force 87–8
    calculating 91
electron affinity 53
electron pair repulsion theory 114
electronic configurations 108–9, 113
electrons
    in 3d energy levels 113
    flow of 92
    release from metals 84–5
electrophilic addition 159
electrophilic substitution 198–9
electrostatics 56
    model 60
elementary reactions 153
empirical formula, calculating 226–7
enantiomers 170–1, 180
    properties 173
end point 35, 37
endothermic reactions 14, 62–3
energy level diagrams 61, 62, 157
enthalpy change 50
    and acid strength 76, 77
    of atomisation 53
    of formation 53–4
    of hydration 59–61
    of hydrogenation 193–4
    of neutralisation 45
    of solution 59, 61

entropy 63
entropy change
    and acid strength 76, 77
    calculating 65–6
    and Gibbs energy 70
    and standard electrode potential 94
    of the surroundings 65, 69
    of the system 64–5
    and temperature 66
    total 64–6, 94
    in various systems 67–9
equilibrium constant K_c 10–11
    calculating 72–3
    and concentration 16–17
    determination 10
    and Gibbs energy 72–3
    and pressure 17
    and standard electrode potential 95
    and temperature 14–15, 73
equilibrium constant K_p 12–13
    and pressure 17
    and temperature 14–15
equilibrium position 62–3
    and temperature 14–15
equivalence point 35–6, 43–4
equivalent protons 243
errors, in pH measurement 34
esterification 185
esters 185
    hydrolysis 188–9
    nomenclature 188
    physical properties 188
ethanedioate ions 135
ethanedioic acid 98–9, 159
ethanoic acid 28, 29, 33, 39–40
ethanoyl chloride 206
exothermic reactions 14, 62–3, 157
experimental lattice energy 56, 57

Fajan's rules 57
Fehling's solution 178–9
filtration 223
first order reaction 145, 148, 151
fossil fuels 50
fractional distillation 221
Friedel–Crafts reaction 164–5
    benzene 196–7, 198–9
fuel cells 78–9, 96–7
functional group isomers 170

gas chromatography (GC) 230–1
    GC-MS 232–3
    limitations 232
gases
    entropy change 64, 67
    greenhouse 46
    measuring volume 142–3
    partial pressure 12–13
    standard electrode potential 88
geometric isomers 170
Gibbs energy 70–1
    applications of 74–7
    and equilibrium 72–3
glutamic acid 210, 211
glycerol 189
glycine 210, 212

265

graphs 251–3
greenhouse gases 46
Grignard reactions 216–17
Grignard reagents 215–16
^{1}H NMR spectroscopy 235, 240–5
    high-resolution 243–5
    interpreting chemical shift 240–4
    low resolution 240–2
Haber process 158
haemocyanin 136–7
haemoglobin 120–1, 213
half-cells 85–9
    reactions 86, 88
half-equivalence point 43–4
half-life of a reaction 148
half-volume method 44
halogen carriers 196
halogenation 185
halogenoalkanes 154–5, 161–3
    amine preparation from 202–3
    amine reactions 206–7
    Grignard reagents from 215–16
    S_N1 and S_N2 mechanisms 174–5
hazard warning symbols 218
hazards 218–19
heating under reflux 220
Henderson–Hasselbalch equation 40
Henry's law 18
Hess cycle 61
Hess's law 54, 61
heterogeneous catalysts 132–3, 158–9
heterogeneous reactions
    K_c 11
    K_p 12–13
high performance liquid chromatography (HPLC) 230, 232
high-resolution ^{1}H NMR 243–5
high resolution mass spectrometry 228–9
histidine 47
homogeneous catalysts 134–5, 157–8
homogeneous reactions
    K_c 10–11
    K_p 12
hydration, enthalpy change of 59–61
hydrochloric acid 76
    cobalt complex reactions 125
    concentrations 218
    dilution effect on pH 33
    manganese(IV) oxide reaction 93–4
hydrofluoric acid 76
hydrogen bonding 177, 183
hydrogen cyanide 180
hydrogen ion concentration 27–9
hydrogen-oxygen fuel cells 78–9, 96–7
hydrogenation 196
    enthalpy change 193–4
hydrogencarbonate 41
hydrolysis
    carboxylic acid preparation 184
    esters 188–9
    halogenoalkanes 154–5, 161–3
    proteins 213
indicators 36–8
    choice of 37–8
    pH range 37

infinite dilution 59
initial-rate method 150
instantaneous reaction rate 149
insulin 213, 214
integration trace 240
inter-ionic distance 52–3, 56
intermediates 154, 161
intermolecular forces 177
interstitial hydrides 158
iodine
    carbonyl compound reaction 179
    propanone reaction 155–6, 157
    titration 100, 143
ion–dipole interactions 60
ionic charge 52–3, 57, 60
ionic product of water K_w 30–1
ionic radius 57, 60
ionic solids
    dissolving in water 67–8
    enthalpy change of solution 59
ionisation 25
    energy 54
iron 109
    as catalyst 158
    complexes 125
iron(II) ions 98, 99, 122, 134
isoelectric points 211
isomerism 170–1

K_a see acid dissociation constant K_a
K_c see equilibrium constant K_c
Kekulé structure 193–4
ketones 176
    reactions 178–9
Kevlar® 209
kinetic stability of a reaction 73
K_p see equilibrium constant K_p

lattice energy 52–5
    definitions 52
    and enthalpy changes of solution and hydration 61
    experimental and theoretical 56–8
    factors affecting magnitude 52–3
ligands
    examples of 111
    exchange reactions 122–3, 124
    multidentate 118–21
    symbols and equations 110
linear complexes 115
liquids, entropy change 64, 67
lithium aluminium hydride 178
lithium fluoride 52
lithium ion batteries 103
lithium oxide 52–3
logarithms 27, 72, 250–1
low resolution ^{1}H NMR 240–2
lysine 210

magnesium chloride 52
magnesium oxide 52–3
magnesium sulfate 74–5
magnetic resonance imaging (MRI) 234
manganate(VII) ions 98, 159
manganese 109
    alkaline-manganese batteries 102–3

manganese(IV) oxide 93–4
mass spectrometry (MS) 228, 247
    GC-MS 232–3
    high resolution 228–9
melting temperatures
    carbonyl compound derivatives 181
    determination of 223
metalloenzymes 164–5
metals
    electron release 84–5
    see also transition metals
methanoic acid 29
methyl orange 36–8
methylamine 204
3-methylbutanoic acid 216
2-methylpropan-2-ol 217
mobile phase 224, 225
molecular formula, calculating 226–7
molecular mass, relative 228–9
molecular models 171
monobasic acids 24–5
monodentate ligands 118
multidentate ligands 118–21
multiplets 243–5

nandrolone 233
neutral solutions 30, 33
neutralisation
    carboxylic acids 184–5
    standard enthalpy change of 45
neutralisation point 35
nickel 109
nickel-cadmium (NiCd) cell 96
nitration, benzene 196, 198
nitric acid 25
nitriles 184, 203
nitrogen monoxide 133
nuclear magnetic resonance spectroscopy (NMR)
    atoms detected by 235
    ^{13}C NMR spectroscopy 235, 236–9
    definition 234
    ^{1}H NMR spectroscopy 235, 240–5
    NMR spectrum 234
    solvents used in 235
nuclei 234
nucleons 234
nucleophilic addition 180–1
nucleophilic substitution 154, 161–3
    S_N2 mechanism 174–5
nylon 209

octahedral complexes 114–15
optical activity 172–3
    amino acids 211
    in nucleophilic addition 180
    and reaction mechanisms 174–5
optical isomerism 170–1
orders of reaction 145–6
    determining 148–52
organic analysis
    empirical and molecular formulae from 226–7
    structural formulae from 227
organic compounds, boiling temperatures 177

organic synthesis 215–17
    extending a carbon chain 215–16
    Grignard reactions 216–17
    planning 215
    practical techniques 220–3
overall order of a reaction 145
oxidation 84
    aldehydes 178–9
    carboxylic acid preparation 184
    ethanedioate ions 135
    ethanedioic acid 159
oxidation number 84, 109
    change in 122
    vanadium 129
oxidising agents 87, 88–9
oxygen transport 120, 136–7
ozone layer 158

paper chromatography 224
paracetamol 206
partial dissociation 26
partial pressure 12–13, 17
peaks
    in ^{13}C NMR 236, 238
    in low resolution ^{1}H NMR 240
    splitting patterns 243–5
pentan-2-ol 217
peptides 212–13
Periodic Table 263
    transition metals in 108
persulfate ion 134
pH
    and activity 27
    aqueous solutions of strong bases 30–1
    blood plasma 22, 41–2
    buffer solutions 39–40
    definition 29
    dibasic acid 28–9
    and hydrogen ion concentration 27, 28
    neutral solutions 30
    range of indicators 37
    seawater 46
pH measurements 32–4
pH meter 35
pH titration curves 35–8
    and buffer solutions 43–4
phenol 200–1
phenol red 37
phenolphthalein 37–8, 201
phenylamine 203, 204–5, 221
phosphorus(V) chloride 185
pK_a 28–9, 77
pK_{in} 37, 38
pK_w 30
plane of polarisation 173
plane-polarised light 172–3
point of inflexion 36
poisoning catalysts 158
polarimetry 172–3
polarisation 57–8
polyamides 208–9
polyesters 190–1
polymerisation
    addition 190
    condensation 190–1

polypeptides 212
positional isomers 170
positive ions 84–5
potassium dichromate(VI) 178, 184
potassium(VII) manganate 98–9, 135, 159
potential difference 87
pressure
    and equilibrium constant 17
    partial 12–13, 17
primary cells 102–3
primary halogenoalkanes 155, 161–3
promoters 158
propan-1-ol 216, 236–7, 240
propan-2-ol 236–7, 240
propanoic acid 29
propanone 155–6, 157
proteins 212–14
proton acceptors 24
proton donors 24
protons, equivalent 243
purity, testing for 223

qualitative analysis 226
quantitative analysis 226–7
quartets 243–5

racemic mixtures 173, 180
radical substitution reactions 154
rate–concentration graph 150–1
rate constants 145
    and temperature 160–3
rate-determining step 146, 154–6
rate equation 145
    determining 148–52
    and mechanisms 153–6
rate of reaction 140
    definition 144
    measuring 142–4
    mechanisms 145–6
    and temperature 160–3
reaction mechanisms 153–6
    and optical activity 174–5
reaction profile diagrams 157, 161–2
reaction quotient Q_c 16–17
recrystallisation 223
red blood cells 120–1
redox reactions 82
    carbonyl compounds 178–9
    electrode potential values and 127–8, 131
    vanadium 129–30
    see also oxidation; reduction
redox titrations 98–101
reducing agents 87, 88–9
reduction 84
    carbonyl compounds 178
    carboxylic acids 184
    dichromate(VI) ions 127–8
reference electrode 85
reflux, heating under 220
relative molecular mass 228–9
retention times 230–1, 232
R_f values 225
risks 218–19
rose oil 246–7

safety 218–19
salt bridge 85
salt solutions, pH 32–3
salts
    from amino acids 211
    solubility 74–6
saponification 189
saturated vapour pressure 13
scandium 109
scents 246–7
seawater, pH 46
second law of thermodynamics 50, 63, 64
    and Gibbs energy 70–1
second order reaction 145, 151
separation techniques 220–3
    see also chromatography
silver chloride 75
simple distillation 220–1
singlets 243–5
six-fold coordination 114
S_N1 mechanism 175
S_N2 mechanism 174–5
sodium chloride
    Born–Haber cycle 53
    lattice energy 52
sodium ethanoate 39–40
sodium fluoride 52–3
sodium hydroxide 30–1
sodium ion 110
sodium nitrate 74–5
sodium thiosulfate 100–1, 143
solids
    enthalpy change of solution 59
    entropy change 64, 67–8
solubility
    carbonyl compounds 177
    carboxylic acids 183
    salts 74–6
solubility product 74
soluble salts 74–5
solvents
    extraction 222
    in NMR 235
sparingly soluble salts 74, 75–6
splitting patterns 243–5
spontaneous processes 63
square planar complexes 116–17
standard conditions 73, 85
standard electrode potential 84–9
    difference between two values 91
    electrochemical cells 90
    making predictions using 92–4
    measuring 85–8
    and redox reactions 127–8, 131
    systems involving gases 88
    and thermodynamics 92–5
standard enthalpy change of atomisation 53
standard enthalpy change of neutralisation 45
standard hydrogen electrode 29, 85
standard lattice energy 52
standard redox potential see standard electrode potential
starch 100
stationary phase 224, 225
steam distillation 221
stereoisomers 170

storage cells 96
strong acids 25–6
    dilution effect on pH 33
    enthalpy changes of neutralisation 45
    hydrogen ion concentration 27
    pH 32
    strong base titrations 35, 37
    weak base titrations 36, 38
strong bases
    pH 30–1, 32
    strong acid titrations 35, 37
    weak acid titrations 36, 37–8, 43–4
structural formulae, determining 227
structural isomers 170
sulfuric acid 28–9
    manufacture 132–3
    zinc and copper reactions 93
surface adsorption theory 132, 158

tartaric acid 173
temperature
    and entropy change 66
    and equilibrium constant 14–15, 73
    and Gibbs energy 71
    and rate constant 160–3
    see also boiling temperatures; melting temperatures
terephthalic acid 190–1
tertiary halogenoalkanes 154, 161–3
Terylene 191
tetrahedral complexes 115
tetramethylsilane 235
theoretical lattice energy 56–8
thermodynamics
    chemical reactions 71, 73
    second law of 50, 63, 64, 70–1
    and standard electrode potential 92–5

thermometric titrations 38
thin layer chromatography 225
third order reaction 145
titanium 109
titrations
    acid–base 35–8, 43–4
    measuring reaction rate 143
    redox 98–101
Tollens' reagent 178–9
total entropy change 64
    calculating 65–6
    and standard electrode potential 94
trans-platin 116, 117
transition metal reactions
    cobalt and iron complexes 124–5
    types of 122–3
transition metals 106
    as catalysts 132–5
    characteristics 108
    definition 108
    electronic configurations 108–9, 113
    in the Periodic Table 108
    symbols and equations 110
    see also complexes; ligands
transition state 157, 161
trichloroethanoic acid 29
triglycerides 189
tripeptides 212
triplets 243–5

unimolecular mechanism 175
unpolarised light 172
urease 213

vanadium 109, 129–30
vanadium(V) oxide 132–3
voltmeter 85–6

volume–time graph 149
washing 222
water
    acyl chloride reactions 186
    freezing of 66, 70
    ionic product 30–1
    ionic solids dissolving in 67–8
    pH 46
    reactions with amines 204–5
    saturated vapour pressure 13
    see also aqueous solutions
water of crystallisation 99
weak acids 25–6
    conjugate base buffers 39–40
    dilution effect on pH 33
    enthalpy changes of neutralisation 45
    hydrogen ion concentration 28
    K_a and pK_a values 29
    pH 32
    strong base titrations 36, 37–8, 43–4
    weak base titrations 36, 38
weak bases
    conjugate acid buffers 41
    pH 32
    strong acid titrations 36, 38
    weak acid titrations 36, 38

xenon tetrafluoride 116

zero order reaction 145, 148, 151
zinc 109
    sulfuric acid reaction 93
zinc-carbon cells 102–3
zwitterions 211